GRASSLANDS: ECOLOGY, MANAGEMENT AND RESTORATION

GRASSLANDS: ECOLOGY, MANAGEMENT AND RESTORATION

HANS G. SCHRÖDER
EDITOR

Nova Science Publishers, Inc.
New York

For permission to use material from this book please contact us:
Telephone 631-231-7269; Fax 631-231-8175
Web Site: http://www.novapublishers.com

LIBRARY OF CONGRESS CATALOGING-IN-PUBLICATION DATA

Schröder, Hans G.
Grasslands : ecology, management & restoration / Hans G. Schröder.
p. cm.
ISBN 978-1-60692-023-7 (hardcover)
1. Grassland ecology. 2. Grassland restoration. 3. Grassland conservation. I. Title.
QH541.5.P7S37 2008
577.4--dc22

2008031076

Published by Nova Science Publishers, Inc. ✢ New York

Contents

PREFACE

Grasslands are areas where the vegetation is dominated by grasses (Poaceae) and other herbaceous (non-woody) plants (forbs). Plants of the sedge (Cyperacae) and rush (Juncaceae) families can also be frequent in grasslands. Grasslands occur naturally on all continents except Antarctica, and in many other areas they have replaced the natural vegetation due to human influence. In temperate latitudes, such as north-west Europe, grasslands are dominated by perennial species, whereas in warmer climates annual species form a greater component of the vegetation.

Grasslands can be found in most terrestrial climates. Grassland vegetation can vary in height from very short, as in chalk downland where the vegetation may be less than 30 cm high, to quite tall, as in the case of North American tallgrass prairie, South American grasslands and African savannah. Woody plants, shrubs or trees may occur on some grasslands — forming wooded, scrubby or semi-wooded grassland, such as the African savannahs or the Iberian dehesa. Such grasslands are sometimes referred to as wood-pasture or Savannah woodland. Grasslands cover nearly fifty percent of the land surface of the continent of Africa.

This new book gathers new and important research from around the world in this field.

Expert Commentaries - It is often perceived that the management practices associated with grass based farming systems are less damaging to biodiversity than are those associated with arable farming systems. Therefore, much scientific attention has been dedicated to the rejuvenation and conservation of field margins within arable systems. However, modern agricultural practices such as sward reseeding, fertilisation and silage cutting are known to be detrimental to the botanical and invertebrate faunal communities associated with grasslands. Despite this, the conservation value of field margins within agriculturally improved grasslands has largely been overlooked by scientists and policy makers to date, notwithstanding that agricultural grassland accounts for a large proportion of agricultural land in many countries, e.g., 79% in Ireland and 67% in the UK.

In this commentary, the authors review the ecological research which has focused on the conservation of field margins and in particular, those within grass-based farming systems. They identify the potential benefits for biodiversity and the ecological services facilitated through the retention/rejuvenation/reintroduction of these habitats within intensively managed grasslands. They also discuss some methods for the rejuvenation and/or establishment of a diverse field margin flora, subsequent management and the challenges associated with each option.

In order to provide reliable projections of future CO_2 emissions from soils, due to global warming, it is important to consider the global distribution of soil BC. Through an initial assessment of the World Soil Archive (http://library.wur.nl/isric/) the authors demonstrate the variability and trends of global soil BC distribution between different climates and soil types and discuss the implications of this chemically recalcitrant form of C on the global C cycle. By doing this, the authors also demonstrate that current methods exist to routinely analyse BC and in the future to develop a global BC map.

Short Communications - Vegetative production is defined as a coupling between crop plant production as the primary product and livestock production as the endproduct in farmlands. The primary product is forage rather than grains that are often the traditional agricultural endproduct. Accordingly, vegetative production requirements for climatic resources are different from traditional grain production. In the first short communication, climatic resources were compared between the north and south China. Traditional (grain) and vegetative (shoot) productions have many production aspects in common and also have differences in physiology, cost inputs, utilization efficiency of solar energy, and vegetative and grain yields. Advantages of vegetative production compared to traditional grain production are presented. Based on the differences of the requirements of irradiance between vegetative and traditional grain productions, enhanced development of vegetative production was proposed through utilization of existing grasslands and winter fallow croplands in south China. This new strategy promises to reduce runoff and loss of topsoil, decrease dust clouds, and restore a good portion of damaged ecosystems in China.

In the second short communication, the authors discuss a central issue in conservation biology and nature management: whether or not characteristic species of a given habitat type could be preserved by fragmented habitat patches or not. The classical theory of island biogeography predicts that the number of species supported by an island increases with the area of the island. However, there is a significant difference between real and habitat islands. In real islands, the surrounding habitat (ocean, sea, lake, river, etc.) is usually inhospitable to organisms occurring on islands. In the case of habitat islands, the bordering habitat (the matrix) is usually less hostile. Consequently, species richness of real islands is not influenced notably by the surrounding habitat. This difference is increasingly emphasized when studying the predictions of island biogeography theory on habitat islands. A clear distinction should be drawn between specialist species that truly perceive the habitat patches as islands and are unable to survive in the surrounding matrix, and those species that occur in both the habitat patch and the matrix (generalist species).

In this case study, the authors demonstrated that depending on the ratio of specialist and generalist species in an assemblage, the species-area relationship may be positive or negative. Ground beetles (*Coleoptera: Carabidae*) of sandy grassland patches were studied in Eastern Hungary (Central Europe). The total number of ground beetle species correlated negatively with grassland area. Based on this result, one can draw the (seriously false) conclusion that it is sufficient to conserve small patches because they support most species. This negative relationship was due to the increasing ratio of generalist species with decreasing patch size. Analyzing the habitat specialist species (open-habitat species associated with sandy soils), the significant negative relationship turned over, and became significantly positive; i.e., the ratio of habitat specialist species increased with patch size, as predicted by the theory of island biogeography.

In the third short communication, the authors discuss a solar chimney power generating technology as a solar thermal technology on a large scale which combines three parts: a collector, a high chimney (reinforced concrete chimney or floating chimney that can stretch up to several thousand meters), and turbines installed at the chimney base. The best locations of these systems are vast desert regions with high solar insolation and large daily range of temperature. As reported by Zhou *et al.* in 2008, special microclimate conditions will result around the solar power generation belts consisting of many commercial solar chimney power plants. Resulting rainfalls will support the growth of different types and quantitities of plants, including herbs, shrubs, and even trees. This will promote restoration of desert land and even create fertile soil and modify the local ecology. Produced plants can be used for biofuel. Furthermore, vegetables can be planted in the periphery of the collector acting as a green house and as a result benefit agricultural production. Other plants can then be used as a source of power as fuelwood or biomass to produce biogas. Many ecological and agricultural benefits are eminent in this respect.

Chapter 1 - The benefits of multi-paddock rotational grazing on commercial livestock enterprises have been evident for many years in many countries. Despite these observations and the results of numerous studies of planned grazing deferment before the mid-1980s that show benefit to species composition, most recent rangelands grazing studies suggest that rotational grazing benefits neither vegetation nor animal production relative to continuous grazing. Detailed comparisons of research methods and practical experiences of successful practitioners of multi-paddock grazing systems identify a number of areas that explain why such different perceptions have arisen. Consistent with producer experience, published data from small paddock trials on both temporal and spatial aspects of grazing management indicates the potential for significantly higher production under multi-paddock rotational grazing relative to continuous grazing and conservative stocking.

While research findings often suggest multi-paddock grazing management is not superior to continuous grazing, researchers have not managed trials to answer practical questions such as: how good is this management option, where is it successful, and what does it take to make it work as well as possible? In contrast, successful ranchers manage strategically to achieve the best possible profitability and ecosystem health. They use basic knowledge of plant physiology and ecology generated by research within an adaptive, goal-oriented management approach to successfully implement planned grazing management.

Published research and experience from ranchers have indicated that the following management factors are the keys to achieving desired goals: (1) planned grazing and financial planning to reduce costs, improve work efficiency and enhance profitability and environmental goals; (2) adjusting animal numbers or having a buffer area available so that animal numbers match forage availability in wet and dry years; (3) grazing grasses and forbs moderately and for short periods during the growing season to allow adequate recovery; (4) timing grazing to mitigate detrimental effects of defoliation at critical points in the life cycle of preferred species inter- and intra-annually; (5) where significant regrowth is likely, grazing the area again before the forage has matured too much; (6) using fire to smudge patch-grazing imprints and manage livestock distribution; and (7) using multiple livestock species. In all these areas, management is the key to success.

Many researchers have failed to sufficiently account for these management factors, either in their treatment applications or in the evaluation of their results. To define the potential impact, researchers must quantify the management strategies for best achieving whole-ranch business and ecosystem results under different grazing management. Conducting research on ranches that have been successfully managed with planned multi-paddock grazing for many years, together with systems-level simulation modeling, offer complementary approaches to traditional small-paddock field research. These methods are particularly applicable where logistics preclude field experimentation, or when assessing impact over decadal time frames. This chapter discusses these points, suggests areas of research that may explain differences in perception among land managers and researchers, and provides information to achieve the full potential of planned multi-paddock grazing management.

Chapter 2 - Desertification models predict that arid grasslands exist in one of two stable states: grassland or desertified shrubland. This prediction is derived from an assumed positive relationship between grass cover and water infiltration rate: once grass cover is reduced, water infiltration rates are insufficient to support perennial grass. These models assume that other factors known to affect water infiltration rate are unimportant. While this alternate stable state view is widely accepted and suggests that restoration of desertified grasslands will be difficult, there have been four recent reports of the reversal of desertification (significant increases in the abundance of perennial grasses) following long-term livestock removal. At one site, recovery of perennial grass inside a long-term grazing exclosure was associated with increased water infiltration rates due to release from soil compaction. To assess the generality of this finding, the author examined water infiltration rates inside and outside five other long-term livestock exclosure sites in southwestern North America that differ in degree of desertification and time since livestock removal. At each site, water infiltration rates were significantly higher inside compared to outside the grazing fence. The relative difference in water infiltration rate across the sites increased with time since livestock removal at a rate of approximately 1.7% per year. These data show that increased water infiltration rate following livestock removal appears to be a general phenomenon in arid grassland systems. In addition, they suggest that the effect of livestock on water infiltration rates should be included in desertification models to better understand the dynamics of vegetation in these systems. Finally, this work suggests that restoration of desertified grasslands may be possible given sufficient time to allow changes in soil properties including release from soil compaction and the concomitant increases in water infiltration rate that follow.

Chapter 3 - At the times of America's discovery in 1492 the temperate part of South America was covered by grasslands. This area includes today Central and Northeastern parts of Argentina, a great deal of Uruguay and Southern Brazil (Figure 1) (Soriano et al., 1991). Among specific features, those grasslands had a unique characteristic: they were developed without high grazing pressures because the region lacked of large herbivores. This differs from the situation found in equivalent African and North American grasslands (Mc Naughton et al., 1988; Sala et al., 1986). The floristic composition of these grasslands varies in accordance with the regional climatic gradients and the local substrate heterogeneity (León et al., 1979; Burkart et al., 1990; Perelman et al., 2001; Soriano et al., 1991). Since European colonization, this vegetation has been progressively modified by the introduction of domestic cattle and the appearance of annual crops, pastures and forestation. These disturbances have led to the invasion and naturalization of several exotic vegetal species, provoking changes in the distribution of the native species (Ghersa and León, 1999) and alterations in the structure and functioning of the vegetal communities (Sala et al., 1986; Rusch and Oesterheld, 1997). Nowadays, only semi-natural grasslands remain in some locations within the region- They are located where the agricultural possibilities of the lands are seriously constrained (Ghersa and

León, 1999). This typically occurs in the so called Flooding Pampa. Two main reasons affect agricultural development in this subregion: the occurrence of frequent ponding and/or floods alternating with summer drought cycles, and the prevalence of halo-hydromorphic soils (León et al., 1975; Batista et al., 1988; Lavado and Taboada, 1988).

Chapter 4 - Here the authors report the main trends observed in the soil seed bank of some species-rich pastures located near Vic (NE Spain), evaluating the relationship between soil seed bank and standing vegetation in two contrasting communities (xerophilous pasture and mesoxerophilous grassland), and the potential role of the soil bank in community regeneration. The authors analyzed 140 soil samples via germination tests and direct observation. The total seed pool (including soil seed bank and standing vegetation) was characterized in terms of morpho-functional traits.

The soil seed content was high and showed strong seasonal variation. The authors found 12,178.8 to 785.7 seeds m^{-2} in the germination tests, and 6,909.1 to 4,882.5 seeds m^{-2} through direct observation. The shallow soil horizon contained much higher densities and species richness than the deep horizon, and higher relative frequencies of elongated or appended seeds. In both communities, 35-45% of the species richness in the soil seed bank corresponded to opportunists, which were rarely found in the established vegetation. In contrast, half of the taxa in the aboveground xerophilous pasture and two-thirds in the mesoxerophilous grassland were not found in the soil bank.

The persistent seeds in the soil bank belonged to a few taxa, mainly not found or rare in the established vegetation. Most of the main species of this vegetation, such as dominant grasses and abundant sub-shrubs, were poorly represented in the soil seed bank. These trends indicate the limited role of this bank in the regeneration of the communities studied, particularly in the mesoxerophilous grassland. The changes in land use occurred during the last half century have strongly increased the fragmentation of semi-natural plant communities, and impair the conservation of plant diversity of this species-rich landscape.

Chapter 5 - During the last century changing patterns of land use in the UK have led to the large scale loss and degradation of species-rich chalk grasslands, a habitat of conservation importance across Europe. In particular, the conversion of chalk grasslands into arable land has represented a major threat to this habitat type. The re-creation of chalk grasslands on ex-arable land therefore has the potential to both increase the area of this threatened habitat, and to provide links between existing remnant fragments. The authors present a four year study investigating the effects of three different methods of introducing plant species during grassland re-creation: 1) a simple grass seed mix, intended to suppress the establishment of pernicious weeds; 2) spreading of hay; and 3) sowing of brush harvested seeds. The latter two methods introduce local provenance seeds collected from species-rich chalk grassland. Using a replicated block design, the authors investigate the efficacy of these management practices in promoting the re-creation of both plant and phytophagous beetle assemblages. They compare re-creation success relative to a target chalk grassland and contrast differences in the responses between these two trophic levels. The introduction of local provenance seeds by hay spreading and brush harvesting is shown to be beneficial during re-creation. However, if used in combination with simple grass-only seed mixtures the establishment of some chalk grassland species may be reduced. The authors discuss the relevance of these findings for the re-creation of plant and phytophagous beetle assemblages of species-rich grasslands on ex-arable land. They also consider the implications of differences in establishment rates between these two taxa for the long term success of habitat creation.

Chapter 6 - Over recent decades, vegetation ecology has increasingly become a predictive functional science that is more oriented towards the physiological properties of the organism and less concerned about the presence or absence of particular organisms. The mechanisms underlying ecosystem functioning result in typical plant community structures and floristic compositions that are frequently seen also in heavily polluted locations. Mathematical models provide useful tools when trying to decipher the links between biodiversity, productivity, stability, decomposition, nutrient acquisition and resilience in ecosystems, with important findings for the direction of natural resource management and applications in remediation technologies. Phytoremediation is the use of vegetation for the *in-situ* treatment of contaminated soils, sediments and water. It is characterized by lower capital inputs and lower costs, as compared to conventional methods. It is best applied at sites with shallow contaminations of metal, nutrient or organic pollutants, that are amenable to one of its applications: rhizofiltration, rhizosphere bioremediation, phytotransformation, phytostabilisation, phytoextraction and, in extreme cases, phytomining. Assigning plant species to functional groups on the basis of their physical and/or ecophysiological characteristics is one of the strategies applied when trying to explain and predict the interactions of plants with their environment. It has become clear however, that no single functional classification is applicable to all situations. Metallophytes, in particular, have received special attention in phytoremediation studies, primarily because of the mechanisms that enable them to resist metal concentrations that are toxic to most plant species, and thus to survive and dominate areas with natural or anthropogenic induced mineralization. The focus of the chapter is on the recent advances in vegetation ecology and their consequences for contemporary phytoremedial applications.

Chapter 7 - In different regions of southern Brazil, often a mosaic of grassland and forest is found, but little is known how these vegetation pattern has been formed. Palaeoecological background information is needed for management and conservation of the highly diverse mosaic of forest and grassland in southern Brazil. Questions on the origin of grasslands, its development, dynamic and stability, its response to environmental change such as climate, and the role of human impact, are essential. Further questions on its natural stage of vegetation or its alteration by pre- and post-Columbian anthropogenic activity are also important. To answer these questions, palaeoecological and palaeoenvironmental data based on pollen and charcoal analysis of radiocarbon dated sedimentary archives from southern Brazil are used to provide an insight into past vegetation changes that allow us to improve our understanding of the modern vegetation and to develop conservation and management strategies for the strongly affected grassland ecosystems in southern Brazil.

Chapter 8 - Agricultural activities are significant producers of nitrous oxide (N_2O) emission to the atmosphere (Johnson et al., 2007; Ugalde et al., 2007). That is, about 58% of total anthropogenic N_2O emissions are caused by agriculture (IPCC, 2007). The main cause of agricultural increases in N_2O is the application of N fertilizers and animal manures. Nitrous oxide is a long-lived greenhouse gas in the atmosphere with 296 times the global warming potential of CO_2. Denitrification is credited as the primary producer of N_2O (Johnson et al., 2005), which is generally favoured in poorly aerated soils with high NO_3^- concentrations (Van Groeningen et al., 2005). Currently, about 32% of the agricultural land in the EU is used for grassland production (Eurostat, 2005). The response of grassland to fertilizer N is greater than that of most other crops due both to its long growing season and to being harvested as vegetative growth (Whitehead, 1995). In the Basque Country, around 36% of agricultural land is occupied by grassland, and there is a high risk of N_2O losses by denitrification due to high rainfall, typically in excess of 1000 $mmyr^{-1}$, and warm temperatures, especially in spring and autumn when fertilizer is applied (Estavillo et al., 1994). Besides, the current dairying model is still developing towards an intensified dairy production system, resulting in increases in generation of animal manure, which may be also a significant source of harmful nutrient emissions into the environment if handled improperly. Also, the mountainous orography characteristic of the area together with the historically parcelled rural soil distribution and the current high industrial pressure has dramatically increased rural soil prices, which does not contribute to enhance farmland availability. Efficient use of nutrients is one of the major keys of sustainable agricultural production systems because inefficient nutrient use not only results in excessive and potentially harmful losses to the environment, it also negatively affects economic performance of production systems (Oenema and Pietrzak 2002).

This review focuses on N_2O emission especially in the Basque Country as related to grassland soil, considering the effect of management and mitigation options.

Chapter 9 - The stages of grassland development on Kunashir and Iturup Islands (Southern Kuril Islands) and Lesser Kuril Ridge are reconstructed based on pollen analysis and radiocarbon dating of soil profiles. There are anthropogenic grasslands and meadows in wind shadows formed during cooling events in Late Holocene; these formations have been relatively stable during last 1500-2000 years. Grasslands on small islands in the south of the Lesser Kuril Ridge have existed last 4000-6000. Floristic components began to change in the Late Pleistocene. Grassland soils have multiple profiles. Grasslands have various floristic compositions. Herbaceously rich meadows are located on terraces and divides, with *Sasa* on some terrace surfaces and slopes. Shrub-grass associations developed on supersaturated substratum; herb-graminoid – on well drained areas. Grass associations typical for swamp accumulative lowlands are developed on low coastal plains.

The lithogenic component of landscapes is most dynamic on the islands and it quickly responds to climatic changes and changes in sea level. For example, substratum renewal has largely been a response to aeolian processes during small amplitude regressions in the Middle and Late Holocene. Sand buildup and the addition of volcanic ash with varied chemical compositions have changed water and mechanical soil characteristics that greatly influenced grass associations. Supersaturated sedge and sedge-herb meadows even on low isthmuses shifted to herb-graminoid and the role of xerophilous species increased.

Chapter 10 - Nitrous oxide (N_2O) emissions from grazed pasture soils represent a significant source of atmospheric N_2O. Nitrous oxide is produced by transformations of nitrogen from cattle excrements, urine and fertilizers entering the soil. Specific conditions are necessary for emissions of N_2O and these emissions are both spatially and temporally highly variable. Spatial, and in some extent temporal variability, is due to the complex effect of nitrogen input, pasture management and environmental conditions (e.g., aeration and water status, pore space and pore size distribution, available carbon and mineral nutrients concentration, pH and other soil chemical properties, as well as microbial communities abundance and diversity) on N_2O emissions and all this makes estimation of emissions very difficult. Several types of environments or events characterized by potentially high N_2O production and emissions (=emission hot spots) can be distinguished in the livestock farming system. These include camping areas, drinking sites, feedlots, shade areas, footpaths, dung and urine patches (where combined effects of nutrients in urine, dung and compaction occur resulting in creating conditions for high nitrification and denitrification rates). Here, the author proposes overwintering areas, that is pasture sites where cattle is located in high stock densities for a relatively long period during winter season (where severe damage of plant cover is common and the effects are typical for the above camping areas other hot spots are even reinforced as another type of hot spots for N_2O emissions. Due to overgrazing effect, damage of vegetation and high stock density, as well as high N-inputs in excrements and lower utilization of deposited N by plants in a cold period, cattle overwintering areas appear to have a large potential for accelerated microbial N transformations and thus gaseous losses. This contribution examines various aspects related to the production of N_2O and its emission from the soil of cattle-grazed pasture sites, aiming to identify its major sources as well as environmental soil conditions favourable for N_2O production (=emission hot spots), as well as to summarize mitigation strategies for N_2O emissions. In particular, it is focused on the cattle overwintering areas, representing specific and often significant emission hot spots. Relevant measures to decrease N_2O fluxes from the grazed grasslands are presented. However, practical options to decrease N_2O emissions seems to be rather limited due to a number of reasons, including a limited knowledge on intrinsic mechanisms regulating N_2O production in the pasture soils.

In: Grasslands: Ecology, Management and Restoration
Editor: Hans G. Schröder
ISBN 978-1-60692-023-7

Expert Commentary A

THE ECOLOGICAL VALUES AND FUNCTIONS OF GRASSLAND FIELD MARGIN HABITATS

H. Sheridan*[*,1], G. Purvis[1] and G. O'Donovan[2]
[1]School of Biology and Environmental Science
Agriculture and Food Science Centre, University College Dublin
Belfield, Dublin 4, Ireland
[2] Broadview Ecological Consultants, Stroud
Gloucestershire, GL6 8PU, UK

ABSTRACT

It is often perceived that the management practices associated with grass based farming systems are less damaging to biodiversity than are those associated with arable farming systems. Therefore, much scientific attention has been dedicated to the rejuvenation and conservation of field margins within arable systems. However, modern agricultural practices such as sward reseeding, fertilisation and silage cutting are known to be detrimental to the botanical and invertebrate faunal communities associated with grasslands. Despite this, the conservation value of field margins within agriculturally improved grasslands has largely been overlooked by scientists and policy makers to date, notwithstanding that agricultural grassland accounts for a large proportion of agricultural land in many countries, e.g., 79% in Ireland and 67% in the UK.

In this commentary, we review the ecological research which has focused on the conservation of field margins and in particular, those within grass-based farming systems. We identify the potential benefits for biodiversity and the ecological services facilitated through the retention/rejuvenation/reintroduction of these habitats within intensively managed grasslands. We also discuss some methods for the rejuvenation and/or establishment of a diverse field margin flora, subsequent management and the challenges associated with each option.

* Corresponding author email: helen.sheridan@ucd.ie

Introduction

Loss of biodiversity in response to the intensification of agricultural systems has been reported worldwide (see Benton *et al.*, 2003). As arable systems are generally heavily input driven and undergo regular disturbance, they are often regarded as being simpler in form and poorer in species than are grass-based farming systems (Altieri, 1994). However, in many areas, the management practices associated with traditional grassland systems have undergone radical changes during the last 50 years e.g. the change from multi-species swards to simple *Lolium-Trifolium* mixtures, increased use of inorganic fertilisers (increased by 100% between 1970 and 1986 in the UK (MAFF *et al.*, 1997)), the shift from hay to silage making and increased stocking rates (Chamberlain *et al.*, 2000). As a result, contemporary grassland management practices are often more similar to those associated with arable systems than with those of traditional, low-input meadows, in terms of the biodiversity which they support.

Currently, semi-natural grassland (i.e. grasslands that have not been reseeded, had applications of artificial fertiliser nor have been subjected to intensive grazing or cutting during the last 45-50 years i.e. 'agriculturally unimproved grasslands' (Vickery *et al.*, 2001)) only account for approximately 200,000ha in the UK (Crofts and Jefferson, 1999). There has been an estimated 80% reduction in their area in Sweden during the period 1870 to 1990 (Berg and Gustafson, 2007), while data for Northern Ireland reveal a 33% increase in the area under improved grassland during the period 1987 and 1998 (Cooper *et al.*, 2003). Changes in land cover classes in the Republic of Ireland between 1990 and 2000, identified through the CORINE Land Cover assessment, include a 25% increase in land area classified within their 'arable land class' which includes land used for silage production (EPA, 2007). The inclusion of silage production areas within the 'arable land class' is notable as it reflects the similarity of management practices which occur under both farming systems. This removal and fragmentation of habitats coupled with intensive grassland management practices has resulted in increased pressure for survival on many taxa.

In recognition of their importance, many of the agri-environment schemes operating in EU now contain measures which focus on the retention and protection of field margin habitats within grassland systems e.g. Rural Environment Protection Scheme (REPS) in Ireland and Environmental Stewardship – Entry and Organic Level Stewardship in the UK. While details of individual schemes' requirements of participating farmers differ, nutrient and pesticide exclusion from these habitats is a common theme.

Definition of Field Margin Structure

The relationship between field boundary and field margin habitats is inextricably linked. However, for the purposes of this manuscript, the term field margin primarily refers to an uncultivated strip of land, typically covered in herbaceous vegetation, which lies adjacent to the field boundary feature (Sheridan, 2005) (Figure I).

These are ecotones which lie between agricultural grass fields and field boundary habitats. Diversity within margins can be relatively high due to the presence of plant species which are common to both habitats. However, this diversity is, to a large extent, dependent on the grassland management practices of the farmer.

Figure I. Uncultivated field margin covered in herbaceous vegetation adjacent to a hedgerow field boundary and an intensively managed *Lolium-Trifolium* sward which has been cut for silage.

Benefits and Services Provided by Field Margin Retention within Intensively Managed Grasslands

Elevated levels of soil fertility resulting from increased use of inorganic fertilisers favour rapidly growing, nitrophilous species (Thomas *et al.*, 2002) which may result in a competitive asymmetry among plant species (Shipley and Keddy, 1994). This generally leads to the exclusion of slower growing and less invasive perennial species (Andrews and Rebane, 1994). Most indigenous grasses and herbs cannot compete with the aggressive sown *L. perenne* under the enhanced soil conditions (Frame, 2000). As a result, field margins can often provide the only areas suitable for the development of a diverse semi-natural grassland flora within intensively managed grassland agricultural systems.

Decreased botanical diversity coupled with management practices such as silage cutting, leads to limited architectural diversity within the sward. This in turn has profoundly negative effects on grassland arthropod populations (Woodcock *et al.*, 2007a; Woodcock *et al.*, 2007b). For example, spider communities at agriculturally improved sites may become dominated by highly invasive, colonising species such as *Erigone dentipalpis*, *Erigone atra* and *Oedothorax fascus*. This may be explained, in part at least, by the management practices which preclude the development of a litter layer, thus resulting in a lack of small litter-inhabiting species such as *Ceratinella brevipes* and *Tiso scopigera* and the linyphiine species such as *Allomengea scopidera* and *Meioneta saxatilis* (Rushton *et al.*, 1989). Carabid beetle community diversity has also been shown to decrease under intensive grassland management practices such that it becomes similar to that found within arable systems, with *Pterostichus niger* being the only large species recorded (Foster *et al.*, 1997). This is most likely due to the inability of the long-lived larvae of the large carabid species to tolerate the sward height reduction found under intensive grazing and silage cutting regimes. Silage making may also involve the immediate removal of vegetation following cutting, resulting in the loss of a much greater proportion of invertebrate biomass than happens during hay making.

Application of organic manure may also negatively effect invertebrate populations through the reduction of the number of suitable refuges or hibernation sites available. For example, Desender (1982) recorded lower abundance of hibernating carabids in areas close to cattle dung than in the remainder of the field. Thus it is likely that large scale dung or slurry spreading could have serious consequences for invertebrate communities and reiterates the importance of excluding these substances from field margin habitats.

Vegetation cover within field margins may be one of the key factors in overwintering site selection made by arthropods. This is due to its buffering capacity which reduces diurnal temperature fluctuations and therefore may diminish invertebrate mortality rates in the uppermost layers of soil (Pfiffner and Luka, 2000; Desender, 1982). The tussock forming grasses e.g. *Dactylis glomerata* and *Holcus lanatus* are known to be particularly important in terms of this (Thomas *et al.*, 1991). They are also known to be particularly favoured by members of the Lepidoptera e.g. the Hesperiidae and Satyridae (Longley *et al.*, 1997).

Avifauna and small mammals are also affected by the vegetation and invertebrate diversity of field margins with approximately 25% of bird species associated with hedgerow habitats actually being dependent on the herbaceous vegetation of the adjacent margin (Hooper, 1970). However, species such as yellowhammer may avoid using hedgerows which are adjacent to improved pastures and silage leys (Bradbury *et al.*, 2000). It is also probable that in a similar manner, as is believed to take place in arable ecosystems, field margins within grasslands act as wildlife corridors, linking together isolated habitat patches and allowing movement of taxa, which would not otherwise take place due to the perceived hostility of the surrounding landscape.

Aside from the intrinsic value of the wildlife which is dependent on field margin habitats to a greater or lesser extent for hibernation, feeding, breeding etc., many taxa also provide important ecological services. For example pollinating insects such as bees and hoverflies are estimated to provide a service equivalent to the monetary value of $112 billion in the U.S. alone (Kearns *et al.*, 1998). In addition to the services provided by the biodiversity which it supports, the actual land area of the field margin may provide an important ecological buffer, preventing the movement of pollutants such as inorganic fertilisers, from areas of application i.e. agricultural fields into adjacent habitats (Marshall and Moonen, 2002). Where a field boundary feature e.g. a hedgerow is also *in situ*, this may increase the effectiveness of this role through the interception of fertiliser granules (Tsiouris and Marshall, 1998).

The botanical diversity of grassland field margins may also provide other agronomic benefits in the form of grazing animal performance. Grazing animals, like humans, appreciate variation in their diet as it enhances palatability. It has been observed that when a small quantity of a second species is present in what otherwise might be termed a monoculture, animals tend to graze it preferentially (Cooper and Morris, 1983; Eliot, 1908). Forage consumption directly influences approximately 90-95% of a grazing animals nutrient and mineral supply (Rogers *et al.*, 1989). Therefore, mineral deficiencies are largely determined by deficiencies in the diet. The majority of essential mineral elements are associated with one or more catalytic functions in the cell (McDonald *et al.*, 2002) and among the conditions known to be caused by mineral and trace element deficiency are: abortion, stillbirth, death of young, infertility, retention of the placenta, scour, increasing susceptibility to infections such as pneumonia, mastitis, stiff joints, muscular weakness and a general lack of thrive (McDonald *et al.*, 2002; Rogers *et al.*, 1989).

Certain herb or wildflower species can be highly beneficial as they add to the nutritive value of the herbage. For example *Plantago lanceolata* (ribwort plantain) has long been recognised as one of the most mineral-rich herbs available. Stapledon (1936) recommended its use where there were insufficient levels of Calcium and Phosphorus in the diet. In addition, *P. lanceolata* and other deep rooting herbs such as *Sanguisorba minor* (salad burnet), *Achillea millefolium* (yarrow), which may be present within the field margins of otherwise intensively managed grassland fields, can retrieve minerals from the lower layers of the soil profile. These minerals can then be made available to the grazing animal. Deep rooting herbs are also known to improve soil structure by increasing aeration and drainage (Culleton *et al.*, 2002; Foster, 1988). Other beneficial species include *Trifolium pratense* (red clover) and *T. repens* (white clover). The importance of these species is attributed to their symbiotic relationship with the nitrogen-fixing bacteria present in their roots.

ESTABLISHMENT/REJUVENATION OF BOTANICAL DIVERSITY WITHIN GRASSLAND FIELD MARGINS

Natural Regeneration

Where the botanical composition of a field margin has become severely degraded, rejuvenation may be possible through the alleviation of detrimental farming practices such as those outlined above. This natural regeneration of the habitat preserves the local flora (Baines *et al.*, 1996). However, it can be a slow and unreliable process. The reasons for this are twofold; firstly there may be a lack of desirable species and a corresponding abundance of long-lived seeds of many 'undesirable' species within the seedbank (Mortimer *et al.*, 1988; Radosevich and Holt, 1984). Seeds of dicotyledonous weed species generally remain viable in the seed bank for a more prolonged period than do those of grass species (Lewis, 1973). Therefore, the success of this method is partially dependent on the availability of 'desirable' seed sources and a corresponding lack of 'undesirable' seed in the vicinity of the margin habitats (Baines *et al.*, 1996). Secondly, even where a desirable seed bank is in place, the lack of seed dispersal pathways may inhibit the rejuvenation of botanical diversity (Muller *et al.*, 1998; Jefferson and Usher, 1989).

Introduction of Seed Mixtures

Where lack and limited dispersal of desirable species propagules is a difficulty, increased botanical diversity in the field margin may be achieved through the introduction of grass and wild flower seed mixtures (Sheridan *et al.*, 2008). This method can have the additional benefit of reducing the abundance of undesirable weed species through their rapid exclusion by the development of the sown perennial species (Sheridan *et al.*, 2008). However, this may be influenced by the diversity of the mixture used, with highly diverse mixtures more likely to suppress undesirable species than low diversity mixtures (van der Putten *et al.*, 2000). The principle difficulty associated with this method of botanically diverse field margin establishment is the poor establishment rate of some herb species (Sheridan *et al.*, 2008; Bokenstrand *et al.*, 2004; Hopkins *et al.*, 1999). In addition, due to the high labour intensity involved in their collection, seed mixtures tend to be relatively expensive. However, these difficulties can at least to some extent, be alleviated through the selection of reputable seed suppliers, careful tailoring of the mixture to soil type and other environmental conditions prevalent in the area, and fine seed bed preparation.

Provenance

Careful consideration should be given to the provenance of any seed or other plant propagules being introduced, as the widespread use of non-native seed can compromise the genetic diversity of any particular species or population (Price, 2003, Smith *et al.*, 2005). In addition, some species may demonstrate the home site advantage hypothesis i.e. propagules of local origin showing enhanced fitness when compared with non-local propagules (Montalvo and Ellstrand, 2000). Hay strewing and brush harvesting of seed from species rich donor sites can be effective methods of introducing desirable propagules of native provenance to a recipient site. Although results from individual species are known to differ depending on their structural location within the sward e.g. colonisation results by lower growing species such as *Lotus corniculatus* (birdsfoot trefoil) and *Prunella vulgaris* (self-heal), tend to be greater with hay strewing compared with brush harvesting techniques. Hay strewing has the additional benefit of not requiring the use of specialised machinery for its adoption by farmers (Edwards *et al.*, 2007).

Management of Grassland Field Margins

Misplacement of Nutrient Inputs

One of the principle constraints to successful establishment and persistence of botanical diversity within field margins is nutrient enrichment due fertiliser and slurry misplacement, or due to dunging of grazing animals (Sheridan, 2005). This can lead to the perception held by some farmers and policy makers that these may act as sources of weeds and pests. However, difficulties associated with weed encroachment into the main sward within pastoral systems usually only arise following a reduction in sward density, which may happen as a result of farming practices such as silage cutting, poaching or overgrazing (Lewis and Hopkins, 2000). The spread distance of some machinery used for land-spreading of fertiliser or slurry i.e. 12 to 15 meters in the case of splash-plate slurry spreaders (Huijsmans, 2003), necessitates wider operating distances from the edge of the field margin to be maintained.

Cutting of Vegetation

Cutting of vegetation can benefit low-growing plant species, particularly due to reduced competition for light and space (Bokenstrand *et al.*, 2004). The removal of the cuttings also has the desirable effect of reducing soil fertility (Berendse *et al.*, 1992). However, heterogeneity in field margin structure is known to be important for numerous taxa in terms of mobility, foraging ability etc. e.g. invertebrates (Anderson *et al.*, 2005; Clausen *et al.*, 2001) and birds (Atkinson *et al.*, 2004). Management practices such as cutting may have conflicting results, favouring the abundance of one taxon, while apparently having no, or negative impacts on the abundance of other taxa (Cole *et al.*, 2007). In addition, although management can influence the biodiversity benefits derived from field margins, it may also influence their suitability to facilitate other ecological services (see Marshall and Moonen, 2002). Therefore, application of generic management techniques to field margins could prove detrimental to particular taxa.

Introduction of Grazing Animals

Grazing is a pivotal issue concerning the management of grasslands and therefore grassland field margin habitats. Large herbivores affect the relative abundance and competitive ability of plants through factors such as defoliation, poaching, dunging and urination (Jensen, 1985). The principle ecological aims of introducing grazing to a site include the prevention of successional change of the herbaceous vegetation to scrub/woodland, and to maintain or increase the structural heterogeneity of the vegetation (Tallowin *et al.*, 2005).

The introduction of herbivores is generally believed to enhance plant diversity (Olff and Ritchie, 1998). However, individual plant responses to grazing are often difficult to predict (Watkinson and Ormerod, 2001). For example Sheridan *et al.* (2008) reported a reduction in plant species diversity in experimental grassland field margins following the introduction of grazing. According to Bullock and Marriott (2000), herb species generally respond positively to grazing while the response of grasses tends to be more species-specific. Management for plant diversity through grazing is strongly related to the type and stocking rate of the herbivores in question (Olff and Ritchie, 1998). For example, due to their grazing habit, swards grazed by sheep tend to be very short and homogenous (Berg *et al.*, 1997), while those grazed by cattle tend to more heterogeneous. Results are also dependent on grazing intensity, with high stocking rate generally believed to lower botanical diversity (Olff and Ritchie, 1998). However, low stocking rates can be equally negative, resulting in the luxuriant growth of woody species such as *Rubus* and/or *Ulex* spp. or problematic, invasive species such as *Pteridium aquilinum* (Sheridan, 2005).

Appropriate grazing intensity is also dependant on initial levels of botanical diversity within the habitat and the objectives of its introduction i.e. the maintenance and/or enhancement of biodiversity. For example Tallowin *et al.* (2005) reported maintenance of initial levels of botanical diversity and enhanced levels of faunal diversity within species rich grassland swards which were grazed to heights of 8 – 12cm over five years. However, no enhancement of positive indicator species was recorded while there was an increase in pernicious weeds.

Width of Margin

Field margins have a high edge to area ratio and are therefore vulnerable to farming practices being carried out in the adjacent grassland sward. Countries across the EU have adopted various different widths of field margins within their individual Agri-Environment Schemes e.g. 1.5-3.0m in Ireland (REPS), 2.0-6.0m in UK (Environmental Stewardship – Entry Level Stewardship). An important role of increased width of field margins is to provide a buffer for the inner area of the margin against agricultural disturbance in the adjacent agricultural area (Marshall *et al.*, 2006). In addition, the probability of persistence and survival of many species increases exponentially with increases in population size (Opdam, 1990). Where field margin width is increased, this is more likely to facilitate the increased population sizes necessary for floral and faunal species persistence in the long term.

Conclusion

While few taxa are found exclusively within field margins, it is likely that many would have reduced ranges or be absent altogether from intensively managed grassland farms without these or other non-cropped habitats. Where the field margin flora has become impoverished over time, this may be rejuvenated or re-established through processes such as natural regeneration, hay-strewing and reseeding. However, the success of any of these methods will to a large extent depend on soil fertility status, nutrient exclusion, the abundance of particular species seeds in the seed bank and subsequent management of the vegetation through cutting and/or grazing.

Acknowledgements

We thank the Department of Agriculture, Food and Rural Development Ireland, for funding provided through the Research Stimulus Fund Programme 2006. Thanks also to Teagasc for previous funding provided to the lead author through the Walsh Fellowship Scheme.

References

Altieri, M.A. (1994). *Biodiversity and Pest Management in Agroecosystems*. Haworth Press, USA.

Anderson, A., Helden, A., Purvis, G., Kirwan, L. and Sheridan, H. (2005). Assessing indicators of biodiversity in agricultural grasslands – some preliminary results from monitoring grassland arthropods and experimental conservation field margins. *Tearmann: Irish Journal of Agri-Environmental Research*, 4, 51-66.

Andrews, J. and Rebane, M. (1994). *Farming and Wildlife. The Royal Society for the Protection of Birds*, Bedfordshire, England.

Atkinson, P.W., Buckingham, D. and Morris, A.J. (2004). What factors determine where invertebrate-feeding birds forage in dry agricultural grasslands? *Ibis,* 146, 99-107.

Baines, M., Feber, R. and Smith, H. (1996). The field margins project – management of arable field margins for sward diversity and weed control. In: MacDonald, D.W., Tattersall, H. (Eds.), *The WildCRU Review: The Tenth Anniversary Report of the Wildlife Conservation Research Unit at Oxford University*, 79-86.

Benton, T.G., Vickery, J.A., and Wilson, J.D. (2003). Farmland biodiversity: is habitat heterogeneity the key? *Trends in Ecology and Evolution*, Vol. 18 No. 4, 182-188.

Berendse, F., Elberse, W.T. and Geerts, R.H.M.E. (1992). Competition and nitrogen loss from plants in grassland ecosystems. *Ecology*, 73, 46-53.

Berg, A. and Gustafson, T. (2007). Meadow management and occurrence of corncrake *Crex crex*. *Agriculture, Ecosystems and Environment*, 120, 139-144.

Bokenstrand, A., Lagerlöf, J. and Redbo Torstensson, P. (2004). Establishment of vegetation in broadened field boundaries in agricultural landscapes. *Agriculture, Ecosystems and Environment*, 101, 21-29.

Bullock, J.M. and Marriott, C.A. (2000). Plant responses to grazing and opportunities for manipulation. In: Rook, A.J. and Penning, P.D. (Eds.) Grazing Management. *British Grassland Society Occasional Symposium* No. 34, 17-26.

Chamberlain, D.E., Fuller, R.J., Bunce, R.G.H., Duckworth, J.C. and Shrubb, M. (2000). Changes in the abundance of farmland birds in relation to the timing of agricultural intensification in England and Wales. *Journal of Applied Ecology*, 37, 771-788.

Clausen, H.D., Holbeck, H.B. and Redderson, J. (2001). Factors influencing abundance of butterflies and burnet moths in the uncultivated habitats of an organic farm in Denmark. *Biological Conservation*, 98, 167-178.

Cole, L.J., McCracken, D.I., Baker, L. and Parish, D. (2007). Grassland conservation headlands: Their impact on invertebrate assemblages in intensively managed grassland. *Agriculture, Ecosystems and Environment*, 122, 252-258

Cooper, A., McCann, T. and Meharg, M.J. (2003). Sampling Broad Habitat change to assess biodiversity conservation action in Northern Ireland. *Journal of Environmental Management,* 67, 283-290.

Cooper, M., and Morris, D.W. (1983). *Grass Farming*. Farming press Ltd. Suffolk, UK.

Crofts, A. and Jefferson, R.G. (1999). *The Lowland Grassland Management Handbook, 2nd Edn*. English Nature, Peterborough, UK.

Culleton, N., Barry, P. and Sheehy, J. (2002). Grass/Clover mixtures. *Principles of Successful Organic Farming*, Teagasc, Dublin, Ireland.

Desender, K. (1982). Ecological and faunal studies on Coleoptera in agricultural land. II Hibernation of Carabidae in agroecosystems. *Pedobiologia*, 23, 295-303.

Edwards, A.R., Mortimer, S.R., Lawson, C.S., Westbury, D.B., Harris, S.J., Woodcock, B.A. and Brown, V.A. (2007). Hay strewing, brush harvesting of seed and soil disturbance as tools for the enhancement of botanical diversity in grasslands. *Biological Conservation*, 134, 372-382.

Eliot, R.H. (1908). In Simpkin, Marshall and Hamilton (Eds.). *The Clifon Park System of Farming* (4th Edn.).

Environmental Protection Agency (EPA) (2007). Land Cover. http://www.epa.ie/environment/biodiversity/landcover. Viewed 04/12/07.

Foster, G.N., Blake, S., Downie, I.S., McCracken, D.I., Ribera, I., Eyre, M.D. and Garside, A. (1997). Biodiversity in agriculture. Beetles in adversity? *Biodiversity and Conservation in Agriculture.* BCPC Symposium Proceedings No. 69. British Crop Protection Council, 53-63.

Foster, L. (1988). Herbs in pastures. Development and research in Britain, 1850-1984. *Biological Agriculture and Horticulture*, 5, 97-133.

Frame, J. (2000). *Improved Grassland Management.* Farming Press.

Hooper, M.D. (1970). Hedges and birds. In: *Hedgerow Birds*, 3, 114-117.

Hopkins, A., Pywell, R.F., Peel, S., Johnson, R.H. and Bowling, P.J. (1999). Enhancement of botanical diversity of permanent grassland and impact on hay production in environmentally sensitive areas in the UK. *Grass and Forage Science*, 54, 163-173.

Huijsmans, J.F.M. (2003). *Manure Application and Ammonia Volatilisation.* Ph.D. Thesis, University of Wageningen, Netherlands.

Jefferson, R.G. and Usher, M.B. (1989). Seed rain dynamics in disused chalk quarries in the Yorkshire Wolds, England with special reference to nature conservation. *Biological Conservation,* 47, 123-136.

Jensen, A. (1985). The effect of cattle and sheep grazing on salt-marsh vegetation at Skallingen, Denmark. *Vegetatio*, 60, 37-48.

Kearns, C.A., Inouye, D.W. and Waser, N.M. (1998). Endangered Mutualisms: The conservation of plant-pollinator interactions. *Annual Review of Ecological Systems*, 29, 83-112.

Lewis, J. (1973). Longevity of crop and weed seed: survival after 20 years in the soil. *Weed Research*, 13, 179-191.

Lewis, G.C. and Hopkins, A. (2000). Weeds, pests and diseases of grassland. In: Hopkins, A., (Ed.) *Grass – Its Production and Utilization*, 3rd edition, British Grassland Society, Blackwell Science.

Longley, M. and Sotherton, N.W. (1997). Factors determining the effects of pesticides upon butterflies inhabiting arable farmland. *Agriculture, Ecosystems and Environment*, 61, 1-12.

McDonald, P., Edwards, R.A., Greenhalgh, J.F.D. and Morgan, C.A. (2002). *Animal Nutrition* (6th Edn.) Harlow, Prentice Hall.

MAFF, The Scottish Office Agriculture, Environment and Fisheries Department, Department of Agriculture for Northern Ireland and Welsh Office (1997). *The Digest of Agricultural Census Statistics*, UK. The Stationery Office, London.

Marshall, E.J.P., West, T.M., Kleijn, D. (2006). Impacts of an agri-environment field margin prescription on the flora and fauna of arable of arable farmland in different landscapes. *Agriculture, Ecosystems and Environment*, 113, 36-44.

Marshall, E.J.P. and Moonen, A.C. (2002). Field margins in northern Europe: their functions and interactions with agriculture. *Agriculture, Ecosystems and Environment*, 89, 5-21.

Montalvo, A.M. and Ellstrand, N. (2000). Transplantation of the subshrub *Lotus scoparius*: Testing the home-site hypothesis. *Conservation Biology*, 14, 1034-1045.

Mortimer, S.R., Hollier, J.A., and Browne, V.K. (1998). Interactions between plant and insect species-richness in the restoration of calcareous grasslands. *Applied Vegetation Science*, 1, 101-114.

Muller, S., Dutoit, T., Alard, D. and Grévilliot, F. (1998). Restoration and rehabilitation of species-rich grassland ecosystems in France: a review. *Restoration Ecology*, 6, 94-101.

Olff, H. and Ritchie, M.E. (1998). Effects of herbivores on grassland plant diversity. *Tree,* Vol. 3, No. 7, 261-265.

Opdam, P. (1990). Dispersal in fragmented populations: the key to survival. In R.G.H. Bunce and D.C. Howards (Eds.). *Species Dispersal in Agricultural Habitats*, 3-17.

Pfiffner, L. and Luka, H. (2000). Overwintering of arthropods in soils of arable fields and adjacent semi-natural habitats. *Agriculture, Ecosystems and Environment*, 78, 215-222.

Price, E.A.C. (2003). Lowland Grassland and Heathland Habitats. *Routledge*, 157-158.

Radosevich, S.R. and Holt, J.S. (1984). *Weed Ecology – Implications for Vegetation Management.* Wiley and Sons.

Rogers, P.A.M., Fleming, G.A. and Gately, T.F. (1989). *Mineral Imbalances and Mineral Supplementation in Cattle*. Teagasc, Dublin, Ireland.

Rushton, S.P., Luff, M.L. and Eyre, M.D. (1989). Effects of pasture improvement and management on the ground beetle and spider communities of upland grasslands. *Journal of Applied Ecology*, 26, 489-503.

Sheridan, H., Finn, J.A., Culleton, N. and O'Donovan, G. (2008). Plant and invertebrate diversity in grassland field margins. *Agriculture, Ecosystems and Environment*, 123, 225-232.

Sheridan, H. (2005). *The Experimental Restoration and Enhancement of Grassland Field Margins within the Irish Rural Environment Protection Scheme* (REPS). Unpublished Thesis, University College Dublin, Ireland.

Shipley B. and Keddy, P.A. (1994). Evaluating the evidence for competitive hierarchies in plant communities. *Oikos*, 69, 340-345.

Smith, B, M., Diaz, A., Winder, L. and Daniels, R. (2005). The effect of provenance on the establishment and performance of *Lotus corniculatus* L. in a re-creation environment. *Biological Conservation* 125, 37-46.

Stapledon, R.G. (1936). *The Land: Now and To-morrow*. Faber and Faber Ltd. London.

Tallowin, J.R.B., Rook, A.J. and Rutter, S.M. (2005). Impact of grazing management on biodiversity of grasslands. *Animal Science*, 81, 193-198.

Thomas, S.R., Noordhuis, R., Holland, J.M. and Goulson, D. (2002). Botanical diversity of beetle banks. Effects of age and comparison with conventional arable field margins in Southern UK. *Agriculture, Ecosystems and Environment*, 93, 403-412.

Thomas, M.B., Wratten, S.D. and Sotherton, N.W. (1991). Creation of 'island' habitats in farmland to manipulate populations of beneficial arthropods: predator densities and emigration. *Journal of Applied Ecology*, 28, 906-917.

Tsiouris, S. and Marshall, E.J.P. (1998). Observations on patterns of granular fertiliser deposition beside hedges and its likely effects on the botanical composition of field margins. *Annals of Applied Biology*, 132, 115-127.

Van der Putten, W.H., Mortimer, S.R., Hedlund, K., van Dijk, C., Brown, J., Leps, J., Rodriguez-Barrueco, C., Roy, J., Len, T.A.D., Gormsen, G.W., Korthals, S., Lavorel, I.S., Smilauer, R. and Smilauer, P. (2000). Plant species diversity as a driver of early succession in abandoned fields: a multi-site approach. *Oecologia*, 124, 91-99.

Vickery, J.A., Tallowin, J.R., Feber, R.E., Asteraki, E.J., Atkinson, P.W., Fuller, R.J. and Brown, V.K. (2001). The management of lowland neutral grasslands in Britain: effects of agricultural practices on birds and their food resources. *Journal of Applied Ecology*, 38, 647-664.

Watkinson, A.R. and Ormerod, S.J. (2001). Grasslands, grazing and biodiversity: editors' introduction. *Journal of Applied Ecology*, 38, 233-237.

Woodcock, B.A., Potts, S.G., Westbury, D.B., Ramsay, A.J., Lambert, M., Harris, S.J. and Brown, V.K. (2007). The importance of sward architectural complexity in structuring predatory and phytophagous invertebrate assemblages. *Ecological Entomology*, 32, 302-311.

Woodcock,B.A., Potts, S.G., Pilgrim, E., Ramsay, A.J., Tscheulin, T., Parkinson, A., Smith, R.E.N., Gundrey, A.L., Brown, V.K. and Tallowin, J.R. (2007). The potential of grass field margin management for enhancing beetle diversity in intensive livestock farms. *Journal of Applied Ecology*, 44, 60-69.

In: Grasslands: Ecology, Management and Restoration
Editor: Hans G. Schröder
ISBN 978-1-60692-023-7

Expert Commentary B

THE GLOBAL EXTENT OF BLACK C IN SOILS: IS IT EVERYWHERE?

Evelyn Krull, Johannes Lehmann*[1]*, Jan Skjemstad, Jeff Baldock*[2] *and Leonie Spouncer

[1]Ph.D. 1997 University of Bayreuth, Germany
Diplom Geoecologist 1993 University of Bayreuth
[2]CSIRO Land and Water Adelaide Laboratory
PMB 2, Glen Osmond SA 5064

THE ROLE OF BLACK CARBON IN GLOBAL CLIMATE MODELS

The latest projections of the Intergovernmental Panel on Climate Change (IPCC) estimate a 3°C increase in global temperatures within the next 100 years (IPCC 4th Assessment Report, 2007), and global warming is seen as a major driver in accelerating decomposition of soil organic matter, resulting in increased production of CO_2 (e.g. Davidson and Janssens, 2006). The 2007 IPCC report recommended coupling models of terrestrial biogeochemical and atmospheric and oceanic processes in order to improve general circulation models and to recognize the quantitative value of soil organic carbon (SOC) in the global carbon cycle.

Estimates of CO_2 emissions from soil rely on predictions of the response of different SOC pools to global warming and correct estimation of the size of these pools. In comparison to the pool representing the most stable and biologically unreactive fraction, commonly referred to as passive or inert organic carbon (IOC), the decomposition of labile C is expected to be faster as a response to temperature increase. IOC is a fraction of the SOC pool that is not readily available for microbial decomposition and has turnover times exceeding 100 years (e.g. Krull et al., 2003).

Black C (BC) is usually considered the most abundant form of IOC and is defined as the 'carbonaceous residue of incomplete combustion of biomass and fossil fuels' (Schmidt and Noack, 2000). BC is important to several biogeochemical processes; for example, BC potentially modifies climate by acting as a potential carbon sink for greenhouse gases (Kuhlbusch, 1998) and leads to increasing solar reflectance of the Earth's atmosphere, but also to a heating of the atmosphere (Crutzen and Andreae, 1990). BC production from fossil fuel combustion contributes to aerosol C, decreasing surface albedo and solar radiation (IPCC 4th Assessment Report). Due to its condensed aromatic structure, BC has a low biochemical reactivity. ^{14}C ages of BC in soils vary between 1160 and 5040 years (e.g. Schmidt et al. 2002).

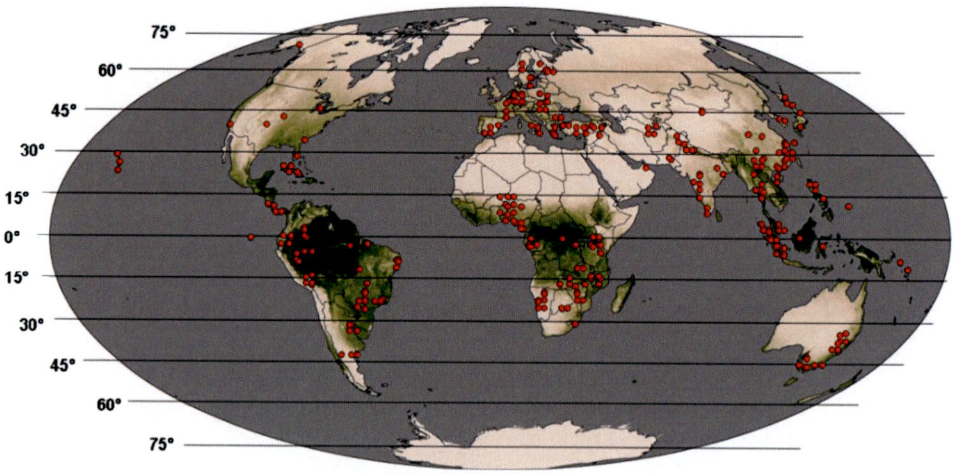

Figure 1. Distribution of soilsused for the prediction of BC in surface soils and global NPP.

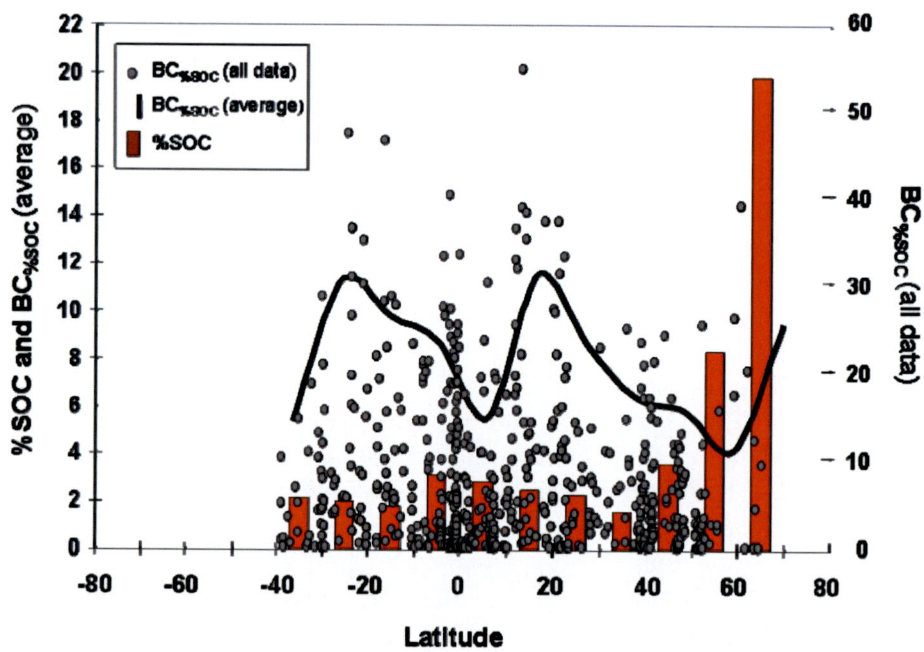

Figure 2. Distribution of $BC_{\%SOC}$ and %SOC in surfaces soils across latitudes.

Baldock (2007) reported that BC constitutes up to 60% of SOC, indicating that BC can make up a significant part of SOC affecting the response of SOC to temperature changes and the overall turnover time of SOC. Thus, the effect of such a large and unreactive C pool must be effectively integrated into global C cycle and climate models. However, the latest IPCC report regards BC only as an aerosol and not as part of SOC, despite an earlier recommendation to the IPCC to "... better gauge the influence of BC on the global carbon cycle" as the result would ensure a "more accurate global black C budget and a better understanding of the role of BC as a potential sink in the global C cycle." (2006 IPCC guidelines for national greenhouse gas inventories; appendix 1). The lack of incorporating soil BC in global climate change assessments may be largely due to the lack of global databases on BC in soils and sediments and the understanding of factors and processes that may influence BC contents in soils (climate, soil texture, primary productivity and fire abundance).

In order to provide reliable projections of future CO_2 emissions from soils, due to global warming, it is important to consider the global distribution of soil BC. Through an initial assessment of the World Soil Archive (http://library.wur.nl/isric/) we demonstrate the variability and trends of global soil BC distribution between different climates and soil types and discuss the implications of this chemically recalcitrant form of C on the global C cycle. By doing this, we also demonstrate that current methods exist to routinely analyse BC and in the future to develop a global BC map.

Are there Routine Methods to Generate a Global BC Map?

We have developed a rapid fourier-transform infrared-based technique that provides predictive capability for major soil properties, including BC content. BC was quantified by a novel mid-infrared (MIR) method coupled with partial least squares (PLS) (Janik et al., 2007) that allowed rapid analysis of a large number of samples, not feasible by other published methods (Hammes et al., 2007). This method was originally developed and calibrated on Australian soils (Janik et al., 2007); however, subsequently, we were able to verify that the predictive capacity of the MIR technique for soils from other parts of the world was robust in most cases.

BC in World Soils: What Drives BC Production and Decomposition?

Our initial assessment took advantage of the large collection of soils that constitute part of the ISRIC – world soil information database (http://library.wur.nl/isric/). We obtained over 400 samples that span all major climate zones and soil types across most parts of the world (Fig. 1). Utilising the MIR/PLS technique, we developed a database that illustrates for the first time the predicted proportion of BC in this large soils dataset. We hope that this initial dataset may become an incentive for the rigorous establishment of a global BC map, data from which may then be incorporated into future IPCC reports and C cycle models.

Figure 2 shows the proportion of BC (as % of total SOC: $BC_{\%SOC}$) as well as total SOC contents across northern and southern hemispheres. These data show that BC in surface soils (A and A/B horizons) is ubiquitous in large parts of the world and occurs in the majority of the sampled locations. The variability was large, resulting in $BC_{\%SOC}$ varying from over 50%

to almost 0%. The soils richest in BC occur in latitudes of 20-30° in both hemispheres, situated mostly in central and South America and southern parts of Africa. These areas correspond to tropical climates with a pronounced dry season in winter ('Aw' in the Koeppen classification). In higher latitudes in the northern hemisphere (60-70°), BC contents were also high and were associated with an increase in total SOC (Fig. 2). These areas are classified as humid-temperate, constantly moist climates (Cf in Koeppen classification). Higher SOC content at high latitudes results from cool climates and low decomposition rates, which promotes accumulation of organic matter.

Despite the high spatial variability of $BC_{\%SOC}$, soils with a high proportion of BC fell largely in the soil group of Vertisols. Vertisols occur worldwide in seasonal climate zones and in lower relief positions (Richardson and Vepraskas, 2000). Such areas of deposition would favour accumulation of BC through erosion and deposition. Another factor contributing to the high amounts of $BC_{\%SOC}$ in Vertisols could be the comparably high proportion of clay (known to stabilize soil organic matter) and the type of clay (smectitic). The MIR-predicted data showed that Vertisols were amongst the soil types with the highest clay contents (average 44%). High clay content may promote retention of fine BC. However, clay alone is not a determining factor for high $BC_{\%SOC}$. Oxisols and Ultisols, having the highest clay content (from MIR prediction) of all analysed soil types (45%), had the lowest $BC_{\%SOC}$ values. Oxisols, and to a lesser degree Ultisols, are mostly found in high-rainfall tropical climates, lacking seasonal rainfall (Af in Koeppen classification) and having a lower fire activity. While fire activity and BC content are both high in the southern hemisphere this relationship is not consistent and does not concur with findings in parts of the northern hemisphere where fire activity is lower, yet $BC_{\%SOC}$ is still high (Carmona-Moreno et al., 2005).

Thus, a combination of factors, such as climate (the necessity of a pronounced dry season to ensure fire occurrence), position in the landscape (areas of accumulation) and mineralogy (high amount of expansive clays) may be instrumental in promoting the formation and/or retention of $BC_{\%SOC}$ in soils. As illustrated in Figure 1, net primary productivity alone did not appear to be a significant factor globally as it is highest in equatorial regions where $BC_{\%SOC}$ was lowest.

While we show here that BC contents in surface soils are often significant, BC can also accumulate deeper in the soil. In addition, soil erosion and transport of BC may result in significant accumulation of BC in rivers, estuaries and off-shore sediments (Krull et al., 2006).Thus, the inclusion of BC in global climate models will require a thorough assessment of BC contents not only in surface but in deeper soil horizons as well as aquatic sediments. The analyses conducted in this study indicate that methods exist to accomplish BC measurements for large areas. However, the high variability of the data indicates that broad empirical measurements and extrapolation over large areas are not sufficient for the aim of producing a global BC map. Thus, the next steps in a comprehensive assessment of global BC stocks and distribution have to include a detailed and consistent sampling format as well as a thorough assessment of the processes that control sources and sinks of BC.

ACKNOWLEDGMENTS

We thank Janine McGowan for assistance in the analyses of the samples and both her and Ryan Farquharson for review of an earlier version of the manuscript. We thank CSIRO Land and Water for supporting the analyses of this work and ISRIC for access to samples from the World Soils Database.

REFERENCES

Baldock, J.A. Composition and cycling of organic carbon in soils (2007), *in* Marschner, P. and Rengel, Z. (eds.), *Nutrient Cycling in Terrestrial Ecosystems.* Springer, Berlin.

Carmona-Moreno, C., Belward, A., Malingreau, J.P., Hartley, A., Garcia-Alegre, M., Antonovskiy, M., Buchshtaber, V., Pivovarov, V. (2005), Characterizing Interannual Variations in Global Fire Calendar Using Data From Earth Observing Satellites, *Global Change Biology 11*, 1537-1555.

Crutzen, P.J. and Andreae, M.O. (1990), Biomass burning in the tropics: impact on atmospheric chemistry and biogeochemical cycles, *Science 250*, 1669-1678.

Davidson, E.A. and Janssens, I.A. (2006), Temperature sensitivity of soil carbon decomposition and feedbacks to climate change, *Nature 440*, 165-173.

Hammes, K. *et al.* Comparison of quantification methods to measure fire-derived (black/elemental) carbon in soils and sediments using reference materials from soil, water, sediment and the atmosphere. *Global Biogeochemical Cycles* **21**, GB3016, doi:10.1029/2006GB002914 (2007).

IPCC 4th Assessment report (2007), Working Group I: The physical science basis of climate change, http://ipcc-wg1.ucar.edu/wg1/wg1-report.html

2006 IPCC Guidelines for National Greenhouse Gas Inventories, vol. 4, Agriculture, Forestry and other Land use, Appendix 1, http://www.ipcc-nggip.iges.or.jp/public/2006gl/vol4.htm

Janik, L.J., Skjemstad, J.O., Shepherd, K.D., and Spouncer, L.R. (2007) The Prediction of Soil Carbon Fractions Using Mid-Infrared-Partial Least Square Analysis. *Australian Journal of Soil Research 45*, 73-81.

Krull, E.S., Baldock, J.A., and Skjemstad, J.O. (2003), Importance of Mechanisms and Processes of the Stabilisation of Soil Organic Matter for Modelling Carbon Turnover, *Functional Plant Biology 30*, 207-222.

Kuhlbusch, T.A.J. (1998), Black carbon and the carbon cycle. *Science 280,* 1903-1904.

Richardson, J.L. and Vepraskas, M.J. (2000), *Wetland Soils: Genesis, Hydrology, Landscapes, and Classification.* CRC Press, Florida.

Schmidt, M.W.I. and Noack, A.G. (2000), Black carbon in soils and sediments: Analysis, distribution, implications, and current challenges, *Global Biogeochemical Cycles 14*, 777-793.

Schmidt, M.W.I., Skjemstad, J.O., and Jäger, C. (2002), Carbon isotope geochemistry and nanomorphology of soil black carbon: Black chernozemic soils in central Europe originate from ancient biomass burning, *Global Biogeochemical Cycles 16*, 1123 doi:10.1029/2002GB001939.

In: Grasslands: Ecology, Management and Restoration
Editor: Hans G. Schröder
ISBN 978-1-60692-023-7

Short Communications

SUSTAINABLE DEVELOPMENT OF PRATACULTURE AND ECOLOGICAL REHABILITATION VIA SHIFT FROM GRAIN PRODUCTION TO VEGETATIVE PRODUCTION IN SOUTH CHINA

Guodong Liu and Yuncong Li
Tropical Research and Education Center
University of Florida, FL 33031, USA

ABSTRACT

Vegetative production is defined as a coupling between crop plant production as the primary product and livestock production as the endproduct in farmlands. The primary product is forage rather than grains that are often the traditional agricultural endproduct. Accordingly, vegetative production requirements for climatic resources are different from traditional grain production. In this study, climatic resources were compared between the north and south China. Traditional (grain) and vegetative (shoot) productions have many production aspects in common and also have differences in physiology, cost inputs, utilization efficiency of solar energy, and vegetative and grain yields. Advantages of vegetative production compared to traditional grain production are presented. Based on the differences of the requirements of irradiance between vegetative and traditional grain productions, enhanced development of vegetative production was proposed through utilization of existing grasslands and winter fallow croplands in south China. This new strategy promises to reduce runoff and loss of topsoil, decrease dust clouds, and restore a good portion of damaged ecosystems in China.

Keywords: vegetative production, traditional grain production, growth curve of crop plants, prataculture, winter fallow croplands

INTRODUCTION

Due to China's rapid economic growth, many interested parties are concerned about environmental sustainability and persistence of China's agriculture in light of the many demands placed upon it. Herein, we discuss sustainability of prataculture and ecosystems in south China. Prataculture generally includes golf course, pasture, range, city grasslands, forage, and livestock productions. This chapter will focus on forage and livestock production systems because this type of prataculture is still relatively small, and has not received much attention yet in the country.

Traditionally, herbivore productions are located in north China. For example, in early 2000s, Hunan Province, a typical southern province, had produced only 130 mL milk per capita per year (Ye et al., 2002). The milk demand requires importation from other provinces such as Inner Mongolia or even overseas. Accordingly, the rural communities do not have much experience in consumption of milk or other herbivore products. The southern provinces in the country have traditionally produced grains such as rice (*Oryza sativa* L.) and wheat (*Triticum aestivum*, L.). However, south China has climatic advantage compared to northern China with respect to bovine production because of southern China's abundant water and heat resources. In recent decades, central and local governments have attempted to regulate and optimize the structure of agricultural industries and diversify food components. As the economy develops, the country has also had to address challenges that threaten its important ecosystems.

To achieve agricultural diversification and adequate food supply in south China, strategic and technical assistance are required at the local level for both government agencies and producers. If prataculture is to continue to be developed, then ecosystem management must also develop to offset unsustainable practices. Sustainability for southern China must include both agricultural improvements and environmental protection.

COMPARISON WITH CLIMATIC CONDITIONS BETWEEN THE NORTH AND SOUTH IN CHINA

China covers the latitudes from about 18 to more than 53 °N. North China is the land area north of the Yellow River. South China is defined as those regions south of the Yangtze River. The area around Huai River and between Yellow River and Yangtze River is called as central China. Simply based upon this large change in latitude, there are differences in climatic conditions between north and south China (Table 1) (Chinese Academy of Meteorology and Nanjing Institute of Meteorology, 1981)˙ Obviously, the south has great advantages in rainfall and heat resources, but disadvantages in light resources compared to the north (Table 1).

Urumqi and Beijing (Figure 1) are both northern cities; whereas, Fuzhou and Guiyang both southern cities. Annual sunshine hours in the northern cities (Figure 1) are approximately two to three times as much as those in the southern cities. The sunshine hours in Urumqi are actually 182% greater than that in Guiyang.

Therefore, Dr. Ying Ding (1888-1964), the first President (1957-1964) of the Chinese Academy of Agricultural Sciences, noted that northern China, particularly the northwest, had a great advantage compared to other parts of the country for producing grains (Dr. Gengling Liu, personal communication, 1995). Ding's insight is certainly correct because reproductive growth (grain production) requirements for climatic resources are remarkably different from vegetative growth as the following section will show.

Table 1. Comparison between agricultural meteorological resources in the north of Yellow River and south of Yangtze River

Agricultural meteorological resources	In the north of Yellow River	In the south of Yangtze River
Total annual radiation (kcal/cm^2)	130-160	80-110
Total annual sunshine hours (hr)	2400-3200	1200-2200
Percentage of sunshine (%)	60-70	30-50
Accumulative temperature above 0°C	2500-5000	5500-8000
Frost-free period (day)	100-200	250-365
Annual rainfall (mm)	100-1000	1200-2400

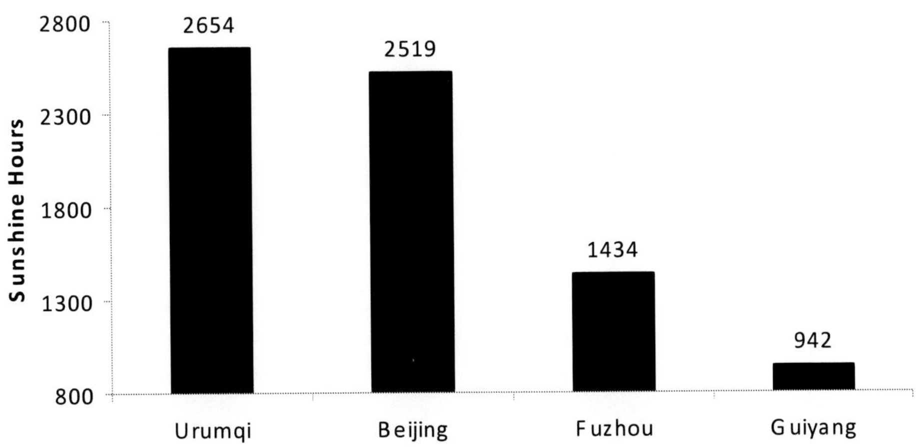

Figure 1. The differences in annual sunshine hours among four cities in China . Urumqi: the capital of Xingjiang Province, a northwestern city; Beijing: the capital of the country, a northern city; Fuzhou: the capital of Fujian Province, a southern city; Guiyang: the capital of Guizhou Province, a southwestern city.

Comparison with Requirements between Traditional and Vegetative Productions

Generally speaking, cereal crops prefer sunny weather during their floral induction, even though there are three essentials for reproductive growth including obligate vegetative growth, specific photoperiodism, and thermoperiodism i.e. vernalization (Gardener et al., 1985). There is a yield loss if there are not enough sunshine hours during floral induction and initiation. For example, winter wheat lost 37%, 60%, and 99% grain yield for 30%, 60%, and

90% shadow during floral development, respectively (Liu, 2000). Lang (1965) reported that floral induction had as many as eight processes including: 1) the high-intensive-light process, 2)pigment conversion, 3) the time-measuring process, 4) florigen synthesis, 5) florigen stabilization, a post-inductive process preparatory to florigen translocation, 7) florigen translocation and 8) floral differentiation. Fewer sunshine hours result in lower grain yield because four of the eight processes of floral induction require intensive sunshine (Lang, 1965). Actually, the first phase for floral induction can NOT start without intensive sunshine. Phases 5, 7, and 8 also require light for proper grain production. These processes need at least three consecutive sunny days. Otherwise, the plant reverts to the previous physiological status. This reversion if light requirements are not met is one of the main reasons why grain yield of winter wheat is only 1500 kg per hectare in Guizhou, Hunan, and other southern provinces in China. Availability of sunlight hours from October to April in the provinces is inadequate for proper grain production.

COMPARISON WITH NUTRIENT YIELDS BETWEEN TRADITIONAL AND VEGETATIVE PRODUCTIONS

Jiang and his coworkers found that protein yield of vegetative production was 420% greater than that of grain production (Jiang et al., 1994), given the same growth and climatic conditions. Similarly, dry matter, fat yield, and mineral gain concentrations such as phosphorus and calcium were much greater for vegetative production than for grain production (Table 2). This large difference can be attributed to weather from late autumn to spring favors vegetative growth in south China where sunshine hours from October to April is only 358 hours for Guiyang and 700 hours for Fuzhou, respectively. Obviously, there is great advantage in vegetative production compare to grain production given this climate.

Table 2. Comparison in harvest of traditional (barley, *Hordeum vulgare* L., grain) and vegetative (ryegrass, *Lolium multiflorum,* Lam., shoot system) agricultures

Organ of harvest	Ryegrass (shoot)	Barley (grain)	Ryegrass/Barley (times)
Times of harvesting	2 - 3	1	2.5
Protein yield (kg/ha)	1755.8	334.5	5.2
Fat yield (kg/ha)	460.5	93.0	5.0
Carbohydrate (kg/ha)	5409.0	1947.0	2.8
Calcium (kg/ha)	83.3	5.3	15.9
Phosphorus (kg/ha)	30.6	6.9	4.4
Yield (kg/ha)	7650.0	3000.0	2.55

Growth Curve (Sigmoid Curve) and Its Role in Forage and Livestock Production

The sigmoid curve is an S-shaped curve (Gardener et al., 1985), which can be used to describe the life-cycle of products, organizations, empires, and even human relationships. As the curve predicts, nearly all of life's endeavors start slowly. The growth curve progresses to an exponential stage before rising to its maximum, after which there is decline (Handy, 1994). The sigmoid curve can be divided into three typical steps: slow-rapid-slow growth stages, or early-middle-late growth stages. The middle growth stage is also called the grand growth period. The time frame may vary from minutes to years depending on the selected organism or organ, but the S-shaped accumulation pattern (Figure 2) typifies every single individual including human beings, every single organ, and even every single cell (Gardener et al., 1985). The growth curve provides us an insight and opportunity to set up our production strategy and planning. For grain production, there is only one grand growth period with early and late growth stages. However, for vegetative production, there can be more than one grand growth period. For example, vegetative growth may contain a short early growth stage with or without a late growth stage.

Vegetative production has two objectives: to produce forage for herbivore consumption; and to produce livestock with its attendant products. During livestock production, the herbivorous animals convert plant crude protein and other constituents into livestock products such as meats and woolen goods. Accordingly, the late plant growth stage should be avoided because lignin (Table 3) is formed during this stage, reducing conversion efficiency from forage to livestock products. During the grand growth period, plants may have more than 7-fold greater photosynthesis efficiency (Jiang et al., 1994). Greater efficiency translates into better and more forage with the potential for conversion into additional livestock products. Based upon the literature, forage crops should be grazed or cut just before or at the flowering stage (Table 3) to avoid the late growth stage and reduce or control lignin content in the forage. Accordingly, one or two more grand growth periods can be gained in the growth season (Table 2) because grain production had only one grand growth period but vegetative production could be harvested two to three times. Every harvest could have one grand growth period. Every increase in number of harvest times could increase one grand growth period for forage production.

Vegetative production is a coupled system linking plant and livestock production. The S-shaped curve is also very important in livestock production for efficient bioconversion from forage to livestock products because the growth of herbivore may also follow a similar function. For example, research results conducted in Inner Mongolia showed that lamb yield per sheep was 13.8, 21.8, 25.8, and 27.6 kg for slaughtering at ages of 10, 24, 48, and 60 months, respectively. The yield at age of 60 months was just as much as two-fold that at age of 10 months. The corresponding feed consumption for the 60 months was as much as 11.2 times that of the 10 months. This observation indicated that the efficiency of the feed conversion of the lambs at 10 months was 4.6 times greater than that of lambs at 60 months. Thus, butchery of 10-month-old lambs can maximize economic return, assuming demand for products from the younger lambs is equivalent to that of 60-month lambs. However, local herdsmen desire to have the maximum lamb yield per sheep because they are not familiar with the S-shaped growth curve; and traditionally the maximum weight yield provides the great feeling of success. In this case, traditional practices led to uneconomical returns, despite the apparent belief that the herdsmen were successfully producing lamb products.

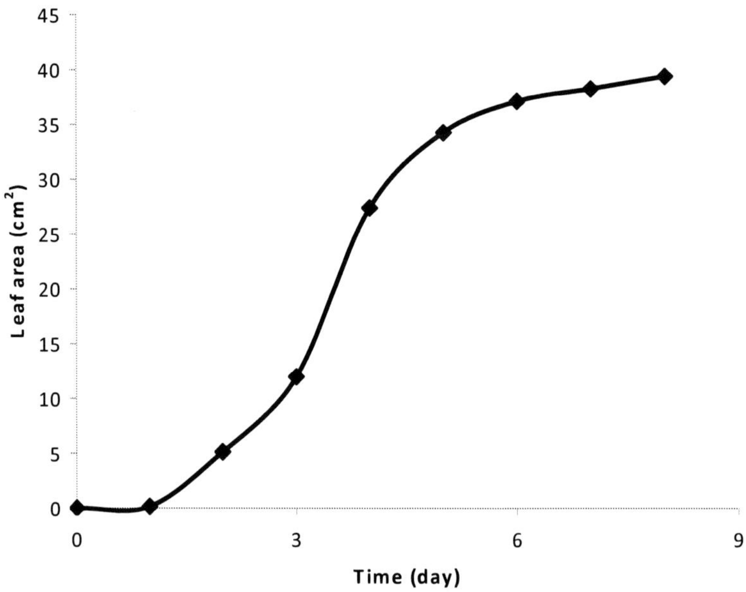

Figure 2. Growth curve of leaves of pinto bean (Ting, 1982).

Table 3. Changes in components (%) of ryegrass (*Lolium* L.) at different stages (Li and Li, 1994)

	Crude protein	Crude fat	Mineral	Nitrogen-free extract	Crude fiber	Lignin
Seedling stage	18.6	3.8	8.1	48.3	21.2	3.6
Before flowering	15.3	3.1	8.5	48.3	24.8	4.6
Flowering	13.8	3.0	7.8	49.6	25.8	5.5
Filling stage	9.7	2.5	5.7	50.9	31.2	7.5

VEGETATIVE PRODUCTION AND ECO-REHABILITATION IN SOUTH CHINA

In the Yangtze River Watershed, much of the grasslands and farmlands are on sloped landscape. Due to a traditional strategy favoring grain production, a great proportion of the sloped grasslands were converted to grain cereal production. For example, vegetation cover in the upper reach of Yangtze River Watershed was about 40% in early 1950s but has decreased to 10% because of reclamation of sloped grasslands. Actually, sloped farmlands have been increased by 50% since 1950s. The total area of sloped farmlands covers more than 8 million hectares. In some regions, the sloped grasslands of 60 degrees were converted to cereal or tuber production (e.g., sweet potato, *Ipomoea batatas*, L.). As an agriculture consultant of the International Foundation of Agricultural Development of the United Nations, the first author visited Huangdao Village, Huangdao Township, Wanshan District, Guizhou Province, in early 1997; and noticed that there many denuded mountains with newly tilled soils planted to cereal crops. This practice of allowing soil on steep slopes to be exposed during the initial phases of cereal production resulted in considerable, damaging soil erosion. Subsequently, cereal production continued to decline as soil was eroded. For example, in 1957, there were 26.4 square kilometers of soil erosion areas that accounted for slightly more than 20% of the total watershed area. The percentage of soil erosion areas was, however, increased to 31.5% in 1990s. The total erosion areas were 57 square kilometers (Cai, 2006).

Frequently tillage of the sloped grasslands (so-called sod busting) results in desertification, especially in karst areas in Guizhou and Guangxi Provinces. In Guizhou, 12.8% of the total area has undergone desertification (Cai, 2006). Clearly, control of soil erosion and rehabilitation of ecosystems in south China are challenges that must be addressed quickly.

Tillage of sloped farmlands, especially slopes greater than 10%, should be avoided. Cereal and tuber production on steeply sloped land is not sustainable due to rapid soil loss. Conversely, tillage can be avoided on the steeper slopes by using the land for vegetative production, which also favors water conservation and a stabilized ecological system. On the other hand, since the southern China terrain contains considerable sloping land, producing vegetative growth would decrease the need for equipment. Likely, most of the cereal harvesting is done by hands, but as the demand for meat and other animal products, mechanization will become popular. The steeper the slope the more problems mechanization will present, especially grain-related equipment. Hay bailing is much simpler on slopes up to 30%. Erosion from bovine traffic on slopes greater than 20% can be severe. Thus, grazing should not be used on these slopes and produced green material should be cut and transported to herbivore instead. Small, short terracing of slopes in the uplands creates a base flow that can increase the number and volume of harvested green matter on the lower slopes. The hydrology of the slope can be used to great benefit. This concept used in Honduras by a German ex-pat (Dr. Edward Hanlon, personal communication, 2008). All the slopes around his property were barren due to erosion of the top soil, but he had done a little terracing each year, starting at the top of the slope, near the summit. He had a year-round supply of water and all his land was highly productive. Once again, traditional strategy that focuses on cereal and/or tuber production must be rethought. Application of non-till techniques and perennial forage crops coupling with livestock production should be prioritized in south China.

Additionally, due to the climatic and landscape diversities in China, some regions are suited for multiple cropping scenarios, allowing the harvesting of grain several times during the applicable growing seasons, assuming suitable crop rotations. In those regions with climatic resources for multiple cropping, fall farmlands have great potential to develop vegetative production to alleviate the shortage of forage during the winter. For example, in cotton areas along Huai River cool season vegetative crops could be planted from October to April. Using planting of this sort in cool seasons may increase 91.2% net income per unit farmland (Jiang et al., 1994), for example ryegrass following cotton. In Guangdong Province, Yang (1994) showed that vegetative crops after harvesting late rice could produce 60,000 kg fresh forage with more than 2250 kg crude protein per hectare. In another example, ryegrass cultivation in winter/fall paddy fields could increase paddy yield by 14% and 7% for early rice and late rice in Guangdong, respectively. Furthermore, development of vegetative production enhances Carbon Sequestration in farmlands, which helps to control global warming.

Conclusions

1. Frequently plowing sloped farmlands, originally grasslands, for cereal production results in considerable soil erosion and resulting serious challenges in ecology and environment in southern China
2. Popularization of perennial forage crops and no-till techniques on sloped farmlands are able to diversify food supplies, and favor healthy development of both forage and livestock production
3. Applications known production functions to both forage and livestock production systems results in improved efficiencies of limited resources and food productivity
4. Integrated use of the above strategies and techniques will favor harmonization of food production and rehabilitation of ecosystems in the region
5. Implementation of these improved practices requires government intervention and farmer education. Without these two components, China will continue to lose its most valuable resource, the soil

Acknowledgements

The authors have appreciated Dr. and Professor Edward Hanlon at the University of Florida and Dr. and Professor Qibing Wang at the Chinese Academy of Sciences for their review of this manuscript and for their valuable suggestions and comments.

References

Cai, Qihua. 2006. Healthy Yangtze River. ISBN: 7-80708-000-0/TV.1 Yangtze River Press, Wuhan, China. Online available in Chinese on June 20, 2007 and verified on April 22, 2008 at http://www.cjw.gov.cn/index/read/detail/20070618/89828.asp.

Chinese Academy of Meteorology and Nanjing Institute of Meteorology. 1981. Agricultural Climatic Resources and Cropping Planning in China. *China Press for Agriculture*. China Agriculture Publishing House 1-54. Beijing, China.

Gardner, Franklin P., Robert Brent Pearce and Roger L. Mitchell. 1985. *Physiology of Crop Plants*. Pp. 187-208. The Iowa State University Press. Ames, Iowa.

Handy, Charles. 1994. *The Empty Raincoat,* London: Random House.

Jiang, Yuming, Hanqing, Ying and Yukun Luo. 1994. New strategy to raise more people in limited farmlands. Sustainable Development of Agriculture in the Area with Red and Yellow Soils in South China (Volume II). *The Research Group for Sustainable Development of Agriculture in the Area with Red and Yellow Soils in South China* (ed.). China Agriculture Publishing House, 58-67. Beijing, China.

Lang, A. 1965. Physiology of flower initiation. pp. 1380-1536. In: Ruhland, W (ed.),. *Encyclopedia of plant physiology*, vol. XV, part 1, Springer Verlag, Berlin.

Li, Fuxin and Xipei Li. 1994. *Encyclopedia of Compound Feed*. p. 295. Ocean University of China Press. Qingdao, China.

Liu, Guodong. 2000. Development of vegetative industry and adjustment of agricultural structure in China. *China Population, Resources and Environments.* 10(1): 71-75.

Liu, Guodong, and Gengling Liu. 1999. Vegetative agriculture and sustainable development of prataculture in the South China. *Acta Prataculturae Sinica*. 8(2): 1-7.

Ren, Jizhou. 1998. Development of Vegetative Industry is an Important Way to escape poverty towards well-off conditions in Mountainous Areas in China. *Proceedings of Annual Meeting of Chinese Academy of Engineering 1998*. 1-5.

Ting, Irwin. P. 1982, *Plant Physiology*. P. 464. Academic Press, New York.

Yang, Zhongyi, Yulin Yu and Huizhi Chen. 1994. Rotation of Ryegrass and Rice: I. Productivity of Italian Introduced Ryegrass under Intensive Farming in South Subtropical Area in China. *Journal of Prataculture*. 3(4): 20-26.

Ye, Zhihuan, Guodong Liu, and Yongzhong Qian. 2002. *Application Prospective and Technical Strategy for Efficient Utilization of Agricultural Resources*. Pp. 63-115. Science Press, Beijing, China.

Yun, Xujiang. 1996. Exploration of Fallow Farmlands and Enhancement of Forage and Herbivore Productions. *Prataculture Science*. 13(3): 1-3.

In: Grasslands: Ecology, Management and Restoration
Editor: Hans G. Schröder
ISBN 978-1-60692-023-7

Short Communications

THE REAL PATTERN IN THE SPECIES-AREA RELATIONSHIP: CASE STUDY USING GROUND BEETLES

***Tibor Magura*[1], *Viktor Ködöböcz*[1] *and Béla Tóthmérész*[2]**
[1]Hortobágy National Park Directorate, H-4002 Debrecen, POB. 216., Hungary
[2]Department of Ecology, University of Debrecen, H-4010 Debrecen, POB. 71., Hungary

ABSTRACT

A central issue in conservation biology and nature management is whether or not characteristic species of a given habitat type could be preserved by fragmented habitat patches or not. The classical theory of island biogeography predicts that the number of species supported by an island increases with the area of the island. However, there is a significant difference between real and habitat islands. In real islands, the surrounding habitat (ocean, sea, lake, river etc.) is usually inhospitable to organisms occurring on islands. In the case of habitat islands, the bordering habitat (the matrix) is usually less hostile. Consequently, species richness of real islands is not influenced notably by the surrounding habitat. This difference is increasingly emphasized when studying the predictions of island biogeography theory on habitat islands. Clear distinction should be drawn between specialist species that truly perceive the habitat patches as islands and are unable to survive in the surrounding matrix, and those species that occur in both the habitat patch and the matrix (generalist species).

In this case study, we demonstrated that depending on the ratio of specialist and generalist species in an assemblage, the species-area relationship may be positive or negative. Ground beetles (*Coleoptera: Carabidae*) of sandy grassland patches were studied in Eastern Hungary (Central Europe). The total number of ground beetle species correlated negatively with grassland area. Based on this result, one can draw the (seriously false) conclusion that it is sufficient to conserve small patches because they support most species. This negative relationship was due to the increasing ratio of generalist species with decreasing patch size. Analyzing the habitat specialist species (open-habitat species associated with sandy soils), the significant negative relationship turned over, and became significantly positive; i.e. the ratio of habitat specialist species increased with patch size, as predicted by the theory of island biogeography.

INTRODUCTION

As a consequence of the increasing agricultural intensification, the area of natural and semi-natural grasslands declined worldwide (Reidsma et al., 2006). This decline created a network of isolated and fragmented grassland habitats. Habitat fragmentation has two main components. First, the total area of the habitat sustaining populations decreases. Secondly, these habitats tend to be more isolated (Saunders et al., 1991). Less mobile arthropod species, like ground beetles (*Coleoptera: Carabidae*), are especially sensitive to habitat loss and isolation (Samways, 2005; Lövei et al., 2006). A central issue in conservation biology and nature management is whether characteristic species of a given habitat type could be preserved by fragmented habitat patches or not. The classical theory of island biogeography predicts that the number of species supported by an island increases with the area of the island (MacArthur and Wilson, 1967). Recently, however, several papers refined the oversimplified original assumptions of the island biogeography theory concerning habitat island, which emphasizes the effects of surrounding habitats on the species richness (Kupfer et al., 2006; Lövei et al., 2006, Magura and Ködöböcz, 2007).

In the present case study evaluating ground beetles in sandy grasslands, we demonstrated that depending on the ratio of specialist and generalist species in an assemblage, the species-area relationship may be positive or negative.

METHODS

Study Area

Eight patches of formerly contagious sandy grasslands located in the Nyírség region (the Great Hungarian Plain, Eastern Hungary) were studied. In the 19th century, this region was covered by natural habitats (marshes, fen meadows, mires, sandy grasslands and sandy oak woods). During the 20th century, as a consequence of the intensification of farming and forestry, these habitats were abolished or became highly fragmented. Today, the fragmented sandy grassland patches are surrounded by arable lands and non-native tree plantations. Recently, the studied grassland patches (prevalent vegetation association was *Potentillo arenariae-Festucetum pseudovinae*) have been lightly grazed with cows and sheep (cattle density was less than 0.25 heads/ha). The matrix habitats surrounded by these patches were similar: non-native deciduous tree plantations (black locust and ennobled poplar species) and croplands (maize and corn). The area of the studied patches varied between 2.3–353.5 ha and the distance between the patches was at least 2 km.

Sampling Design

Ground beetles of the eight sandy grasslands were collected during three years (2001-2003) using unbaited pitfall traps, consisting of plastic cups (diameter 100mm, volume 500ml) with 70% ethylene glycol as a killing and preserving solution. There were 10 traps, scattered randomly within the individual patches (at least 100m from the grassland edges).

Traps were checked fortnightly from the end of March to the end of October in every year. Ground beetles were identified to species using the keys of Hůrka (1996). To ensure a more complete species inventory in the studied grassland patches, beetles caught from the three trapping years were pooled.

Data Evaluation

Based on the literature data (Hůrka, 1996), considering local conditions, collected ground beetle species were divided into two ecological groups according to their habitat preference: habitat specialist species (open-habitat species associated with sandy soils) and generalist species (species occurring likewise in both the closed and open canopy habitats and are not associated with sandy soils). The relationship between the patch area and the number of ground beetle species was examined by linear regression analysis (Kutner et al., 1996). We analyzed separately the total number of ground beetle species collected in the fragment, as well as the ratio of habitat specialists and generalists to the total number of species. Analyzing the ratio (or relative frequency) instead of the actual numbers removes the inherent differences in species richness among the studied patches. The distribution of data used in the linear regression analyses was normal (tested by the Kolmogorov-Smirnov test, Sokal and Rohlf, 1995).

RESULTS

From the eight studied sandy grassland patches 8,620 ground beetles belonging to 67 species were sampled during the three-year collecting period. The majority of the trapped beetles (7,469 individuals belonging to 31 species) were habitat specialist species, as they were open-habitat species associated with sandy soils.

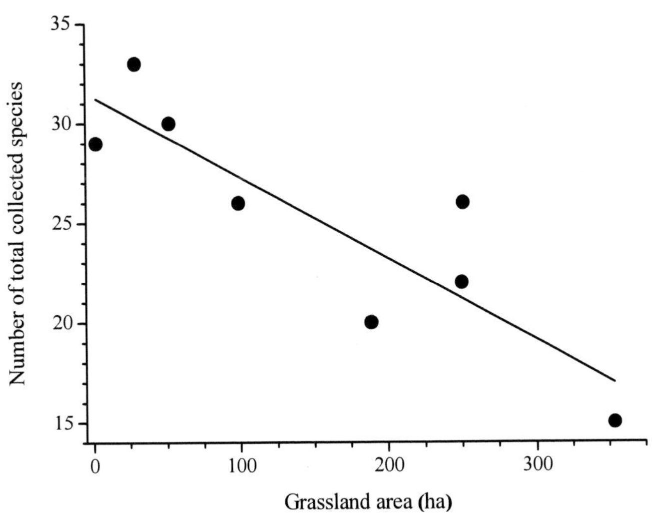

Figure 1. Relationship between the sandy grassland patch area and the total number of ground beetle species collected.

A significant negative relationship was found between the total number of ground beetle species and the area of the sandy grassland patch by linear regression analysis ($F= 19.3805$; d.f.= 1,6; $p= 0.0046$; $R= -0.8738$; Figure 1).

The area of grassland patch and the relative frequency of generalist species (expressed by the ratio of the generalist species to the total number of species) also showed a significant negative relationship ($F= 8.7794$; d.f.= 1,6; $p= 0.0252$; $R= -0.7707$); a decreasing area resulted in an increasing generalist species ratio (Figure 2a).

A significant positive relationship was found between the relative frequency of habitat specialist species (open-habitat species associated with sandy soils) and the grassland area ($F= 8.7794$; d.f.= 1,6; $p= 0.0252$; $R= 0.7707$), indicating the importance of habitat specialist species with increasing patch size (Figure 2b).

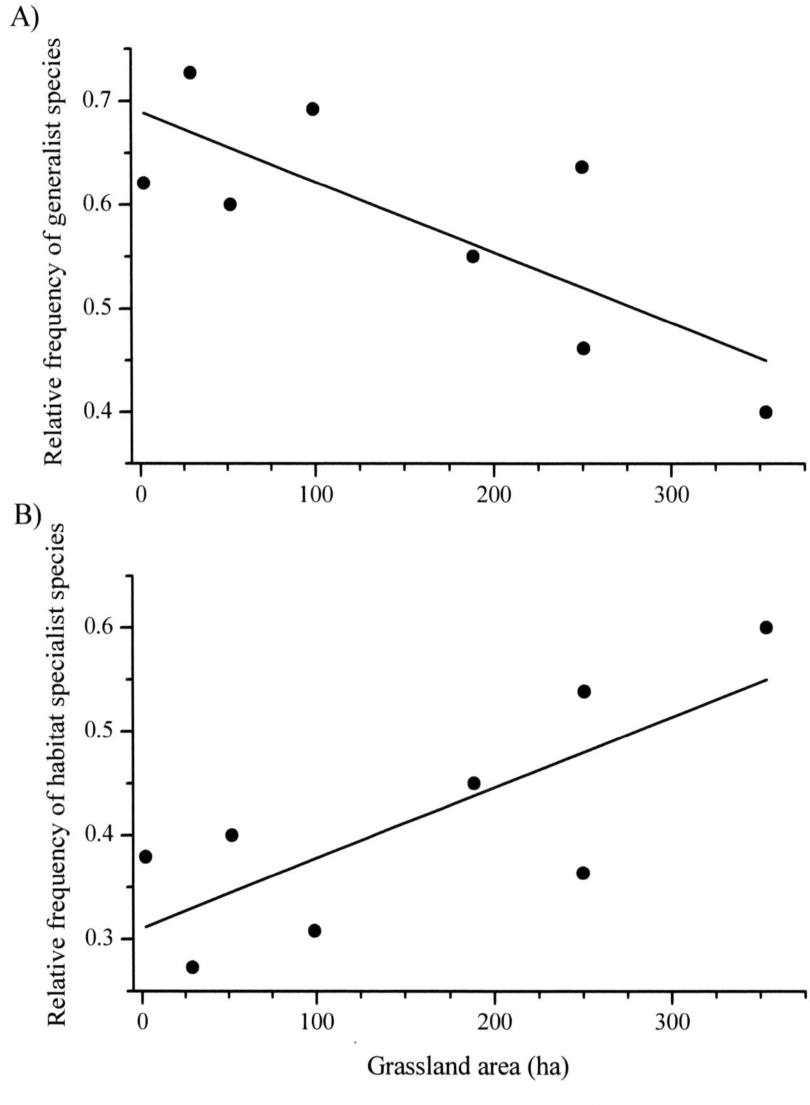

Figure 2. Relationship between the sandy grassland patch area and the relative frequency of generalist species (A) and the relative frequency of habitat specialist species (B).

CONCLUSION

Results of the published studies regarding the relationship between the area of habitat island and the number of observed animal species were rather inconsistent. Several authors, in accordance with the prediction of the classical theory of island biogeography, reported a significant positive correlation (Faeth and Kane, 1978; Mader, 1980; Nilsson et al., 1988; McCoy and Mushinsky, 1994; Bolger et al., 1997; Abensperg-Traun and Smith, 1999; Peintinger et al., 2003; Hovestadt et al., 2005; Watson et al., 2005). Others, contrary to the prediction, described a significant negative relationship (Bauer, 1989; Estades and Temple, 1999; Magura et al., 2001; Lövei et al., 2006). Furthermore, some studies found that overall animal species richness was unrelated to the habitat area (Ås, 1993; Hopkins and Webb, 1984; Davies and Margules, 1998; Brose, 2003; Gandhi et al., 2004; Juliao et al., 2004).

The above inconsistency could arise from the fact that the original theory of island biogeography considered real islands. There is a significant difference between real and habitat islands. In real islands, the surrounding habitat (ocean, sea, lake, river etc.) is usually inhospitable to organisms occurring on islands. In the case of habitat islands, the bordering habitat (the matrix) is usually less hostile. Consequently, species richness of real islands is not notably influenced by the surrounding habitat, while habitat islands could be inhabited by colonists from the matrix: "species can colonize the islands from the sea" (Cook et al., 2002). The above difference is increasingly emphasized when studying the predictions of island biogeography theory on habitat islands. Clear distinction should be drawn between those species that occur in both the habitat patch and the matrix (e.g. generalist species) and the specialist species that truly perceive the habitat patches as islands (Bauer, 1989; Magura et al., 2001, 2008; Cook et al., 2002; Lövei et al., 2006). The specialist species are unable to survive in the surrounding matrix.

In this case study, we demonstrated that depending on the ratio of habitat specialist and generalist species in an assemblage, the species-area relationship may be positive or negative. Based on the significant negative correlation between the total number of ground beetle species and the grassland area one can easily draw the (seriously false) conclusion that it is sufficient to conserve the small patches because they support most species. In fact, however, this was due to the increasing ratio of generalist species with decreasing patch size. Previous studies on ground beetles also emphasized that generalist species from the neighboring matrix and from the edge may cause increased overall species richness in habitat patches with limited size and/or high degree of isolation (Bauer, 1989; Halme and Niemelä, 1993; Desender et al., 1999; Magura et al., 2001; Lövei et al., 2006; Magura and Ködöböcz, 2007). Removing the non-habitat specialist species from the assemblages and analyzing the importance of only habitat specialist species (open-habitat species associated with sandy soils), the significant negative relationship turned over and became significant positive as predicted by the theory of island biogeography. This duality in the species-area relationship concerning ground beetles is not a special case, as several studies reported similar results in different habitats. Bauer (1989) also found that the relationship between the size of limestone outcrop and the overall ground beetle species richness was significantly negative, while the number of limestone specialist species and the area of outcrop showed a significant positive relation. Species richness of farm woodland ground beetles also correlated positively with the area of woodland, while the total number of species did not (Usher et al., 1993). Similar patterns exist for conifer forest patches, where only the forest specialist species showed a positive species-area relationship (Halme and Niemelä, 1993). De Vries (1994) also reported that the relationship between the area of heath fragments and the species richness of heath specialist ground beetles was significantly positive, while the area-total species correlation was statistically not significant. The species-area relationship in deciduous forest patches was also significantly negative when all species were considered, while it was significantly positive for the forest specialist species (Magura et al., 2001; Lövei et al., 2006).

ACKNOWLEDGEMENTS

This study was conducted in the frame of the National Biodiversity Monitoring System in Hungary funded by the Ministry of Environment and Water. We are especially grateful to Zoltán Elek and Tivadar Molnár for field assistance.

REFERENCES

Abensperg-Traun, M. and Smith, G. T. (1999). How small is too small for small animals? Four terrestrial arthropod species in different-sized remnant woodlands in agricultural Western Australia. *Biodiversity and Conservation, 8,* 709-726.

Ås, S. (1993). Are habitat islands islands - Woodliving beetles (Coleoptera) in deciduous forest fragments in boreal forest. *Ecography, 16,* 219-228.

Bauer, L. J. (1989). Moorland beetle communities on limestone 'habitat islands'. I. Isolation, invasion and local species diversity in carabids and staphylinids. *Journal of Animal Ecology, 58,* 1077-1098.

Bolger, D. T., Alberts, A. C., Sauvajot, R. M., Potenza, P., McCalvin, C., Tran, D., Mazzoni, S. and Soule, M. E. (1997). Response of rodents to habitat fragmentation in coastal southern California. *Ecological Applications, 7,* 552-563.

Brose, U. (2003). Island biogeography of temporary wetland carabid beetle communities. *Journal of Biogeography, 30,* 879-888.

Cook, W. M., Lane, K. T., Foster, B. L. and Holt, R. D. (2002). Island theory, matrix effects and species richness patterns in habitat fragments. *Ecology Letters, 5,* 619-623.

Davies, K. F. and Margules, C. R. (1998). Effects of habitat fragmentation on carabid beetles: experimental evidence. *Journal of Animal Ecology, 67,* 460-471.

Desender, K., Ervynck, A. and Tack, G. (1999). Beetle diversity and historical ecology of woodlands in Flanders. *Belgian Journal of Zoology, 129,* 139-155.

De Vries, H. H. (1994). Size of habitat and presence of ground beetle species. In K. Desender, M. Dufrêne, M. Loreau, M. L. Luff and J-P. Maelfait (Eds.), *Carabid beetles: Ecology and evolution* (pp. 253-259.). Dordrecht: Kluwer Academic Publishers.

Estades, C. F. and Temple, S.A. (1999). Deciduous-forest bird communities in a fragmented landscape dominated by exotic pine plantations. *Ecological Applications, 9,* 573-585.

Faeth, S. H. and Kane, T. H. (1978). Urban biogeography-city parks as islands for Diptera and Coleoptera. *Oecologia, 32,* 127-133.

Gandhi, K. J. K., Spence, J. R., Langor, D. W., Morgantini, L. E. and Cryer, K. J. (2004). Harvest retention patches are insufficient as stand analogues of fire residuals for litter-dwelling beetles in northern coniferous forests. *Canadian Journal of Forest Research, 34,* 1319-1331.

Halme, E. and Niemelä, J. (1993). Carabid beetles in fragments of coniferous forest. *Annales Zoologici Fennici, 30,* 17-30.

Hopkins, P. J. and Webb, N. R. (1984). The composition of the beetle and spider faunas on fragmented heathlands. *Journal of Applied Ecology, 21,* 935-946.

Hovestadt, T., Poethke, H. J. and Linsenmair, K. E. (2005). Spatial patterns in species-area relationships and species distribution in a West African forest-savanna mosaic. *Journal of Biogeography, 32,* 677-684.

Hůrka, K. (1996). *Carabidae of the Czech and Slovak Republics.* Zlin: Kabourek.

Juliao, G. R., Amaral, M. E. C., Fernandes, G. W. and Oliveira, E. G. (2004). Edge effect and species-area relationships in the gall-forming insect fauna of natural forest patches in the Brazilian Pantanal. *Biodiversity and Conservation, 13,* 2055-2066.

Kupfer, J. A., Malanson, G. P. and Franklin, S. B. (2006). Not seeing the ocean for the islands: the mediating influence of matrix-based processes on forest fragmentation effects. *Global Ecology and Biogeography, 15,* 8-20.

Kutner, M., Nachtschiem, C., Wasserman, W. and Neter, J. (1996). *Applied linear statistical models.* Boston: McGraw-Hill.

Lövei, G. L., Magura, T., Tóthmérész, B. and Ködöböcz, V. (2006). The influence of matrix and edges on species richness patterns of ground beetles (Coleoptera, Carabidae) in habitat islands. *Global Ecology and Biogeography, 15,* 283-289.

MacArthur, R. H. and Wilson, E. O. (1967). *The theory of island biogeography.* Princeton, NJ: Princeton University Press.

Mader, H-J. (1980). Die Verinselung der Landschaft aus tierökologischer Sicht. *Natur und Landschaft, 55,* 91-96.

Magura, T. and Ködöböcz, V. (2007). Carabid assemblages in fragmented sandy grasslands. *Agriculture, Ecosystems and Environment, 119,* 396-400.

Magura, T., Báldi, A. and Horváth, R. (2008). Breakdown of the species-area relationship in exotic but not in native forest patches. *Acta Oecologica, 33,* 272-279.

Magura, T., Ködöböcz, V. and Tóthmérész, B. (2001). Effects of habitat fragmentation on carabids in forest patches. *Journal of Biogeography, 28,* 129-138.

McCoy, E. D. and Mushinsky, H. R. (1994). Effects of fragmentation on the richness of vertebrates in the Florida scrub habitat. *Ecology, 75,* 446-457.

Nilsson, S. G., Bengtsson, J. and Ås, S. (1988). Habitat diversity or area per se? Species richness of woody plants, carabid beetles and land snails on islands. *Journal of Animal Ecology, 57,* 685-704.

Peintinger, M., Bergamini, A. and Schmid, B. (2003). Species-area relationships and nestedness of four taxonomic groups in fragmented wetlands. *Basic and Applied Ecology, 4,* 385-394.

Reidsma, P., Tekelenburg, T., van den Berg, M. and Alkemade, R. (2006). Impacts of land-use change on biodiversity: An assessment of agricultural biodiversity in the European Union. *Agriculture, Ecosystems and Environment, 114,* 86-102.

Samways, M. J. (2005). *Insect diversity conservation.* Cambridge: Cambridge University Press.

Saunders, D. A., Hobbs, R. J. and Margules, C. R. (1991). Biological consequences of ecosystem fragmentation: a review. *Conservation Biology, 5,* 18-32.

Sokal, R. R. and Rohlf, F. J. (1995). *Biometry* (2nd edition). New York: W. H Freeman and Company.

Usher, M. B., Field, J. P. and Bedford, S. E. (1993). Biogeography and diversity of ground-dwelling arthropods in farm woodlands. *Biodiversity Letters, 1,* 54-62.

Watson, J. E. M., Whittaker, R. J. and Freudenberger, D. (2005). Bird community responses to habitat fragmentation: how consistent are they across landscapes? *Journal of Biogeography, 32,* 1353-1370.

In: Grasslands: Ecology, Management and Restoration ISBN 978-1-60692-023-7
Editor: Hans G. Schröder

Short Communications

SOLAR POWER GENERATION AND USE OF GRASSLAND IN ECOLOGY MANAGEMENT

Xinping Zhou, Jiakuan Yang and Bo Xiao
School of Environmental Science and Engineering, Huazhong University of Science and Technology (HUST), 1037 Luoyu Road Wuhan, Hubei 430074, P.R. China

ABSTRACT

Solar chimney power generating technology is a solar thermal technology on a large scale which combines three parts: a collector, a high chimney (reinforced concrete chimney or floating chimney that can stretch up to several thousand meters), and turbines installed at the chimney base. The best locations of these systems are vast desert regions with high solar insolation and large daily range of temperature. As reported by Zhou *et al.* in 2008, special microclimate conditions will result around the solar power generation belts consisting of many commercial solar chimney power plants. Resulting rainfalls will support the growth of different types and quantitities of plants, including herbs, shrubs, and even trees. This will promote restoration of desert land and even create fertile soil and modify the local ecology. Produced plants can be used for biofuel. Furthermore, vegetables can be planted in the periphery of the collector acting as a green house and as a result benefit agricultural production. Other plants can then be used as a source of power as fuelwood or biomass to produce biogas. Many ecological and agricultural benefits are eminent in this respect.

Keywords: solar energy, power generation, grassland, ecology management

Deserts and desertified lands are vast accounting for one third of land area on the surface of the earth. The area of deserts and desertified lands increase year by year [1]. Desert and desertified lands [3] in China cover 171 million hectares, accounting for 18 % of the Chinese land area. They comprise a vast desert belt, 4500 km from east to west, 600 km from south to north and stretching west to the Tarim Basin and east to the Song Nen Plain, which enlarges by 2460 km^2 per year. The deserts and desertified lands are gradually degraded from grasslands or forest lands due to bad natural climate condition and unreasonable human activity such as overgrazing [2]. These deserts or desertified lands are not available for agricultural and residential use. The ecology in these lands where precipitations are less than the sum of evaporations and percolation to the underground through sands and the plants including herbs, shrubs, and trees are extinct is increasingly destroyed. The destruction poses a great threat to human life by producing sandstorm downstream.

There exist high solar insolation and large daily ranges of temperatures in these vast lands. Increasing needs to satisfy the improvement of better standard of life in the recent years has put a big constraint on fossil fuels, the main energy source, which is being exhausted at a fast rate, hence greenhouse effect being is being aggravated [4], It is therefore, urgent to develop technologies utilizing renewable and clean energy source to solve the above problems for energy needs in future. Solar chimney power generating technology (Figure 1) is a promising large-scale power generating technology [5-9]. The best locations for these systems are vast desert and desertified regions where the land may be free [10] with high solar insolation and large daily ranges of temperatures. The technology was first described in a publication by Günter in 1931 and tested with the 50 kW Manzanares prototype plant in the 1980s [4,11]. The solar chimney power system combines three familiar components: an air collector, a chimney situated in the center of the collector, and turbine generators installed at the chimney base. It works on the principle that the turbines are driven by airflow produced by buoyancy derived from hot air heated inside the collector. Additional tight water-filled heat storage placed under the inner collector roof [12] or solar-pond thermal storage system [6] supports continuous 24 hour operation of turbines. For commercial power plants producing energy economically, not only a large collector area is necessary for collecting a large amount of solar energy, but also a very high gigantic chimney. The height is needed only for the larger driving force that is proportional to the height and also to make the best use of the heat available. The height combined with a large collector surface area will reduce the cost of energy production.

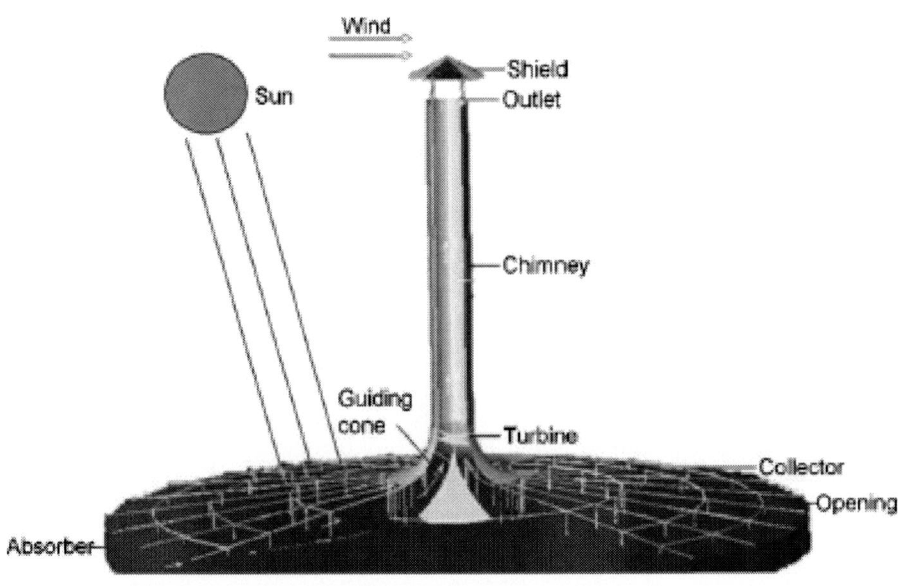

Figure 1. Schematic diagram of solar chimney power system.

The gigantic chimney may be reinforced concrete chimney that can stretch more than 1000 meters within the technological restrictions [5] or floating chimney that can stretch up to several thousand meters [8]. Recently, the Australian government decided to support the construction of a prototype solar chimney power plant with a 1 km high solar chimney. The proposed solar plant in Mildura, Australia, can produce 200 MW of electricity [13]. The Chinese local government is supporting the construction of a 600 m high experimental floating solar chimney prototype in a desertified land in Gansu Province.

As reported by Zhou *et al.* [10,14], special microclimate conditions will result around the solar power generation belts consisting of many commercial solar chimney power plants resulting from the volume of warm air that carries many tiny granules as effective condensation nuclei of moisture transferred from the ground level to a high altitude.

The special microclimate conditions accompanied by the operation of solar chimney will occur 24 h. In the free atmosphere, the velocity and direction of wind at a high height usually vary little [15]. This can produce continuous rainfalls above a long and narrow region which in turn can produce abundant water, which is the crucial factor of restoration of desert lands [1]. This is also the reason why oases are usually formed at the foot of high mountains where precipitations in the high-mountain environment are more than that of deserts, with one part staying in solid phase on the mountain top and the other flowing down to irrigate the oases at the foot of the mountain. In the restoration of deserts and desertified lands to grassland, the types of pioneer plants grown are selected according to the conditions of the local lands, the wind, and water supports including rainfalls and underground water [1] and phytocoenosis is evolved from herbs to shrubs to trees. Rainfalls resulting from the special microclimate conditions will support the restoration of deserts and desertified lands by supporting the growth of different types and quantitities of plants.

The desertified lands can be finally reformed to fertile soils, which have strong water-holding capacity and reduce the evaporation of water, thus increasing surface runoff. The surface runoff can be used in the restoration of the desertified lands away from the rainfall regions and the lands under the outer collector roof. This will help in modifying the local ecology.

Vegetables or fruits can be planted in the periphery of the collector acting as a green house [16,17] and as a result benefit agricultural production. On the other hand, a wet cultivated ground is often darker than a dry flat one, so that this albedo effect generates a synergy among agricultural and power productions.

Other plants can then be used as a source of power as fuelwood or biomass to produce biogas. Many ecological and agricultural benefits are eminent in this respect by using grassland around huge solar chimney belt.

REFERENCES

[1] Zhang, K. B.; Zou, S. Y. *Theory and development of desertification control;* Forestry Publishing House: Beijing, China, 1990.

[2] Cao, H. X. *Climate and desertification*; China Meteorological Press: Beijing, China, 1988.

[3] Zhou, X. P.; Yang, J. K. *Heat Transfer Engineering*. 2007, in acceptance.

[4] Zhou, X. P; Xiao, B.; Ochieng, R. M.; Yang, J. K. *Renewable and Sustainable Energy Reviews*. 2007, doi:10.1016/j.rser.2007.10.003. Utilization of carbon-negative biofuels from low-input high-diversity grassland biomass for energy in China.

[5] Schlaich, J. *The Solar Chimney: Electricity from the sun;* Edition Axel Menges: Stuttgart, Germany, 1995.

[6] Dennis, C. *Nature*. 2006, vol. 443:23–4.

[7] Woody, T. *Business 2.0*. 2006, *vol.* 7, 94–102.

[8] Papageorgiou, C. D. In Proc. ISES 2005 Solar World Congress Conference, Orlando, Florida, USA, August 2005.

[9] Zhou, X. P.; Yang, J. K.; Wang, F.; Xiao, B. *Renewable and Sustainable Energy Reviews*. 2008, in press.

[10] Zhou, X. P.; Yang, J. K.; Xiao, B.; Hou, G. X.; Shi, X.Y. *J Energy Engineering*. 2007, in press.

[11] Ketlogetswe, C; Fiszdon, J. K.; Seabe, O. O. *Renewable and Sustainable Energy Reviews*. 2007, doi:10.1016/j.rser.2007.03.009, in press.

[12] Kreetz, H. *Theoretische Untersuchungen und Auslegung eines temporären Wasserspeichers für das Aufwindkraftwerk*; diploma thesis, Technical University Berlin, Berlin, 1997.

[13] Available at http://www.enviromission.com.au/faqs/faqs.htm.

[14] Zhou, X. P; Yang, J. K; Ochieng, R. M.; Li, X. M.; Xiao, B. *Atmospheric research*. 2008, in submission.

[15] Zhou, S. Z.; Zhang, R. Y.; Zhang, C. *Meteorology and meteorology*; 3rd edn; Higher Education Press: Beijing, China, 1997.

[16] Dai, Y. J.; Huang, H. B.; Wang, R. Z. *Renewable Energy*. 2003, *vol.* 28, 1295–1304.

[17] Private communication with Bonnelle, D. Feb, 2008..

Reviewed by

Dr Reccab M. Ochieng

Department of Physics, Maseno University, P.O. Box 333, Maseno, Kenya

Email: reccabo@yahoo.com

Dr Reccab M. Ochieng is the specialist in both solar power generating and biomass.

In: Grasslands: Ecology, Management and Restoration ISBN 978-1-60692-023-7
Editor: Hans G. Schröder

Chapter 1

BENEFITS OF MULTI-PADDOCK GRAZING MANAGEMENT ON RANGELANDS: LIMITATIONS OF EXPERIMENTAL GRAZING RESEARCH AND KNOWLEDGE GAPS

Richard Teague[1], Fred Provenza[2], Brien Norton[3], Tim Steffens[4], Matthew Barnes[5], Mort Kothmann[6] and Roy Roath[7]

[1]Texas AgriLife Research, Texas A&M System, Vernon TX 76384
[2]Department of Wildland Resources, Utah State University, Logan, UT 84322-52305
[3]Centre for the Management of Arid Environments
Locked Bag 22, Kalgoorlie, WA 6433, Australia
[4]United States Department of Agriculture-Natural Resources Conservation Service
Springfield, CO 81073
[5] United States Department of Agriculture-Natural Resources Conservation Service
Kremmling, CO 80459
[6]Department of Ecosystem Science and Management, Texas A&M University
College Station, TX 77843
[7]Forest, Rangeland and Watershed Stewardship Department, Colorado State University
Fort Collins, CO 80524

ABSTRACT

The benefits of multi-paddock rotational grazing on commercial livestock enterprises have been evident for many years in many countries. Despite these observations and the results of numerous studies of planned grazing deferment before the mid-1980s that show benefit to species composition, most recent rangelands grazing studies suggest that rotational grazing benefits neither vegetation nor animal production relative to continuous grazing. Detailed comparisons of research methods and practical experiences of successful practitioners of multi-paddock grazing systems identify a number of areas that explain why such different perceptions have arisen. Consistent with producer experience, published data from small paddock trials on both temporal and spatial aspects of grazing management indicates the potential for significantly higher production under multi-paddock rotational grazing relative to continuous grazing and conservative stocking.

While research findings often suggest multi-paddock grazing management is not superior to continuous grazing, researchers have not managed trials to answer practical questions such as: how good is this management option, where is it successful, and what does it take to make it work as well as possible? In contrast, successful ranchers manage strategically to achieve the best possible profitability and ecosystem health. They use basic knowledge of plant physiology and ecology generated by research within an adaptive, goal-oriented management approach to successfully implement planned grazing management.

Published research and experience from ranchers have indicated that the following management factors are the keys to achieving desired goals: (1) Planned grazing and financial planning to reduce costs, improve work efficiency and enhance profitability and environmental goals; (2) Adjusting animal numbers or having a buffer area available so that animal numbers match forage availability in wet and dry years; (3) Grazing grasses and forbs moderately and for short periods during the growing season to allow adequate recovery; (4) Timing grazing to mitigate detrimental effects of defoliation at critical points in the life cycle of preferred species inter- and intra-annually; (5) Where significant regrowth is likely, grazing the area again before the forage has matured too much; (6) Using fire to smudge patch-grazing imprints and manage livestock distribution; and (7) Using multiple livestock species. In all these areas, management is the key to success.

Many researchers have failed to sufficiently account for these management factors, either in their treatment applications or in the evaluation of their results. To define the potential impact, researchers must quantify the management strategies for best achieving whole-ranch business and ecosystem results under different grazing management. Conducting research on ranches that have been successfully managed with planned multi-paddock grazing for many years, together with systems-level simulation modeling, offer complementary approaches to traditional small-paddock field research. These methods are particularly applicable where logistics preclude field experimentation, or when assessing impact over decadal time frames. This chapter discusses these points, suggests areas of research that may explain differences in perception among land managers and researchers, and provides information to achieve the full potential of planned multi-paddock grazing management.

INTRODUCTION

Many ranchers who have practiced multi-paddock grazing management for decades are very satisfied with the economic results and improvement to the ecosystem, as well as the change in management lifestyle and social environment of their ranch businesses. Such ranchers regularly win conservation awards from the ranching industry and natural resource professional organizations. In contrast, many grazing researchers have concluded that multi-paddock grazing offers no significant benefit over continuous grazing (Holechek et al. 1999, 2000; Briske et al. 2008), but their studies have been largely small-scale trials focussed on the technical questions of ecological impacts and livestock production conducted in a relatively limited scope of fairly resilient landscapes. In addition, research plots are designed to reduce or eliminate variability, while ranch managers must manage in the environment with all the inherent variability of the landscape. The relevance of such research to a commercial ranch operation is questionable because many studies do not address critical information needed by ranch managers to achieve desired outcomes.

Ranchers have a vested interest in managing for the best result in terms of production, profitability and sustaining natural resources. Achieving management goals requires integrating knowledge from numerous biological, management and economic disciplines and correctly adjusting management actions to changing conditions. Research scientists, in contrast, almost invariably work only within the narrow confines of a single discipline, be that soils, plants, or herbivores. Consequently, research on grazing systems has not considered the multi-disciplinary and broadly integrative ramifications of treatments. Generally multi-paddock research treatments have been applied to assess esoteric goals that may be academically interesting but are of little benefit to ranch managers.

Questions relevant to managers are: (1) what are the advantages of this management option; (2) where is it likely to be successful; (3) how does it need to be managed to make it work as well as possible; and (4) what are the biological interactions and principles that can provide guidance for adjustments to changing conditions or unanticipated outcomes? Ultimately, every situation is unique in time and space, and while we can understand principles and processes, it is the answers to these site-specific questions that enable managers to assess the benefits of adopting various practices as they pertain to the biological, financial and social aspects of their business to achieve desired environmental, economic and social goals (Provenza 2000).

In this chapter, we first discuss the relationship between ecological science and applied management science. We then provide an ecological and physiological framework that underpins management decisions to achieve desired outcomes. We explain concepts critical for implementing successful planned grazing, then discuss how they can be used to manage for desired outcomes, and finally we provide a list of guidelines for planning successful grazing management programs. We use these guidelines as a framework to examine and explain the differences between much of the scientific literature and management experience and suggest areas of research to provide information for managers to achieve the full potential of planned multi-paddock grazing management.

The Relationship between Ecological and Applied Management Sciences

Basic ecological and applied management sciences have different but complementary functions (Provenza 1991). The aim of ecological research is to understand principles and processes as they pertain to the interrelationships among organisms and their environments, and not to examine how ecological information can be used effectively by managers. Applied management sciences bridge the gap between ecological information and the achievement of desired management goals by integrating knowledge from diverse disciplines. They evaluate management consequences within a research-based theoretical framework of ecological processes and how they affect ecological, economic and social factors important to management.

The rangeland discipline is distinguished from ecological disciplines in that we work in a management framework. If we fail to address management-related issues important to land managers, we have abrogated our responsibilities and minimized our impact. We are an essential bridge between ecologists and managers of rangeland, but to be relevant to managers, reductionist studies of processes must, at the very least, be rooted in potential for application to managing landscapes within a systems framework. Knowledge of ecological processes can help managers be more effective in implementing strategies that enhance positive responses, reduce or mitigate negative responses, and benefit the financial efficiency and human relationships of a ranch business.

ECOLOGICAL AND PHYSIOLOGICAL BACKGROUND

Fire and grazing by large ungulates are considered integral to the persistence of grassland ecosystems and species diversity, and their importance for rejuvenating landscapes will only become more important as the costs of fossil fuels rise (Provenza 2008). Natural rangeland communities are constantly responding to the effects of the most recent disturbance, in most cases never achieving a steady-state or climax stage. The absence of these disturbances in grassland ecosystems results in a decline in species diversity and deterioration of physical structure (Vogl 1974; Rice and Parenti 1978; Picket and White 1985; Hulbert 1969, 1988). Grazing by large ungulates has been an integral part of prairie ecosystems from the late Mesozoic, and ecosystems have a high inherent sustainability as a direct result of this long co-evolutionary history (Frank and McNaughton 2002). This is remarkable as they support more herbivore biomass and sustain considerably higher levels of herbivory than any other terrestrial habitat. Nomadic pastoral systems that mimic these grazing patterns also seem to have less detrimental effects on vegetation (Danckwerts et al. 1993).

One key stabilizing element of native grazing ecosystems is that they are characterized by high spatial and temporal variation in forage supply, due to vegetation response to ever changing topography, edaphic effects, and climate. Second, they are dominated by large, migratory ungulate herbivores that are constantly on the move, and although herbivores often graze intensely at any particular site, such grazing never lasts long and defoliated plants are afforded time and usually suitable conditions to regrow (McNaughton et al. 1989). Herbivores move for a variety of reasons including need for nutrients, satiating on nutrient and secondary compounds, fire, predators, movements by a herder, and rotations in a grazing system (Provenza 2003b; Provenza et al. 2003; Bailey and Provenza 2008). Grazers are important regulators of ecosystem processes (Frank and Groffman 1998). Through their impacts, herbivores link their health with that of soils, plants and ultimately people (Provenza 2008). Grazers can increase forage nutrient concentrations and aboveground plant production (Frank and McNaughton 2002). Grazers also enhance mineral availability for soil microbial and rhizospheric processes that ultimately feed back positively to plant nutrition and photosynthesis (Hamilton and Frank 2001), in addition to increasing nutrient cycling within patches of their urine and excrement (Holland et al. 1992). Frank and Groffman (1998) found that grazer control of carbon and nitrogen processes was as important as landscape effects of topography, catenal position and different soils. By increasing resource availability locally, they can also reduce the adverse impacts of secondary compounds in plants (Bryant et al. 1983; Coley et al 1985). However, the positive feedbacks from grazers on the ecosystem are contingent on suitable climatic conditions. During drought these feedbacks are diminished (Wallace et al. 1984; Coughenour et al. 1985; Louda et al. 1990).

Grazing management systems were developed in an attempt to manage grazers and grazing lands in a manner that maintains or improves ecosystem structure and function while achieving social and economic goals (Heitschmidt and Taylor 1991). In stark contrast to native grazing ecosystems, however, the replacement of free-ranging wild herbivores with livestock has generally resulted in overgrazing and degradation of rangelands (Provenza, 2003b). Generally, extensive or poorly managed rotational grazing of domestic animals by humans does not emulate the movements of wild ungulates, and managed herds during dry seasons can be held at stocking rates higher than the land can support. Maintenance of artificially high animal numbers with supplementary feed during less productive periods promotes degradation (Oesterheld et al. 1992; Milchunas and Lauenroth 1993). The more sedentary and concentrated animal use of the vegetation under human management removes the key revitalizing element of periodic deferment and natural response to climate variation. Indeed, in the recent past we have largely restrained the movements of domestic animals and in the process inadvertently trained herbivores to become sedentary, largely with the use of fences in continuous and conventional rotational grazing systems, and with the suppression of fire and large predators (Provenza 2003a).

While the foregoing discussion emphasizes grazer effects on landscapes, defoliation by grazers also significantly affects individual plants morphologically and physiologically. This in turn affects their vigor and productivity, as well as recruitment and survival through the indirect effects on competitive relationships among plants (Briske 1991). The detrimental effects of defoliation are increased with greater intensity or frequency of defoliation (Briske 1991) and can lead to mortality of plants, particularly if environmental conditions deteriorate. Seedlings and juveniles of palatable species are particularly vulnerable. McIvor (2007) found lower survival rates of plants grazed at > 60% utilization as opposed to <60% utilization under continuously grazed management; drought exacerbated these differences. In an early and classic study, Crider (1955) found that a single defoliation removing 50% or more of the shoot volume retarded root growth in 7 of 8 perennial species examined. This observation, among others, prompted the often used term "take half - leave half" as an aphorism for grazing management that emphasizes stocking rate. However, Hormay (1956) observed that preferred plants in preferred sites are utilized closely and repeatedly even when the entire management unit is lightly or moderately stocked on average.

Preferred plants have probably always been severely grazed when encountered, but much of the time the intermittent nature of the severe grazing described in the second paragraph of this section prevented this condition being chronic (McNaughton et al. 1989). The elements of timing, when and for how long, and adequate recovery distinguish severe grazing from overgrazing, and cannot be overemphasized in any discussion of grazing management. Overgrazing occurs on individual plants (Roshier and Nicol 1998) as a result of multiple, severe defoliations without sufficient physiological recovery between defoliations. It is site, stock density, and time-specific, and diet selection of grazing animals can put palatable and actively growing plants in preferred areas at a disadvantage (Earl and Jones 1996). Stocking rate affects only the proportion of plants likely to be used heavily. Therefore, while conservative stocking is an important first step in sustainable management, it must be applied in conjunction with other management practices like short grazing periods at high stock density (O'Connor 1992) and periodic deferment to mitigate the effects of selective grazing

(O'Reagain et al. 2003). Increasing differences in palatability and abundance among different plants in a pasture, decreasing stock density, or increasing the graze period will tend to increase the likelihood of overgrazing the more palatable plants (Earl and Jones 1996).

Vegetation dynamics on a landscape emerge from interactions among plant autecology, community processes, climate, and disturbance, as modified by grazing animal preferences and distribution in response to plant species, topographic and ecological site diversity (Walker 1988). Ash and Stafford-Smith (1996) provide an excellent explanation of how selective use of plants and landscape components can cause a gradually widening area of degradation under continuous grazing, even at light to moderate stocking rates.

Earlier, Hormay (1956) asserted that close and frequent cropping (overgrazing) in preferred areas could not be prevented by regulating only stocking rate, and said that stocking rate determined only the size of the overgrazed areas, and therefore, the spatial scale of degradation if allowed to continue over an extended period of time. He explained that for a given kind of livestock, there are only four factors that can be manipulated to influence desired management goals on rangelands: stocking rate, season of grazing, livestock distribution, and frequency of grazing. These factors are generally acknowledged by many range managers and scientists alike (e.g. Heitschmidt and Walker 1996; Briske et al. 2008). These four management factors have been further reduced to frequency of defoliation, the intensity of defoliation, and regrowth opportunity (Reed et al. 1999), but should also include spatial distribution (Barnes et al. 2008).

For communities to move from one stable state (assemblage of species occupying a site) to another, some external force is required (Walker 1988; Westoby et al. 1989; Danckwerts et al. 1993). Management of grazing to enhance soil moisture and seed production can "condition" the resource to take fuller advantage of episodic events like drought, fire, and unusually wet periods, particularly if the event is of a marginal magnitude, so that a community might move or be prevented from moving from one state to another more readily (Watson et al. 1996; Gerrish 2004). In any management situation the key issue is to be aware of what stable state or states have the greatest chance of fulfilling management objectives, and what combination of events and management is required to cause or prevent movement from one state to another (Westoby et al. 1989; Danckwerts et al. 1993).

The Real Impact of Continuous Grazing

Continuous grazing in large paddocks is usually associated with patch grazing and resource deterioration in localized areas. Grazing under enclosed conditions does not occur uniformly over time or over a landscape (Willms et al.1988; O'Connor 1992; Ash and Stafford-Smith, 1996; Bailey et al. 1996; Gerrish 2004; Witten et al. 2005). Livestock grazing large paddocks exhibit spatial patterns of repetitive use, heavily using preferred patches and avoiding or lightly using others. The process of patch-selective grazing results in the effective stocking rate on heavily used patches being much higher than that intended for the area as a whole. The resulting heterogeneity in landscapes may be desirable or undesirable depending on desired outcomes and scale of disturbance. Any positive feedbacks on the ecosystem from grazers noted in the previous section are contingent on suitable climatic conditions, and are weakly expressed during low-rainfall years (Wallace et al. 1984; Coughenour et al. 1985;

Louda et al. 1990). If threshold amounts of biomass and litter are not maintained, a degradation spiral is initiated (Thurow 1991; Ash and Stafford-Smith, 1996), with heavily used patches as foci (Fuls 1992; O'Connor 1992; Teague et al. 2004).

Consequently, even at light stocking levels the heavily grazed patches and preferred species are subject to excessive grazing pressure. Root biomass (or density) and rooting depth decline as: (1) a greater proportion of leaf material is removed (Crider 1955; Cook et al. 1958; Singh and Mall 1976; Hodgkinson and Baas Becking 1977; Stroud et al. 1985), especially during the growing season (Ganskopp 1988; Engel et al. 1998); (2) the interval between defoliation events declines (Harradine and Whalley 1981; Motazedian and Sharrow 1987; Danckwerts and Nel 1989); or (3) a function of long-term grazing impacts (Weaver 1950; Tomanek and Albertson 1957; Schuster 1964; Blydenstein 1966). This reduces plant vigor and causes desirable plants to die in overgrazed patches during droughts, enabling less desirable species or invading weeds to occupy the vacated space and expand into surrounding vegetation.

A continuously grazed paddock must be large enough to supply enough forage for an entire year or climatic grazing season. Paddock size is driven up by herd size big enough for efficient ranch operation and by the cost and logistics of water supply and fencing. Paddock size will also be influenced by declining forage production per hectare with decreasing average rainfall in semi-arid environments. Landscape heterogeneity increases with the size of a grazing unit (Senft et al. 1985; Stuth, 1991; Bailey et al. 1996), resulting in heavier impact on preferred areas and a greater proportion of the paddock receiving light utilization or total neglect. Selection is affected a little by small-scale heterogeneity at the feeding station level but is profoundly affected by large-scale heterogeneity at the landscape level (WallisDeVries and Schippers 1994; Barnes 2002). Both the spatial arrangement of grazing patches and the scale of patchiness are major determinants of selectivity during grazing (WallisDeVries et al. 1999). In addition, patterns of herbivory on the landscape are controlled by the spatial distributions of topography, water, cover, minerals and inter- and intra-specific social interactions (Coughenour 1991). Spatial and temporal variability in primary production localizes and intensifies herbivore impacts (Illius and O'Connor 1999). Historically, herbivores were encouraged to move across landscapes, sometimes in ways that were predictable, such as seasonal movements related to topography and elevation, and sometimes in ways that were not predictable, due to precipitation, fires, and hunting by predators, including humans. These factors compound over time to create long-term impacts on the environment and on primary and secondary production (Coughenour 1991; Fuls 1992; Kellner and Bosch 1992).

Patch selective over-grazing and resource deterioration in localized areas have a number of profound consequences for the interpretation of experimental results and for managing rangeland for sustainable goals. Usually experimental paddocks have been less than 25 ha and often less than 5 ha each (Norton 1998), which is considerably smaller than those of commercial ranches in the same environment. Small herds of livestock continuously grazing in such small experimental paddocks have no problem exploring and accessing the entire paddock one or more times each day. Relatively small research paddocks grazed continuously to compare with rotational grazing do not mimic the continuous grazing of large paddocks. They reduce the manifestation of a critical element of continuous grazing, namely uneven utilization and patches of extreme use over the landscape, as they ignore the documented patterns of selection and resulting effects that occur in large paddocks (Norton 1998; Teague and Dowhower 2003). The conclusions of such research have been extrapolated to all pastoral situations, regardless of paddock size. Small-scale experiments are carried out as though paddock size doesn't matter, and when the paddocks are only 20 ha or less, it doesn't. Unless the issues of scale and spatial heterogeneity are included as treatments, experiments at only small scales do not represent what happens at the scales of commercial ranches. Great care must be taken in offering advice without due consideration of the effects of extrapolating results from small-scale experiments to larger scales.

It is incorrect to assume that at low stocking rates, grazing pressure will be low across a landscape. This occurs only for a short time after a fire has burned a management area. As time progresses after a fire, the pattern of area- and patch-selective grazing resumes, so that the pattern-removal effect of fire disappears completely after several years (Archibald et al. 2005). On average, grazing pressure might be low but it is heavy on some areas and very low or zero on other areas. The misconception that low stocking rates result in no overgrazing contradicts much of the published research relating to the process of uneven utilization in landscapes (Ash and Stafford-Smith, 1996). All that can be truthfully said is that all other things being equal, at lower stocking rates grazing pressure will be lower *on average*. With low stocking rates, a relatively small proportion of the primary productivity is harvested by grazing animals, while the majority senesces and decomposes without being ingested by herbivores. If the damaging effects of area and patch selection are minimised by limiting the number or severity of defoliations (e.g. Derner et al, 1994) through short graze periods and providing *adequate* recovery after grazing, conservation and production goals will be achieved.

The published evidence illustrates unequivocally that patch-selective overgrazing happens under continuous grazing on ranches even if small plot research does not confirm this or underestimates it, as outlined above in this chapter. The implementation of planned grazing management allows recovery of heavily grazed patches and the regulation of grazing intensity and frequency.

Managing for Desired Outcomes

Successful grazing managers must optimize several ecological goals to attain sustainable production goals (Heitschmidt and Taylor 1991; Briske et al. 2008). These include: (1) Planned grazing and financial planning to reduce costs, improve work efficiency, enhance profitability, and achieve environmental goals; (2) Providing sufficient growing season deferment to maintain or improve range condition; (3) Grazing grasses and forbs moderately during the growing season for a short period to allow adequate recovery; (4) Timing grazing to mitigate detrimental effects of defoliation at critical points in the life cycle of preferred species inter- and intra-annually; (5) Where significant regrowth is likely, grazing the area again before the forage has matured too much; (6) Flexible stocking to match forage availability and animal numbers in wet and dry years or having a buffer areas that can be grazed; (7) Using fire and other tools to manage livestock distribution and increase the total plants harvested; and (8) Using multiple livestock species. These goals cannot be accomplished with continuous, season-long grazing in environments that receive enough moisture to have growing periods of more than a few days.

The means of achieving these goals are discussed below. Because this discussion is focused on biological and ecological relationships, we deal primarily with factors 2-6, which are influenced by factors 7 and 8.

Importance of Planned Recovery Periods

Maintenance or improvement of plant species composition and productivity is of paramount importance to achieving sustainable use. Since even at low stocking rates, patch and area overgrazing occurs, management must provide adequate periods of recovery to maintain or improve the range resource. Rangeland improves if the benefits of recovery exceed the damage of grazing, and degradation occurs if the benefits of recovery are less than the damage caused by grazing (Merrill 1954; Thurow et al. 1988; Tainton et al. 1999; Müller et al. 2006).

Significant range improvement in the form of increased proportion of desired species and increased plant vigor has been demonstrated in many parts of the world following growing season deferment. This allows adequate recovery periods at similar or higher stocking rates compared to season-long grazing or regimens that provide shorter recovery periods (Smith 1895; Sampson 1913; Rogler 1951; Scott 1953; Matthews 1954; Merrill 1954; Hormay 1956; Hormay and Evanko 1958; Hormay and Talbot 1961; Hormay 1970; Reardon and Merrill, 1976; Booysen and Tainton 1978; Taylor et al. 1980; Thurow et al. 1988; Taylor et al. 1993; Tainton et al. 1999; Snyman 1998; Teague et al. 2004; Müller et al. 2007).

Ranch-scale research indicates that plant vigor, ecological condition, and carrying capacity benefit from grazing systems at appropriate stocking rates with only one grazing period during the growing season and the application of full growing-season deferment to each pasture once every 3 to 4 years (Danckwerts 1984; Tainton et al. 1999; Müller et al. 2007). Even when treatments incorporating a growing season rest were more heavily stocked than a continuously grazed treatment there were benefits to species composition in the rested treatment (Thurow et al. 1988).

A theoretical analysis by Müller et al. (2007) indicated destocking in times of drought and restocking post-drought was not always adequate to maintain long-term forage productivity in an area with 177 mm of mean annual rainfall; it is the variability in rainfall, not the mean that makes all the difference. This study, which was corroborated with 40 years of data from a 30,000 ha ranch, showed that rest periods during the normal growing season were indispensable to maintaining productivity of semi-arid rangeland.

It is important to stress that periods of deferment will allow recovery only if adequate growing conditions are experienced during the period of deferment. Maintaining or increasing more productive and preferred species depends on the availability of water and nutrient resources (Lee and Bazzaz 1980; Wallace et al. 1984; Coughenour et al. 1985; Polley and Detling 1989), and a positive response to rest is possible only if sufficient resources are available (Louda et al. 1990). In addition to the respite from grazing afforded by rotational grazing, favorable long-term climatic conditions are necessary for the recovery of more productive and palatable grasses. For this reason more arid rangelands require longer recovery periods (Heitschmidt and Taylor 1991), often a year or more (Bradford 1998; Howell 2006).

Defoliation events during the growing period can have remarkable negative effects on plant vigor and total stored carbohydrates compared to comparable levels of grazing at other times of the year in certain mid and tall grass species (Mullahey et al. 1990; Mullahey et al. 1991; Reece et al. 1996; Cullan et al. 1999). Multiple defoliations during this period had a more serious effect than single defoliations of the same intensity. Effects also varied with growing season precipitation. Effects of excessive defoliation or defoliation during this critical period may also be manifest in the following year (Hendrickson et al. 2000). These results indicate that if a goal of management is to increase the frequency of these species, a recovery period of at least 2 months following a mid-summer graze period in at least a significant proportion of years would be advantageous, while multiple defoliation events during this time of year would decrease the frequency of these species in an environment receiving around 260 mm of growing season precipitation (Reece et al. 1996). From this discussion, it is evident that continuous growing season grazing or long graze periods would make management for these taller species extremely difficult at best.

The question of what constitutes adequate recovery for species in semi-arid rangeland environments is one of the most important that can be answered in the field of range science, yet it has not been investigated as a principal subject of inquiry. Certainly, it varies with inter- and intra-annual variations in climate, as well as with the intensity and frequency of defoliation of those target species in the preceding graze period. Trlica et al. (1977) found that herbage yields of some species required more than 26 months of recovery following severe defoliation. Dowling et al. (1996) asserted that recovery periods need to be timed to give an advantage to the desired species. They found that deferment of grazing for 3 months during the summer of a mesic environment increased the proportion of palatable perennial grasses compared to a continuously grazed treatment. The timing of the graze period affects the recovery period needed to benefit the desired species, as the length of the remaining growing season and likelihood of precipitation in the intervening planned recovery period will determine if sufficient regrowth occurs. The species of concern will also have a bearing, as growth rates and such critical physiological characteristics as when the apical meristem is elevated will be affected by temperatures and day length.

Some reasonable "rules of thumb" can be developed by evaluating some of the scientific literature and common practices in the rangeland and agronomic fields. The considerable body of work that has been done in Texas at the Sonora experiment station (e.g. Merrill 1954; Thurow et al. 1988; Reardon and Merrill 1976) indicates that in that environment with a variable precipitation averaging 438 mm, 4 months seems usually to be adequate, but 50 days is too short to maintain higher producing warm-season mid-grasses. Yet, in several studies with similar or even drier climates (e.g. Derner and Hart 2007; Hart et al. 1988), a similar recovery period was used but no differences in performance of animals or vegetation between treatments were observed. They correctly assert that rest and deferment during periods of minimal plant growth and low soil moisture or temperature extremes limit the potential for positive vegetation responses. Therefore, recommendations asserting no benefits of periodic recovery in these circumstances must be suspect.

In many areas where cool season, irrigated, perennial grasses are harvested for forage, one, or at most two, cuttings are harvested in a year, with perhaps some dormant-season grazing of aftermath. Where perennial warm-season grasses are irrigated with relatively long growing seasons, a common interim between cuttings is about 30 days. Therefore, it might be reasonable to assume that at least 30 days of optimum growing conditions are necessary for adequate recovery from a similar level of defoliation. Torrell et al. (2008) found that for central New Mexico native rangeland, natural rainfall provided only about 30 days in a growing season with soil moisture above 30%. Given the discussion in the previous paragraphs, in many of the arid to semi-arid environments characteristic of western North America, in all but the most favorable years it seems reasonable to assume that plants require most of a growing season to adequately recover from a defoliation. If there are both desirable cool- and warm-season plants in the community, one short, early growing season graze period followed by deferment for the remainder of the cool part of the year when these species are growing should normally benefit that component of the sward. Another graze period, later in the season to use warm-season species, followed by recovery until the dormant season, would likely benefit palatable warm-season species. Grazing management programs that provide less recovery than that would not be expected to show reliable, measurable benefits in most years.

In summary, by providing periodic, adequate growing season deferment, multi-paddock grazing management can minimize the detrimental effects of patch and area-selective overgrazing. While adequate deferment can be applied in pauci-paddock systems, multi-paddock systems enable more refined management as discussed below.

Importance of a Short Grazing Period

There are important benefits to both plants and animals by having short grazing periods. The nutritional regime offered to livestock in a rangeland situation is often highly variable, and animals cannot necessarily meet their nutritional needs for high-demand functions like conception, late pregnancy and lactation. This can be turned into a benefit using rotational grazing in which the manager plans his stock movements to place animals in paddocks with the best chance of meeting those higher nutritional requirements. Multi-paddock management can positively or negatively influence both forage productivity and quality. However, management needs to be different in wet and dry areas because there are different factors to take into account due to climatic differences.

One of the purposes of subdividing a management unit into paddocks is to better control the length of time and place where livestock eat plants. This commonly results in shorter graze periods and more even distribution of the animals when considered over the course of a year or grazing season (Barnes et al. 2008), but much more concentrated distribution when considered over shorter periods of time, for example, within a given day or for the length of a graze period. This has the advantages of: (1) decreasing plant selectivity, so that preferred species are not placed at a strong competitive disadvantage with their neighbors; (2) if grazing periods are short, more of the forage consumed consists of "first bites" from the plants on offer (Derner et al, 1994), and one defoliation event not only favors nutritional selectivity but also favors production from residual photosynthetic material; and (3) intake distribution can be spread over a larger part of the total landscape.

Enhancing Plant Composition and Productivity

In more humid and sub-tropical rangeland, forage matures quickly. The taller grass species lignify as they mature and periods of growth are longer than in more xeric areas. In higher rainfall regions, rotational grazing can be used to maintain plants in a vegetative phase of growth that results in higher forage growth rates and presents the animals with higher quality forage. A long rest period or a low grazing pressure allows plant tissues to mature and forage quality to decrease compared to more frequent grazing intervals. Multi-paddock grazing systems management must specifically address the goals of animal productivity and vegetation stability or improvement. As grazing intensity increases to an optimal level, primary production increases above that of ungrazed vegetation, followed by a decrease at greater grazing intensities, provided adequate moisture and nutrients are available for plants to regrow (McNaughton 1979). Evidence supports the grazing optimization hypothesis at both the plant and community level (Belsky1986; Milchunas and Lauenroth 1993), and several potential mechanisms within plants have been hypothesized and tested (McNaughton 1983). Critically, plants respond along continua of water and nutrient availability, and they produce best when they have the resources necessary to regrow following defoliation. Management-intensive grazing can enhance the likelihood of these conditions occurring (Gerrish, 2004).

Under season-long continuous grazing, herbivory can be too intensive and frequent in preferred grazing areas for the occurrence of compensatory growth. Multi-paddock grazing management can regulate the frequency of defoliation of preferred plants and preferred areas in systems where regrowth during a graze period is likely, and afford adequate recovery before regrazing if grazing periods are short enough (Derner et al, 1994). The grazing pattern required to increase primary production mimics migratory herbivores because there is a period of intensive grazing, often early in the season when resources for regrowth typically are available, followed by a long period of little or no grazing (Frank and McNaughton 1993). An absolute increase in plant growth occurs under intensive grazing systems only if plants are not subject to chronic defoliation and have the time and resources to recover following defoliation. Continuous grazing does not allow for recovery on heavily grazed patches (Teague and Dowhower, 2003).

Plants growing in xeric environments with erratic precipitation and areas dominated by temperate C_3 grasses face a much greater challenge from herbivory compared to plants growing in more mesic environments. Under these circumstances, flexibility in the duration and timing of grazing is critical. These environments do not have a well-defined rainy season; growth rates are relatively low, of very short duration, and sporadic through the growing season. Forage quality does not decline as precipitously with maturity. The benefits of shorter graze periods may be relatively small compared to the advantage of longer recovery periods. In these environments the provision of planned recovery periods, as outlined in the previous section, are vital for maintaining or improving range condition, and rotational grazing has a significant role to play in reversing or avoiding the damage caused by patch and area-selective overgrazing that occurs with continuous grazing (Fuls 1992; Snyman 1998; Tainton et al. 1999; Teague et al. 2004; Barnes et al. 2008).

Enhancing Animal Nutrition

Plant health and animal nutrition are linked. To maximise animal performance in more mesic areas, leaf biomass must not be too low or too high, so relative growth rate (RGR) and digestibility are both kept high (Voisin 1959; Booysen 1966; Booysen and Tainton 1978). At low leaf biomass, and high leaf biomass as plants reach the reproductive phase, RGR is low. Maximum RGR occurs at intermediate levels of biomass when plants are vegetative and kept in a leafy condition. Forage digestibility is high at low biomass and decreases as biomass increases. Therefore, managing for intermediate levels of biomass increases forage RGR and ensures high levels of forage digestibility. This principle was applied in the rangeland rotational grazing management systems of Booysen (1969) and Tainton et al. (1999).

Under continuous grazing, the whole area has a low mean RGR because both heavily grazed patches and avoided patches have low RGR. Grazing systems with a number of pastures can be managed for much of the area to exhibit high RGR. Rotational grazing in productive environments where water and nutrients are adequate for regrowth, such as tallgrass prairie, can maintain plants in a vegetative state for a longer time and reduce the probability of plants reaching a reproductive stage (Burke et al. 1998). The importance of length of grazing period and length of rest are well illustrated by a rotational grazing study in southern African rangeland of Zimbabwe where graze periods of 5, 10, or 20 days were combined with recovery periods of 15, 30, 35, 60, 70 and 140 days (Denny and Barnes 1977; Barnes and Denny 1991 cited by Norton 2003). Animal performance declined linearly from ~70 to ~50 kg/head/season with increasing length of grazing period and from ~70 to ~40 kg/head/season as the recovery period increased from 15 to 140 days. The more positive results were obtained by grazing moderately for a short period and regrazing before the maturing grasses had decreased in digestibility. At both high and low stocking rates, livestock production increased as grazing periods became shorter and utilization was correspondingly less severe. Length of the grazing period appeared to have more effect than length of the rest period. In a similar study in South Africa, a trend toward higher forage production at moderate utilization under short grazing periods occurred consistently as the grazing period was reduced from 20 to 10 to 2 days (Tainton *et al.* 1977; Morris and Tainton 1991). The response to shorter grazing periods was greater than the response to longer rest periods, which varied from 20 to 60 days in their 24-year study (Norton 2003).

One of the primary reasons for a short graze period, often overlooked in grazing studies, is to allow an adequate opportunity for animals to select a high quality diet from a mixed sward or landscape with a diversity of topographic, edaphic, and vegetational features. Allowing this selectivity can promote diversity of vegetation structure at different temporal and spatial scales matching or exceeding that in continuously grazed situations (Derner et al, 1994). Because of their diet preferences, as animals graze a paddock they will progressively deplete the palatable plants that contain higher quality levels to meet their dietary requirements. This would, in effect, increase the proportion of low-quality forage available to them as the graze period progresses. In the first few days of a fairly long graze period, the animals will harvest the highest quality components of the plant community. As time goes on, they will be forced each day to take a higher proportion of the lower quality forage. Consuming the lower quality forage will lower performance. Practiced consistently over time, "eat the best and leave the rest" degrades soils and plants as well as animal performance and landscapes (Provenza 2003a; b).

One misunderstanding of many scientists and land managers unfamiliar with sound grazing management guidelines is that the purpose of "intensive grazing management systems" is always to reduce the selectivity that can be expressed by livestock and increase the uniformity of grazing (e.g. Briske et al. 2008), creating a "mowed lawn" effect. While animal distribution can be remarkably influenced by planned grazing management, animals still express selectivity in their diet selection and, therefore, exhibit patchy grazing to a greater or lesser degree (Hunt et al. 2007). In a continuously grazed situation, with a given stocking rate and some opportunity for regrowth, the animals will not be forced to reduce selectivity to the same extent, because they will continuously re-graze higher quality regrowth. This causes overgrazing of the preferred plants and degradation in grazed patches. In a planned grazing situation where herbage allowance is sufficient the animals will still be able to meet their requirements. As animals better learn which plants to mix in their diets a greater proportion of the total available vegetation would meet their requirements if consumed in the right temporal framework, so there is often a period of time when performance lags behind that of continuously grazed animals (Provenza, 2003a). Therefore, grazing periods should be kept short enough so that the animals can maintain sufficient diet quality to meet performance goals.

Heitschmidt and Taylor (1991) argue that converting a 1000 ha paddock into a 4-paddock rotation will depress livestock production. They base this thesis on the premise that a rise in grazing pressure (forage demand relative to forage available) causes greater competition for the smaller forage resource in a rotation paddock, and therefore less opportunity for dietary selection and a decline in value of nutritional intake. The flaw in this reasoning is to assume that all forage in a paddock of 1000 ha is equally available to the livestock, and that utilization is spatially even (Norton 2003). On the contrary, patterns of use always emerge with heavy grazing in patches and many areas overlooked. These patterns tend to be reinforced through time via a network of established tracks and behavioral expressions of territoriality. The forage consumed in a large paddock continuously grazed is not necessarily of higher nutritional value than forage consumed in a rotation paddock stocked at higher density. From another perspective, the subdivision of a 1000 ha paddock into four 250 ha paddocks will almost certainly place livestock in parts of the landscape that were previously neglected, and grazing those locations creates a *de facto* increase in forage available over that encountered prior to subdivision. A rotation will also ensure a definite period of recovery for all grazed plants, which is precluded under a continuous grazing regime. By using grazing management to better control where and when livestock graze the manager can achieve modest ecological impacts by shortening the grazing period. At the same time he can favor animal production by improving distribution and increasing exposure to forage resources over the landscape, and by preventing exposure to heavily used areas with depleted forage. The degree of such control over timing of occupancy of any part of the ranch, and the potential for production benefits, is a function of the number of paddocks at the manager's disposal for an individual rotation cycle.

Herbivore Learning and Diet Mixing

Grazing management influences what animals learn: continuous grazing at low stock densities encourages selective foraging, whereas management-intensive and short-duration grazing at high stock densities encourages animals to learn to mix their diets. The relationship between learning and diet mixing in grazing management plays a critical role. Herbivores learn to optimize intake of foods in a manner consistent with their previous experiences with the mix of foods offered (Provenza et al. 2003). When they learn to eat only a small subset of the more "palatable" foods that provide adequate nutrition, animals are unlikely to learn about the possible benefits of mixing different foods, especially those high in secondary compounds (Provenza, 2003b), nor over generations will they become locally adapted to the areas where they must make their livings (Provenza 2008). Over time, such selective foraging on pastures and rangelands will change the mix of plants on offer, further reducing opportunities to learn. However, herbivores encouraged to learn to eat all plants on a landscape are more likely to learn to eat mixtures of foods that enhance health and nutrition, while mitigating toxicity, assuming appropriate choices are available. Experience and the availability of alternatives both influence food choice (Villalba et al. 2004) and animals can learn to eat unfamiliar food (Shaw et al. 2006). This requires careful management during the familiarization and adaptation periods and ongoing monitoring to achieve the objective of getting the animals to consume a greater variety of plants in the landscape.

When a transition is made from continuous grazing at low stocking rates to multi-paddock management involving short-duration grazing at high stock densities there is an immediate negative effect on animal performance until the animals learn to mix their diets and expand the species they consume, and this can last for up to 3 years in mature animals (Provenza 2003a). Such management, which if implemented with high utilization is known as high intensity, low frequency rotational grazing (HILF; Heitschmidt and Taylor 1991), has been shown in experiments to improve plant communities but have negative impact on animal nutrition and performance (Kothmann 1980; 1984). However, with perseverance such negative results can be overcome, as illustrated by Ray Banister who manages 7,200 acres of hardscrabble rangeland in eastern Montana (Provenza 2003a). His management style evolved over 40 years from reliance on rotational grazing that involved relatively short periods of grazing and rest to boom-bust management that consists of intensive periods of grazing followed by a 2-year period of rest. Banister's boom-bust grazing management stresses soils, plants and herbivores with infrequent intensive grazing pressure, and then allows them to recover. Occasional disturbance, followed by rest, creates and maintains a diversity of micro and macro habitats. Heavy use of all plant species keeps undesirable plants in check. Abundant plant cover in the uplands and riparian areas mitigates soil erosion, which leads to clean water and great habitat for fish, waterfowl, and terrestrial species of wildlife.

The change to boom-bust grazing challenged the cattle on his ranch because they were no longer allowed to eat only the most palatable plants as they had under rotational grazing. Instead, they were forced to eat all of the plants. Under the new management procedures, Banister monitors the least palatable plant species, shrubs like sagebrush and snowberry and various weeds, as indicators of when to move the cattle to a new pasture. Cattle are allowed to move only after their use of the unpalatable species reaches high levels. In so doing, Banister reduces the competitive advantage unpalatable plants have over more palatable species. Heavily grazed plants are at a disadvantage when competing with ungrazed plants for moisture and nutrients. It took Banister's cows 3 years to adapt to the boom-bust style of management. During that time, the weaning weights of calves plunged from robust animals well over 230 kg to scrawny individuals that weighed closer to 160 kg, and then rebounded back to over 230 kg.

Under boom-bust management, adapted cattle begin to eat formerly unpalatable species like snowberry and sagebrush as soon as they enter a new pasture. The cows evidently have learned how to mix their diets in ways that better enable them to eat both the palatable and the unpalatable species. Once the older cows made the transition to a new way of behaving, the young calves were able to learn from their mothers how to thrive under boom-bust management. The calves that Banister keeps as replacements never have to make the harsh transition. They were trained by their mothers that all plants are food at Banister's place. Such learned patterns of foraging behavior are transferred culturally from one generation to the next (Provenza 2003a; 2008). We do not advocate this form of management but it illustrates well how animals can learn and modify their behavior.

Flexibility for Variable Weather

Management is not difficult when forage quality is reasonably good all the time, and when resources for plant growth are abundant and predictable. What makes rangelands different, and managerially difficult, is that management skills must be applied to compensate for poor seasons, poor quality forage, and droughts, all with a high degree of uncertainty. In South Africa, protocols were developed for managing both pauci- and multi-paddock grazing systems in climatically variable semi-arid rangeland to get the best animal performance and improve the vegetation while avoiding the negative consequences to range resources from continuous grazing (Figure 1; Venter and Drewes 1969, cited by Tainton et al. 1999). The approach regulates the level of defoliation and provides growing season recovery after each grazing period, as well as a drought buffer. Animal performance is maintained by keeping highly digestible, leafy forage in front of livestock, and not grazing too heavily (rotational but not HILF grazing).

This approach is flexible enough to work in wet or dry seasons. In years of average or above-average rainfall, animals are rotated through the non-rested areas to achieve moderate to low defoliation and experience recovery until the forage is ready to graze again. The moderate level of defoliation keeps both the forage quality and plant growth rate high. In dry years, the same principles of moderate defoliation and recovery before regrazing are applied. However, as growth is slower in dry years, more paddocks are grazed before the first grazed paddock is ready to be grazed again. In very dry years, this can be achieved only by grazing the area that was planned to be rested. Thus, the rested area serves as a forage buffer during drought. If consecutive dry years are experienced then animal numbers can be reduced or other pastures can be rented.

In more arid areas of southern Africa, rotational resting of a third to a fourth of the grazing area during the growing season provides a forage buffer (Danckwerts 1984; Tainton et al. 1999; Müller et al. 2007), which stabilizes animal numbers and cash flow from year to year. A low-intensity system with 3 to 4 paddocks per herd provides a rotational rest every 3 to 4 years. For greater production, a more intensive, multi-paddock system (> 12 or 16 paddocks per herd) can be implemented with rotational grazing conducted in the paddocks not being rested each year as discussed by Müller et al. (2007) and Beukes et al. (2002).

In the USA, where it is relatively easy to change livestock numbers, successful managers can regulate stocking rate as an alternative. They either reduce the number of animals grazed on contract for other enterprises or keep no more than half the livestock biomass as breeding stock so they can adjust stocking rate quickly before large numbers of people have to sell and markets drop, and minimize the loss of valuable genetics. Both options are less costly and damaging to the range resource.

In any rotational grazing management program, forage within a paddock will be depleted at a more rapid and noticeable rate per unit of time than in a similarly stocked continuously grazed system. If forage becomes depleted or is more abundant at, for instance, the end of the rainy season, then experience indicates that it would provide a much earlier warning to a manager that a stocking rate adjustment may be needed. These more timely responses in the form of stocking rate adjustments and changes in the grazing schedule are a risk management aspect of properly planned and managed grazing that is difficult to quantify scientifically, but is extremely important regardless.

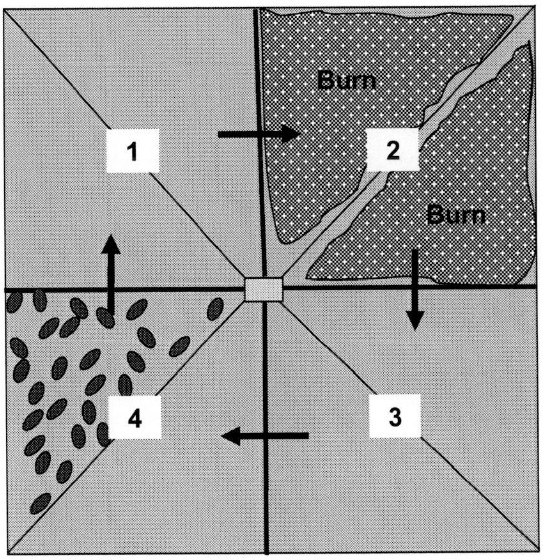

Block 1. Rest for a full growing season but use as emergency drought grazing reserve

Block 2. Apply spring burn followed by RG when ready to graze. 1st grazing priority, graze again whenever ready

Block 3. 2nd in priority for grazing after block 2 that has recently been burned

Block 4. 3rd priority grazing when neither the 1st or 2nd priority blocks are ready to regraze

Year 1. Following a spring burn to remove residual low-quality material, grazing should be closely controlled and selective grazing minimized using Rotational Grazing (RG). The most recently burned paddock is grazed whenever it has recovered sufficiently from the previous grazing, but before the forage has matured and become unacceptable to animals.

Year 2. In the second season this paddock is rated second in priority, the newly burned paddock assuming the priority position. Selective grazing should be controlled as effectively as possible during this second season using RG but, in practice, unacceptable material invariably begins to accumulate because of the reduced grazing control. The now maturing less preferred (Increaser) plants cannot be grazed without damage to the more preferred (Decreaser) species in the sward, and so the development of tuftiness becomes unavoidable.

Year 3. During the third and subsequent seasons (where there are more than four paddocks) selective grazing becomes progressively worse as the more recently burned paddocks receive priority treatment. At the beginning of the growing season when forage growth is slow, and in dry years, most of the paddocks will be grazed before the 1st priority paddock is ready to graze again. In wet years, and from the fourth year onwards, the paddocks may remain ungrazed during the peak growth period and they will largely be used as buffer grazing areas during periods of slow growth when the preference paddocks cannot supply sufficient grazing.

Year 4. The paddock is eventually rested for a full growing season and burned (where there is a large quantity of residual material, burning may also often be advisable in the spring prior to the paddock entering rest). The paddock then enters the cycle once more as the first preference paddock.

Figure 1. Open Camp Grazing management (Venter and Drewes 1969; Tainton et al. 1999).

Importance of Fire

In less arid rangelands, fire is important for both continuous and rotational grazing management, as it alleviates the effects of repeated selective grazing of the same areas year after year. Nutritious regrowth following a burn attracts herbivores, and the ensuing heavily grazed patches are maintained by their continued attractiveness to grazers. When a new area is burned, the new regrowth attracts grazers. As a consequence, grazing pressure is spread over the whole burned area for a period, reducing the pressure on the previously heavily grazed patches. In time the selective grazing of patches and species resumes, so that the patch-masking effect of fire disappears completely after about 3-4 years (Archibald et al. 2005).

In systems with frequent fires, however, these intensively utilized patches do not persist and the landscape can become dominated by tall, fast-growing grasses intolerant of grazing but highly flammable. Managing with fire can, therefore, also reduce the herbivore impact on intensively grazed patches that naturally develop in the landscape by spreading grazing over a larger area for a short period, reducing the grazing pressure on previously heavily grazed areas (Archibald et al. 2005). In addition, patch burning is a relatively new management technique being implemented where a portion of a pasture is treated to increase pasture heterogeneity (Fuhlendorf and Engle 2004). By burning different patches in subsequent years, grazing pressure is reduced on preferred areas while the animals concentrate on the recently burned patches. This natural rotation driven by animal behavior offers some rest and recovery while also providing structural heterogeneity that can improve wildlife habitat. By controlling distribution and providing adequate recovery for heavily utilized areas on the landscape, the need for fire in this context may be less pressing.

Accommodating Extensive and Intensive Management Philosophies

Not all managers are comfortable with managing very intensively using pauci- or multi-paddock systems. Risk-averse behavior, resistance to change, and differences in capabilities of managers, as well as the financial constraints associated with infrastructure development in drier regions, dictate that different levels of complexity need to be developed to provide a choice of management strategies to obtain satisfactory productivity and sustain or improve the range resource.

Keeping risk at an acceptable level is most important for rangeland management. Stocking rate should match forage availability in both wet and dry years by allowing for adequate plant residual biomass to enable rapid regrowth following grazing, and by having buffer areas available. Keeping areas of the ranch dedicated as buffers to accept stock during a shift in stocking rate enhances managerial flexibility and is a short-term strategy preferable to marketing animals. Rotational resting and rotational grazing should ensure improved forage plant composition and productive potential so the effects of drought are decreased and there is speedy recovery after drought (Teague et al. 2004). Conversely, grazing management strategies that facilitate patch degradation increase pressure on desirable plants already weakened by heavy use (Norton 1998) and slow recovery post drought.

At the low-intensity end, pauci-paddock systems allow sufficient growing season deferment at relatively low stocking rates. Systems such as the Merrill 4-paddock 3-herd system for ranges grazed year-long have given excellent results with less intensive management (Heitschmidt and Taylor 1991; Taylor et al. 1993) as have the 3 or 4 paddocks per herd systems mentioned above incorporating growing season rest with or without rotational grazing (Danckwerts 1984; Tainton et al. 1999; Müller et al. 2007).

Those aiming for maximum sustainable yield generally use the more intensive multi-paddock systems with > 16 to 30 paddocks per herd and usually stock at heavier rates (Norton 1998). This increases operational risk significantly as maximum sustainable yield is likely to be at a critical point, the location of which is not exactly known and variable between seasons given the unpredictability of nature. A slightly higher stocking rate or weather change can drive the forage base to unacceptably low levels so it is a problem determining how conservative management should be, a problem encountered when managing all natural resource stocks (Walters 1986). However, producers who manage systems with 30 paddocks or more often find that there is forage to spare and actually less risk, which is due to a combination of better grazing distribution, quite short grazing periods and long rest periods that increase heterogeneity of landscapes. Some producers have found that the multi-paddock system reduces stress and frees up time, which is partly due to the greater facility for better planning of forage resources. The management activities and management decisions may be more intensive, but overall the intensive rotation incurs lower ecological risk.

The Case of High Stock Density Grazing

The most management-intensive form of rotational grazing involves using small paddocks to achieve high stock density with relatively short grazing periods to prevent heavy use. A large number of paddocks per herd are preferable, usually 30 or more, and grazing periods are 7-10 days or even less, so rest periods are automatically long. Having a large number of paddocks per herd gives a manager a lot more flexibility to achieve plant and animal production goals using the guidelines discussed above. We discuss below the implications of using relatively few (4 to 8) compared to a larger number (16 to 50 or more) of paddocks per herd. It is very important to remember that no matter how many paddocks are involved or how sophisticated the management, if the stocking rate is too high for the amount of forage available there will be negative consequences for animal performance, natural resources and profits.

The need for recovery is a function of the severity of defoliation; the more severe the loss of leaf, the more recovery time is required to restore herbage yield, for all the reasons explained above. For simple rotations involving 4-8 paddocks, relatively long grazing periods occur, and paddocks on a commercial operation are large enough for patch grazing to develop to a damaging degree. In such circumstances, the length of the rest period becomes critical for protection of target species. As paddock numbers increase, length of graze period shortens considerably, more of the forage harvested is consumed with one bite from plants on offer, species selectivity declines, severity of grazing becomes less of an issue, and recovery times take care of themselves. As O'Connor (1992) demonstrated, at high stock density species identity has a minor role to play in determining whether any plant is grazed or not.

In multi-paddock systems operating only 2 to 3 paddocks in a deferred rotation, there is little risk to grazing one paddock at a time apart from the fact that periods of occupation are long. If feed supply runs short, the rancher can move the animals a few weeks earlier than planned without causing any negative repercussions. The grazing and rest periods are long enough, and the paddocks large enough, to accommodate a built-in buffer of reserve forage accessible under higher (but not too high) grazing pressure even if patch selection is not reduced. On the other hand, with only 4-12 paddocks in the rotation, managerial flexibility is considerably reduced. Small errors in judgment can cause negative consequences for plant utilization or animal performance. There is only a small built-in forage buffer to accommodate an increase in rate of rotation, and dropping a paddock out of a rotation cycle is scarcely an option. Most of the research trials of rotational grazing have been done in this range of 4-12 paddocks, a relatively inflexible and ecologically risky strategy with little latitude for avoiding undesirable effects of climatic uncertainty.

At the other end of the scale, in multi-paddock grazing with, for example, 30 paddocks, the system is also less risky to manage. A paddock is being grazed for only 12 days of the year (maybe 2 grazing periods of 6 days each on average) and rested for at least 350 days a year. If two paddocks contain insufficient feed, or present a toxic plant problem, you can skip them in a rotation cycle and the rest period for the grazed paddocks in the cycle drops from 25 weeks to 23 weeks, not enough to really depreciate the benefit of a long rest period. Meanwhile, grazing period and stocking density in the grazed paddocks may remain unchanged. Even if the manager decides to reduce a 6-day grazing period to 5 days grazing in all 30 paddocks, the rest period is still about 4 months, which should be adequate for recovery after only 5 days of grazing. This is still only 12 days of grazing per paddock per year.

Rotations involving 20, 30 or even 50 paddocks are not uncommon on commercial ranches. With an increasing number of paddocks, the manager's anxiety should decrease. The risk aspect can be considered from the point of view that if the ranch can be managed so that the same number of stock are taking their annual feed requirements from more of the landscape than was used before, and from more of the plants that grow on the ranch than were grazed before, then the consequence has to be more benign ecological impact and more productive pasture. If there appears to be excess forage on the ranch as a result of this strategy, then stock numbers can be judiciously increased to harvest the surplus, but on the proviso that if conditions deteriorate, some stock must be removed from the ranch or shunted into buffer paddocks.

The impact of high-density grazing under intensive rotational grazing is ecologically benign due to the combination of low selectivity, a small percentage of preferred species experiencing repeat defoliations, and the inherent long rest periods. Because the grazing impact becomes more benign as the number of paddocks in the rotation increases, the need for a long recovery period declines. A factor reducing the need for a long recovery time is the botanical character of palatable species. They persist in grazed vegetation because they invest in fast-growing photosynthetic tissue rather than high-energy-demanding secondary compounds and physical discouragements to herbivory (Bryant et al. 1983; Coley et al. 1985). If selectivity is being reduced and all species in the pasture are experiencing some level of defoliation, the palatable species are at a competitive advantage to out-perform their neighbors in regrowth.

For the doubters who believe that intensive multi-paddock management is damaging to the natural resource base, Norton (1998) listed 9 examples of grazing trials (from Canada, USA, Zimbabwe, Australia and New Zealand) in studies that ran for 5-35 years. These trials reported no adverse ecological effects of either continuous or rotational grazing treatments even though experimental stocking rates were maintained at 40% to 200% above those recommended for commercial properties in the area. He asserted that when using small paddocks, forage availability is not limited by grazing animal distribution. This supports the belief that it is beneficial to use a high number of smaller paddocks.

Increased stocking rates do have negative effects on soil properties (increased bulk density, disruption of biotic crust, reduced aggregate stability and aggregate size distribution) and are positively correlated with the distribution and frequency of animal trampling (Warren et al. 1986; Gerrish 2004). However, Thurow (1991) reported that surface hydrology attributes, such as infiltration and sediment production, are related almost entirely to the amount and distribution of plant cover and are largely unresponsive to rotational grazing, except when stocking rates and cover removal exceed recommended levels, or precipitation events follow too closely on heavy grazing events. The same study indicated that managing for grassland dominated by high seral plants improves hydrological function. High stocking rates in this study showed strongly reduced infiltration rates due to physical effects on the soil and changes in the vegetation towards dominance by lower seral plants. At a heavy stocking rate, infiltration was much higher in an intensively run, multi-paddock rotational grazing system than in a continuously grazed treatment *at the same stocking rate.* This improved hydrologic function may be associated with the improvement noted between trampling events seen by Warren et al. (1986). Similarly, Pluhar et al. (1987) found that infiltration increased and sediment production declined as vegetation standing crop and cover increased. They found no differences in infiltration and sediment yield among a rotationally grazed treatment at a heavy stocking rate and the moderately stocked, deferred rotation and continuously grazed treatments.

A misconception among many opponents of planned multi-paddock grazing is that they think it necessarily decreases the diversity of plant species and the heterogeneity of landscapes. Depending on the circumstances, just the opposite can occur. Smaller paddocks can improve distribution of animals across a landscape. This can increase or decrease diversity, depending on how well animals were distributed before. As discussed, forage demand compared to forage available determines the proportion of plants that will likely be consumed by the animals while they are in a paddock. Then, depending on the manager's goals for structural and species diversity, the greater control and flexibility of animal placement and movement allows them to come back and regraze plants sooner, or allows more regrowth and more structural similarity within a paddock. Small paddocks will allow a "checker board" effect where some paddocks can be closely grazed and neighboring rested paddocks could have tall, untouched growth, creating heterogeneity across the landscape. Thus the manager can decide exactly how best to juxtapose these components to achieve specific plant, livestock or wildlife goals. Likewise, depending on placement of fences and diversity of topography, aspect, soils, and plant communities within paddocks, grazing can be more or less uniform within a paddock for a given graze period.

LIMITATIONS OF RESEARCH ON MULTI-PADDOCK GRAZING

The results of much experimental research are inconsistent with the demonstrated effectiveness on commercial properties in many countries of planned multi-paddock grazing management. The potential for significantly higher production under planned multi-paddock grazing management, consistent with producer experience, can be justified with published research data, using scientific arguments focusing on the temporal and spatial aspects of grazing management and their physiological effects on plant and animal production and complementary relationships. Many grazing system research projects have been conducted with no ecological or production goals and have often ignored relevant ecological research or practical knowledge when choosing and implementing treatments. This lack of treatment goals or consideration of their possible efficacy allows limited opportunity for understanding mechanisms behind observed results. Consequently, the relevance of much small-scale research on grazing systems is of questionable value to commercial ranch managers.

We offer the following reasons to explain these differences in perception: (1) The manner in which researchers have managed multi-paddock grazing treatments has been sub-optimal for providing the best possible vegetation or animal production results; (2) The notion that rotational grazing can control frequency of defoliation within a grazing period is flawed *at the scale of rotations employed in research trials* (Norton 1998); (3) Grazing systems comparisons in small-paddock trials fail to address the problem that continuous grazing in large paddocks causes patch grazing and localized pasture degradation and underuse elsewhere, resulting in low growth rate relative to moderately and more uniformly grazed vegetation; and (4) The omissions in grazing research of a spatial dimension and consideration of the effect of grazing management on animal activities have created a communication gap between scientists and commercial producers, for whom landscape features and herd behavior are a prominent aspect of their production system.

Research experiments are structured to minimize variability when testing hypotheses to promote understanding of ecological processes. Yet, in nature individuality and variability are the rule, not the exception to the rule. Thus, an effectively designed grazing experiment may not necessarily represent the most flexible and hence effective way to manage a ranch enterprise for production or conservation goals. Grazing treatments may be intentionally severe to induce ecological contrasts, and they are usually applied more rigidly than the adaptive management practiced by many grazing managers. Effective management must be adaptive to changing conditions and environments, and timely monitoring information is crucial to this process (Danckwerts et al. 1993). Consequently, research results may be inapplicable or misleading if not placed within an adaptive management framework managing for sustainable use at the landscape scale.

Researchers have rarely taken into account how the manner of conducting experiments has contributed to the results obtained. Any experiment is merely an inflection in time and space of biophysical processes that link soils, plants, herbivores and people. If the same treatments were applied while managing for the best ecological, social and economic outcomes, how would the results differ? For example, as mentioned previously, there was confounding in many experiments between continuous grazing at moderate stocking rates and much heavier (often double) stocking rates with intensive rotational grazing treatments (Briske et al. 2008). The vast majority of grazing studies cited above compared responses to continuous season-long grazing at recommended stocking rates that optimize livestock gains per head or per unit land area (e.g., Bement 1969) with responses to the short-duration grazing at a higher stocking rate, often 1.5 to 2 times greater. Stocking rate, in any given circumstance, has greater effects on animal and vegetation responses than grazing system (Van Poollen and Lacey 1979; Heitschmidt and Taylor 1991; Manley et al. 1997; Gillen et al. 1998; McCollum et al. 1999).

This confounding with different stocking rates between systems being compared is further exacerbated by the variable weather in semi-arid grazing ecosystems. Accepted good management during droughts commonly includes reducing stock numbers early to reduce damage to the vegetation, animal condition and profitability. These patterns at best mimic natural systems in which the number of animals decreases during a drought and then gradually increases as a drought diminishes. Experiments have very rarely done so. If they did, the higher stocking rate treatments might not incur detrimental effects and would be consistent with accepted good management practices. The currently perceived "correct" management of an experiment is often not the best way to manage in a production ranch setting. It is very important to take these differences into account when giving management advice. Whether managers use continuous, pauci- or multi-camp systems they need to know how to manage their ranches to get the best results and avoid the problems associated with each system. To do this they need to know the benefits of each management system if it were managed to give the best result. Most experiments comparing continuous with rotational grazing systems set the rotations, rather than manage them for best animal or plant responses within the constraints of the system.

As outlined above, the small spatial and temporal scales at which most experimental grazing systems research is conducted has produced contrasting viewpoints between research scientists and successful grazing managers. These scale effects are largely ignored when extrapolating experimental results to large-scale ranch operations. The short time frames of most studies, and the lack of awareness of adaptation troughs for soils, plants, herbivores and people, ignore realities of systems as they transition to different modes of use (Provenza 2003a). The majority of experimental paddocks have been considerably smaller (usually < 25 ha and often < 5 ha) than pasture sizes of commercial ranches (500 to 5,000 ha or more; Norton 1998). This juxtaposition of scales has had profound consequences for the translation of experimental results to the management of grazed ecosystems. Small research paddocks that are grazed continuously do not always mimic the continuous grazing of large paddocks because they preclude evaluation of uneven landscape utilization (Norton 1998; Teague and Dowhower 2003; Teague et al. 2004). Smaller experimental paddocks usually result in a more uniform distribution of grazing pressure (Barnes et al. 2008), which restricts ability to evaluate how grazing animals use landscapes. There should be greater awareness of the problems in extrapolating results from small scale experiments to grazed ecosystems at larger scales. For example, the work of Gammon and Roberts (1978) and O'Reagain and Turner (1992) reported that defoliation is not always controlled more effectively in intensive than in continuous grazing systems, and that forage quality and quantity are not consistently and substantially increased in intensive systems compared to continuous grazing. However, these interpretations may have been different if pasture size had been hundreds or thousands of hectares rather than a few hectares. Similarly, work reported by Derner and Hart (2007) and Hart et al. (1988) suggested there were no differences in performance of animals or vegetation between continuous grazing and a treatment including deferment. It is probable that in this circumstance, as in other small-scale experiments (24 ha), the continuously grazed paddocks would have received a more uniform defoliation than would occur in the larger paddocks encountered on commercial ranches that also varied in biophysical characteristics. Thus the patch and area-selective grazing that causes deterioration under continuous grazing would have been underestimated.

Grazing system research generally has not acknowledged other research on the effects of grazing at a large scale (see Coughenour et al. 1985; Fuls 1992; O'Connor 1992; Bailey et al.1996; Teague et al. 2004; Archibald et al. 2005), which demonstrates the often detrimental effects of patch and area-selective grazing in the landscape. Teague et al. (2004) found that in large pastures, 1500-2000 ha in size, resting provided by rotational grazing reduced deterioration and facilitated improvement of shortgrass and midgrass patches associated with patch and area-selective grazing. A phenomenon such as enhanced productivity based on individual plant and community response to defoliation, that is not detected at a small scale, is often likely to emerge at a larger scale (Levin 1993; Bissonette 1997; Turner et al. 2001), particularly if there are interactions between grazers and plants that affect productivity at larger scales as has been shown by Frank and Groffman (1998).

Failure to consider the rate and scale of ecosystems response to changes in management activities may further minimize the value of translating experimental research to management situations. Such effects are unlikely to be detected in small-scale experiments conducted over 2 to 3 years. It is very difficult to determine treatment differences in rangeland ecosystems because of the slow or erratic response times that may be triggered by reactions to stochastic events such as climatic fluctuations that interact with management actions (Danckwerts et al. 1993), but these effects are critical to determining what is sustainable and what is not. Watson et al. (1996) and Walker (1988) emphasize that, even though climate is the most important driver in rangeland ecosystems, management is still critical to sustainable management. Watson et al. (1996) further state that emphasizing climatic or other event-driven or episodic drivers may de-emphasize the importance of management that can better take advantage of these events within a relatively stable plant community. Management directed towards soil moisture, seed production, etc., can "condition" the resource to take fuller advantage of these episodic events when they occur, particularly if the event is of a marginal magnitude (Gerrish 2004). One interesting question that comes from the discussion of Watson et al. (1996) whether such conditioning can move a plant community from one stable state to another when some advantageous circumstance occurs (e.g. a particularly wet spring following a drought or wildfire). Time in the order of decades may be needed to evaluate these types of changes on rangelands (Burke et al. 1998).

Advocates of rotational grazing have long contended that timing of grazing and recovery periods according to plant growth rates is of central importance to their success (Savory 1983; McCosker 1994), and that many failures may have been due to not slowing down the rate of rotation during slow or no growth to provide for a non-grazing interval long enough for complete recovery (Savory with Butterfield 1999). The benefits of rotational grazing in many studies may have been reduced or lost due to calendar-based, multiple-cycle rotations (e.g., Kirby et al. 1986; Burboa-Cabrera et al. 2003; Derner and Hart 2007; Hart et al. 1988; Hart et al. 1993). Recommendations asserting no benefits from periodic recovery in these circumstances must be suspect because the benefits are dependent on timing relative to conditions adequate for plant growth (Mullahey et al. 1990; Mullahey et al. 1991; Reese et al. 1996; Cullan et al. 1999). Many grazing studies have used insufficient recovery periods (e.g., 2 months during dormancy) even when the rotation was flexible (e.g., Bryant et al. 1989; Walker et al. 1989).

Another significant factor that can influence research results is the time the different elements in a system take to change from previous pre-experimental management. A shift from continuous grazing at low stocking rates to intensively managed rotational grazing at the same or higher stocking rates will affect many ecosystem variables, including soils, vegetation, livestock and associated herbivores at different temporal and spatial scales, and it may take several years after making consistent, substantial, correct management changes for the system to adapt to these new conditions (Provenza 2003a; 2008), and decades for changes to be measurable at the landscape level. However, they may respond relatively rapidly in preferred grazing areas or those receiving run-on or sub-surface moisture (e.g. Teague et al. 2004). Therefore, some means of stratified measurement of vegetation responses for grazing research studies and of stratified monitoring to obtain timely monitoring information to facilitate management decisions as advocated by Danckwerts et al. (1993) would be desirable. Developing protocols that are time efficient and repeatable is ripe for scientific investigation, as currently being pursued by people such as Gregg Simonds (personal communication). Animals accustomed to low-density continuous stocking can be trained to increase harvest speed and efficiency, but this takes time and a few never adjust (Provenza 2003a; 2008). Research experiments that operate for short periods following treatment imposition may capture the period of system adaptation and underestimate the potential of long-term intensive grazing systems. When well managed, long-term intensive grazing programs improve after the adaptation phase (Merrill 1954; Provenza 2003a). These factors have largely been ignored in research projects, or there have been constraints to taking them into account.

Future Research

Attempts to study grazing systems are complicated by many issues, not the least of which is having adequate resources. With limited land and funds, researchers have been forced to conduct experiments on small areas of land. The necessity for comparable land attributes between replicates only exacerbates this problem. As changes occur relatively slowly in rangeland ecosystems it is unreasonable to expect meaningful results from research conducted over less than 10-year periods, for reasons outlined by Burke et al. (1998). Indeed, given the dynamics of landscapes, it is reasonable to ask if soils, plants and herbivores ever reach any sort of equilibrium (Provenza 2003b; 2008). While certain questions can be answered from small-scale, short-time-frame operations, the applicability of such research should not be overextended. These limitations may be overcome by monitoring biological processes related to soils, plants and herbivores on ranches managed successfully for many years, often decades, and using systems-level simulation modeling. Comparing the experiences of successful and unsuccessful managers will provide insights unlikely to emerge on research stations.

There are a number of advantages to monitoring ranches with and without track records of successful multi-paddock grazing management. A case study by Jacobo et al. (2006) compared adjacent ranches that used either continuous or rotational grazing, and similarly, Earl and Jones (1996) studied producer-managed rotational grazing at the ranch scale. The strength of this approach is that researchers could evaluate the entire ranch enterprises within the constraints of respective grazing regimes, including the capacity to adaptively manage for the best possible outcomes. This approach simultaneously evaluates ecological and managerial responses. A number of ranches world-wide have been successfully managed at different levels of sophistication using pauci- and multi-paddock systems. Some have operated successfully for nearly three decades. The importance of data gathered from such management over these long periods of time should not be ignored.

The belief that intensive grazing systems involving many paddocks per herd (>20) will necessarily increase profits has largely not been substantiated, even though an ecological economic evaluation of a ranch in South Africa does not refute this claim (Beukes et al 2002). However, successful ranch businesses that employ such management, for example the Richard's Ranch in Jack County Texas, have demonstrated superior profitability and return on capital investment. The Richard's Ranch has demonstrated superior performance and economic results when evaluated in the US National Cattlemen's Beef Association-Standardized Performance Analysis program (Bevers pers. Comm.[1]). This particular ranch has also won conservation awards from the Society of Range Management, Texas Parks and Wildlife Department, and the National Cattleman's Beef Association. It is important that research examines and documents this topic as well as the impact of such management on the natural resources.

Systems-level simulation modeling can complement both small paddock and ranch-based research as the influence of treatments can be explored without the space, variability, time or cost limitations of traditional small-scale or ranch-based research. Simulation modeling would also develop a sound theoretical base for understanding of processes and linking biophysical processes with observed results, essential elements that have so far been lacking. Developing understanding from large data sets requires theory, and theory often requires models to test understanding (Woodward 2005). Modelling of grazing systems has included such topics as ecological economics (Beukes et al. 2002), stock number management strategies (Hahn et al.1999; Diaz-Solis et al. 2003), spatial issues (Witten et al. 2005; Müller et al. 2006) and achieving a better understanding of field experiments (Teague and Foy 2004).

Alternative Research Hypotheses and Paradigms

The original paradigm of rangeland management was based on the widespread observation that degradation of the range resource was largely due to excessive numbers of livestock. In this paradigm the solution was to reduce stocking rates while allowing season-long continuous grazing to continue. Subsequently, another paradigm was developed following the experience of pioneer rancher conservationists and scientists, who had achieved significant range improvement using growing season deferment to allow recovery periods (Smith 1895; Sampson 1913; Scott 1953; Matthews 1954; Merrill 1954; Hormay 1956; Hormay and Evanko 1958; Hormay and Talbot 1961; Hormay 1970; Müller et al. 2006; Booysen and Tainton 1978; Tainton et al. 1999). A number of earlier researchers confirmed the success of using growing season deferment, often in conjunction with rotational grazing (Rogler, 1951; Merrill, 1954; Reardon and Merrill, 1976; Smith and Owensby, 1978; Daines 1980; Danckwerts et al. 1993; Taylor et al. 1993; Kirkman and Moore 1995).

A third, more radical paradigm was developed in the early 1970s based on earlier writings (Voisin 1959; Acocks 1966) which inspired people such as Savory and colleagues (Savory 1978, 1983; Savory and Parsons 1980; Savory and Butterfield 1999) and Gerrish and colleagues (Gerrish 2004) to explore the merits of multi-paddock, high-density rotational grazing in rangeland ecosystems using grazing periods that were unconventionally short and stocking rates that were considered irresponsibly high. Since then, many ranchers have substantially increased stocking rates while simultaneously improving range vegetation composition using these methods (Goodloe 1969; Tainton et al. 1977; Cumming 1989; McCosker 1994; Earl and Jones 1996; Stinner et al. 1997; Norton 1998, 2003; Sayre 2000; Berton 2001; Gordon 2002). Many in the rangeland science discipline have totally rejected this alternative paradigm, even in the face of much anecdotal evidence (Holechek et al. 1999, 2000; Galt et al. 2000; Briske et al. 2008).

[1] Stan Bevers, Extension Economist, Texas AgriLife Extension, P.O. Box 1658, Vernon, Texas 76385.

It is most difficult in science, as in other fields, to shake off accepted views (Dubin 1978). Many scientists feel threatened professionally when an innovative and nontraditional way of thinking is introduced. That is especially true if: (1) a new way of thinking involves a major shift in the scientific paradigm; (2) acceptance of the new theory implies that currently used practices are inadequate or inappropriate; or (3) the new theory threatens the assumptions of the established paradigm. Obsolescence of knowledge threatens the professional integrity of proponents of that knowledge, or the assumptions of the new paradigm appear so contradictory to the assumptions of the accepted paradigm that it is rejected outright. An example from the medical profession is the systematic and long-term intransigence of the established medical profession to prevent legal acceptance of the chiropractic profession, which is now widely accepted and operates with full legal authority (Lisa 1986).

Traditionally, disciplines operate on the tenets of a single major paradigm (Kuhn 1970), which produces valuable but incomplete understanding. All paradigms are a narrow view of the multifaceted nature of most fields of study (Burrell and Morgan 1979; Frost 1980; Provenza 2000). Different paradigms are grounded in fundamentally different assumptions and produce markedly different ways of approaching and building a theoretical base for any discipline (Gioia and Pitre 1990). Considering and comparing more than one paradigm can generate more complete knowledge than is possible with any single paradigm. A broader approach that accounts for differing paradigmatic assumptions yields a more comprehensive understanding of the processes of nature, and their constantly changing manifestations.

It is important to remember when assessing any hypothesis that a single refutation is sufficient to illustrate that the hypothesis being tested should probably be revised to accommodate what has been learned by such a refutation (Kuhn 1970). The numerous instances from research studies outlined in this document and evidence from scores of ranchers around the world provide solid reasons to modify the hypothesis expressed by Briske et al. (2008) that there is no reason to favor multi-paddock rotational grazing over continuous grazing and conservative stocking. Because hypotheses cannot be proved, only rejected, the role of science is to test alternative hypotheses or paradigms and specifically try to refute them. Consequently, we need to expand our methods of enquiry to include ranch-based research and simulation models to develop and test theories, and constantly check conclusions for any inconsistencies between them and evidence from other sources.

To do so, we must focus not only on comparisons of grazing systems, but on the relationships between biophysical processes and management. While it is certainly possible to understand the processes of nature, and much is known about soils, plants and herbivores, the variation inherent in the manifestation of processes in time and space precludes direct comparisons of grazing systems *per se* in experimental analyses. All the physical and biological variables in the various processes are in constant flux, as influenced by history, necessity and chance, and therefore their manifestations become unique in time and space (Provenza 2000). Managers must work with physical and biological processes to manage landscapes. Optimally, this involves knowledge of processes combined with flexibility to respond to ever-changing environments, and that can't be studied with classical grazing studies. Flexibility in the face of unending change is what plants, herbivores and people are about, and that involves ongoing interactions among genes, environments and chance (Lewontin 2000).

Conclusions

A large body of evidence from controlled experimentation before the mid-1980s has shown effects of defoliation by grazing animals on plants and the benefits of adequate recovery following defoliation. The benefits of multi-paddock rotational grazing on commercial livestock enterprises have been evident for many years in many countries. However, despite these observations and the benefit to species composition found in numerous studies of planned grazing deferment, most recent rangelands grazing studies suggest that rotational grazing improves neither vegetation nor animal production relative to continuous grazing. Detailed comparisons of research methods and practical experience of successful practitioners of multi-paddock grazing management have identified a number of areas that explain why such different perceptions have arisen. The uneven distribution of livestock in continuously grazed large paddocks leads to localised pasture degradation, which has not been accommodated in the design of most research studies comparing continuous grazing to rotational grazing. This oversight also assumes spatial homogeneity of forage availability and utilization, which is refuted by a large body of observations at larger scales.

This failure to take into account plant and animal processes at appropriate temporal and spatial scales has resulted in incorrect interpretations for rangeland management. Research at a small scale diminishes the degree of selective use and impact that animals have over the landscape. This has resulted in many researchers interpreting the herbivore as an amorphous, diffuse defoliator, that plucks forage in random fashion or like a harvesting machine blanketing the pasture, and even when defoliating selectively does so in a spatially uniform way as implied by Briske et al. (2008). In fact the herbivore is an animal with a point-sampling defoliation apparatus, that moves in forward motion and normally walks long distances, that responds to visual and tactile cues and reacts to its surroundings in various ways, that engages in activities other than defoliation, that is a social creature influenced by history, necessity and chance, that has biological limits to bite size and energy expenditure, and that develops patterns of behavior in response to its environment and companions. Grazing ungulates have an entirely different impact on the landscape than that implied by Briske et al. (2008), as is well documented by work at the landscape scale we have outlined earlier in this chapter. This points to an entirely different and more meaningful way of designing and interpreting grazing trials.

Another reason for mixed results is that researchers have often applied treatments that did not adequately consider physiological effects, complementary relationships among soils, plants, animal behavior, preferences and selectivity, and ecological processes like water and mineral cycles. As a result, they often do not address nor provide valid answers to practical questions such as: how good is this management option; where is it successful; and what does it take to make it work as well as possible? Consequently, interpretation of grazing trials by some researchers has incorrectly concluded that planned grazing benefits neither vegetation nor animal production relative to continuous grazing. As we have indicated in this document, unless experiments have been conducted in a manner that aims at achieving the best plant and animal responses, the results will probably be misleading in defining the potential of an experimental treatment. Similarly, when reviewing the literature to draw general conclusions (Holechek et al. 1999, 2000, 2004; Briske et al. 2008), each experiment needs to be examined to see how it was conducted and if the objective was such that the study results could be extrapolated to practical ranch situations. If it was not conducted in a manner that current understanding would define as the potential of the treatment, then the interpretation of the experiment will be spurious and misleading. In addition, if such reviews use only references that support a particular viewpoint and do not relate to what a manager needs to know, understanding of the subject will be clouded and not enhanced. Thus it is essential to address and test alternative hypotheses with equal vigor using comparable management goals.

In contrast to the conclusions of many researchers, numerous commercial livestock enterprises in many countries have used a basic knowledge of plant and animal physiology and ecology within an adaptive, goal-oriented management approach to implement successful planned grazing management programs. When evaluated as a body, comparisons of research methods and results and practical experiences of successful planned grazing practitioners identify a number of areas that explain why such different perceptions have arisen. When evaluated using a paradigm encompassing basic ecological and biological principles, these results provide insights that allow the formulation of guidelines for implementing planned grazing management programs that can more effectively meet vegetation, production and financial goals in variable environments relative to continuous grazing and conservative stocking.

Managers need to know how to work adaptively within their operations to produce the best results and minimize inherent problems. Successful ranchers modify their management to achieve the best possible outcomes in terms of profitability and enhancing or maintaining ecosystem health. Researchers have much to learn by working with successful ranchers. Examples of this research approach have compared continuous grazing with an intensive grazing system on commercial ranches (Earl and Jones 1996; Jacobo et al. 2006). The ranches were adaptively managed for the best possible outcomes within the constraints of each system. Using this approach, many of the constraints inherent in the way some grazing systems research has been conducted could be avoided. Monitoring ranches that have been successfully operating intensive grazing management for many years, often decades, might also be the only way we can address the pertinent question raised by Burke et al. (1998) on the much neglected subject of time needed to register changes in rangeland ecosystems. Simulation modeling represents an additional and complementary research approach where cost and logistics preclude field experimentation over large spatial and temporal scales (e.g., Hahn et al. 1999; Beukes et al. 2002; Diaz-Solis et al. 2003; Teague and Foy 2004). This approach is well suited to evaluating the managerial and ecological components of grazing systems, both independently and in combination.

Published research and experience from ranchers has indicated that the following management factors are the keys to achieving desired goals: (1) Careful grazing and financial planning to reduce costs, improve work efficiency, enhance profitability, and achieve environmental goals; (2) Providing sufficient growing season deferment to maintain or improve range condition; (3) Grazing grasses and forbs moderately during the growing season for a short period and allowing adequate recovery; (4) Timing grazing to mitigate detrimental effects of defoliation at critical points in the life cycle of preferred species inter- and intra-annually; (5) Where significant regrowth is likely, grazing the area again before the forage has matured too much; (6) Flexible stocking to match forage availability and animal numbers in wet and dry years, or having a buffer grazing area available; (7) Using fire to manage livestock distribution; and (8) Using multiple livestock species. These can be achieved with more control in multi-paddock systems but the same principles can be applied in pauci-paddock systems as practiced by many ranchers in many countries.

The benefits of properly implemented, planned grazing management, as well as the results of poorly implemented programs have been evident for many years on commercial livestock enterprises in many countries, and are also evident from research trials. For those managers who wish to use simple, less management-intensive operations, various pauci-paddock systems can be employed to plan recovery periods during the growing season with or without using planned rotational grazing. The outlined management guidelines will maximize benefits and minimize potentially negative results. More intensive management with appropriate use of multi-paddock systems can increase productivity and improve rangeland health if managed appropriately using the guidelines above. The key to sustainability using these high-intensity systems is high stock density with short grazing periods and moderate utilization, followed by recovery periods to maintain forage nutritional status and productivity. More even animal distribution is automatically achieved by such a system, and the benefit of this to livestock production is already evident from research studies involving small paddocks and to wild and domestic animals on large ranches. In the variable climate associated with all range ecosystems, management needs to be flexible so animal numbers match forage amounts and animals are presented with high quality material in both wet and dry years. As each ranch and rancher is different we have carefully avoided suggesting whether less or more-intensive management is better. We have throughout concentrated on providing information that will aid in improving management for any level of management intensity.

Managing grazing does not necessarily involve more fencing. Fire can be used to spread grazing pressure and minimize the negative effects of overgrazing on more heavily used patches and areas in a grazing unit and enhance vegetation structural heterogeneity and wildlife habitat. Rotational grazing may also be partly implemented through methods other than intensive fencing, including rotating access to water sources (Martin and Ward 1970), strategic supplementation (Bailey and Welling 2007), herding (Bradford 1998; Coughenour 1991; Butler 2000; Bailey 2005; Bailey et al. 2008), and manipulating animal behavior (Provenza 2003a; Launchbaugh and Howery 2005).

Science is a tool to help people understand the processes of nature (Provenza 2000). With regard to grazing management, researchers have used this device primarily to understand interrelationships among physical and biological processes that link soils, plants and herbivores. They have not, as Briske et al. (2008) point out repeatedly, focused on the most important feature of the system, namely the human element of management. Understanding processes is of little value without the flexibility to continually create in the face of uncertainty, and that is what the human element at its best brings to the table in the form of management. Thinking in terms of grazing systems is far less important than understanding processes and determining how to achieve management goals using that knowledge. What matters is feedback from constantly monitoring and continually adjusting the movements of herbivores to ensure the nutrition and health of soils, plants, herbivores and ultimately people. All of that depends upon animals frequently moving across landscapes, whether driven by their needs for nutrients, a herder, rotations through fenced paddocks, fire, or predators (Provenza 2003b; Provenza et al. 2003). People are the glue that links soils, plants and herbivores in grazing systems, and if we really want to understand the innovation and integration essential to the successes of those relationships, we must understand what the best managers do (Provenza 2003a).

REFERENCES

Acocks, J.P.H. 1966. Non-selective grazing as a means of veld reclamation. *Proceedings of the Grassland Society of South Africa.* 1:33-40.

Archibald, S., W. J. Bond, W. D. Stock, and D. H. K. Fairbanks. 2005. Shaping the landscape: fire–grazer interactions in an African savanna. *Ecological Applications* 15: 96–109.

Ash, A.J. and D.M. Stafford-Smith. 1996. Evaluating stocking rate impacts in rangelands: animals don't practice what we preach. *Rangeland Journal* 18(2): 216-243.

Bailey, D.W. 2005. Identification and creation of optimum habitat conditions for livestock. *Rangeland Ecology and Management* 58:109-118.

Bailey D.W., J.E. Gross, E.A. Laca, L.R. Rittenhouse, M.B. Coughenour, D.M. Swift and P.L. Sims. 1996. Mechanisms that result in large herbivore grazing patterns. *Journal of Range Management* 49: 386-400.

Bailey, D.W. and F.D. Provenza. 2008. Mechanisms determining large-herbivore distribution. Pages 7-28 in H.T.T. Prins and F. van Langevelde (eds.) *Resource Ecology: Spatial and Temporal Dynamics of Foraging*. Springer. Dordrecht, Netherlands.

Bailey, D.W., H.C. VanWagoner, R. Weinmeister, and D. Jensen. 2008. Evaluation of low-stress herding and supplement placement for managing cattle grazing in riparian and upland areas. *Rangeland Ecology and Management* 61:26-37.

Bailey, D.W. and G.R. Welling. 2007. Evaluation of low-moisture blocks and conventional dry mixes for supplementing minerals and modifying cattle grazing patterns. *Rangeland Ecology and Management* 60:54-64.

Barnes, D.L., and R.P. Denny. 1991. A comparison of continuous and rotational grazing on veld at two stocking rates. *Journal of the Grassland Society of Southern Africa* 8: 168-173.

Barnes, M.K. 2002. *Spatial distribution of grazing on a gradient of paddock size and stocking density* [MSc. thesis]. Logan, UT: Utah State University. 63p.

Barnes, M.K., B.E. Norton, M. Maeno, and J.C. Malechek. 2008. Paddock size and stocking density affect spatial heterogeneity of grazing. *Rangeland Ecology and Management* in press.

Belsky, A.J. 1986. Does herbivory benefit plants? A review of the evidence. *American Naturalist* 127:870-892.

Berton, V. (ed.). 2001. The new American farmer: profiles of agricultural innovation. *Sustainable Agriculture Research and Education Program*, USDA, Washington, D.C.

Beukes, P.C., R.M. Cowling and S.I. Higgins. 2002. An ecological economic simulation model of a non-selective grazing system in the Nama Karoo, South Africa. *Ecological Economics* 42:221-242.

Bissonette, J.A.. 1997. Scale-sensitive ecological properties pp 3-31. In *Wildlife and landscape ecology*. Ed. J.A. Bissonette. Springer, New York.

Blydenstein, J. 1966. Root systems of four desert grassland species on grazed and protected sites. *Journal of Range Management* 19, 93-95.

Booysen, P. deV. 1966. A physiological approach to research in pasture management *Proceedings of the Grassland Society of Southern Africa*. 4:77-96.

Booysen, P. deV. 1969. An evaluation of the fundamentals of grazing systems. *Proceedings of the Grassland Society of Southern Africa* 4:84-91.

Booysen, P. deV. and N.M. Tainton. 1978. Grassland management: principles and practice in South Africa. *Proceedings 1st International Rangeland Congress*. Denver Colorado, 551-554.

Bradford, D. 1998. Holistic resource management in the West Elks – why it works. *Rangelands* 20:6-9.

Briske, D. D. 1991. Developmental morphology and physiology of grasses. *In:* R.K. Heitschmidt and J.W. Stuth (Editors.) *Grazing Management: An Ecological Perspective*. Timber Press, Portland OR USA. pp. 85-108.

Briske, D.D., J.D. Derner, J.R. Brown, S.D. Fuhlendorf, W.R. Teague, K.M. Havstad, R.L. Gillen, A.J. Ash, and W.D. Willms. 2008. Rotational grazing on rangelands: reconciliation of perception and experimental evidence. *Rangeland Ecology and Management* 61:3-17.

Bryant, F.C., B.E. Dahl, R.D. Pettit, and C.M. Britton. 1989. Does short-duration grazing work in arid and semiarid regions? *Journal of Soil and Water Conservation* 44:290-296.

Bryant, J.P., F.S. Chapin III, and D.R. Kline. 1983. Carbon/nutrient balance of boreal plants in relation to vertebrate herbivory. *Oikos* 40:357-368.

Burboa-Cabrera, F.R., W.H. Schacht, and B.E. Anderson. 2003. Utilization and grazing distribution of cattle at 4 stocking densities. *Journal of Range Management* 56:328-333.

Burke, I.C., W.K. Lauenroth, M.A. Vinton, P.B. Hook, R.H. Kelly, H.E. Epstein, M.R. Aguiar, M.D. Robles, M.O. Aguilera, K.L. Murphy, and R.A. Gill. 1998. Plant-soil interactions in temperate grasslands. *Biogeochemistry* 42:121-143.

Burrell, G. and Morgan, G. 1979. *Sociological paradigms and organizational analysis*. Helnemann, London.

Butler, P.J. 2000. Cattle distribution under intensive herded management. *Rangelands* 22:21-23.

Coley, P.D., J.P. Bryant, and F.S. Chapin III. 1985. Resource availability and plant antiherbivore defense. *Science* 230:895-899.

Cook, C.W., Stoddart, L.A. and Kinsinger, F.E. 1958.Responses of crested wheatgrass to various clipping treatments. *Ecological Monographs* 28:237-272.

Coughenour, M.B. 1991. Spatial components of plant-herbivore interactions in pastoral, ranching, and native ungulate ecosystems. *Journal of Range Management* 44:530-542.

Coughenour M.B., S.J. McNaughton and L.L. Wallace. 1985. Responses of an African graminoid (*Themeda triandra* Forsk.) to frequent defoliation, nitrogen and water: A limit of the adaptation to herbivory. *Oecologia* 68:105-111.

Crider, F.J. 1955. Root-growth stoppage resulting from defoliation of grass. *USDA Tech. Bull.* 1102.

Cullan, A.P., P.E. Reece, and W.H. Schacht. 1999. Early summer grazing effects on defoliation and tiller demography of prairie sandreed. *Journal of Range Management* 52: 447-453.

Cumming, K.J. 1989. Lehmann lovegrass and simple time control grazing. *Rangelands* 11:150-153.

Daines, T. 1980. The use of grazing pattern in the management of Döhne Sourveld. *Proceedings of the Grassland Society of southern Africa* 15: 185-188.

Danckwerts, J.E. 1984. Towards improving livestock production off sweet grassveld. PhD. thesis. University of Natal, Pietermaritzburg, South Africa.

Danckwerts, J.E. and L.O. Nel. 1989. The effect of frequency of defoliation on Themeda triandra in the False Thornveld of the Eastern Cape. *Journal of the Grassland Society of Southern Africa* 6, 32-36.

Danckwerts, J.E., P.J. O'Reagain, and T.G. O'Connor. 1993. Range management in a changing environment: a southern African perspective. *Rangeland Journal* 15: 133-144.

Denny, R.P. and Barnes, D.L. 1977. Trials of multi-paddock grazing systems on veld 3. A comparison of six grazing procedures at two stocking rates. *Rhodesian Journal of Agricultural Research* 15: 129-142.

Derner, J.D., R.L. Gillen, F.T. McCollum, and K.W. Tate. 1994. Little bluestem tiller defoliation patterns under continuous and rotational grazing. *Journal of Range Management* 47:220-225.

Derner, J.D. and R. H. Hart. 2007. Grazing-induced modifications to peak standing crop in northern mixed grass prairie. *Rangeland Ecology and Management*. 60: 270-276.

Diaz-Solis, H., M.M. Kothmann, W.T. Hamilton and W.E. Grant. 2003. A simple ecological sustainability simulator (SESS) for stocking rate management on semi-arid grazinglands. *Agricultural Systems* 76:655-680.

Dowling, P.M., D.R. Kemp, D.L. Milchalk, T.A. Klein and G.D. Millar. 1996. Perennial grass response to seasonal rests in naturalized pastures of central New South Wales. *Rangeland Journal* 18: 309-326.

Dubin, R. 1978. *Theory building*. Free Press, New York.

Earl, J.M. and C.E. Jones. 1996. The need for a new approach to grazing management – is cell grazing the answer? *Rangeland Journal* 18:327-350.

Engel, R.K., Nichols, J.T., Dodd, J.L. and Brummer, J.E. 1998. Root and shoot responses of sand bluestem to defoliation. *Journal of Range Management* 51:42-46.

Frank D.A. and P.A. Groffman 1998. Ungulate vs. landscape control of soil C and N processes in grasslands of Yellowstone National Park. *Ecology* 79:2229-2241.

Frank, D.A., and S.J. McNaughton. 1993. Evidence for the promotion of aboveground grassland production by native large herbivores in Yellowstone National Park. *Oecologia* 96:157-161.

Frank D.A. and SJ McNaughton. 2002.The Ecology of the Earth's Grazing Ecosystems. *BioScience* 48:513-521.

Frost, P. 1980. Toward a radical framework for practicing organizational science. *Academy of Management Review* 5:501-507.

Fuhlendorf, S.D. and D.M. Engle. 2004. Application of the fire-grazing interaction to restore a shifting mosaic on tallgrass prairie. *Journal of Applied Ecology* 41:604-614.

Fuls, E.R. 1992. Semi-arid and arid rangelands: A resource under siege due to patch-selective grazing. *Journal of Arid Environments* 22:191–193.

Galt, D., F. Molinar, J. Navarro, J. Joseph, and J. Holechek. 2000. Grazing capacity and stocking rate. *Rangelands* 22:7-11.

Gammon, D.M., and B.R. Roberts. 1978. Patterns of defoliation during continuous and rotational grazing of the Matopos Sandveld of Rhodesia. 1. Selectivity of grazing. *Rhodesia Journal of Agricultural Research* 16:117-131.

Ganskopp, D. 1988. Defoliation of thurber needlegrass: herbage and root responses. *Journal of Range Management* 41:472-476.

Gerrish, J. 2004. *Management-Intensive Grazing: The Grassroots of Grass Farming*. Green Park Press, Ridgeland, MS.

Gillen, R.L., F.T. McCollum III, K.W. Tate, and M.E. Hodges. 1998. Tallgrass prairie response to grazing system and stocking rate. Journal of Range Management 51:139-146.

Gioia D.A. and Pitre, E. 1990. Multiparadigm perspectives on theory building. *The Academy of Management Review* 15:584-602.

Goodloe, S. 1969. Short-duration grazing in Rhodesia. *Journal of Range Management* 22:369-373.

Gordon, K. 2002. An eye for grass: SRM bestows Oklahoma couple with "Outstanding Grazing Lands Managers" award. *Rangelands* 24(3):3-5.

Hahn, B.D., F.D. Richardson and A.M. Starfield. 1999. Frame-based modeling as a method of simulating rangeland production systems in the long term. *Agricultural systems* 62:29-49.

Hamilton E.W. and D.A. Frank. 2001. Can plants stimulate soil microbes and their own nutrient supply? Evidence from a grazing tolerant grass. *Ecology* 82:2397-2402.

Harradine, A.R and Whalley, R-D.B. 1981. A comparison of the root growth, root morphology and root response to defoliation of *Aristida ramosa* R. Br. and *Danthonia linkii* Kunth. *Australian Journal of Agricultural Research* 32:565-574.

Hart, R.H., M.J. Samuel, P.S. Test, and M.A. Smith. 1988. Cattle, vegetation, and economic responses to grazing systems and grazing pressure. *Journal of Range Management* 41: 282-286.

Heitschmidt, R.K., and C.A. Taylor, Jr. 1991. Livestock production. *In:* R.K. Heitschmidt and J.W. Stuth (Editors.) *Grazing Management: An Ecological Perspective*. Timber Press, Portland OR USA. p. 161-177.

Heitschmidt, R.K. and J.W. Walker. 1996. Grazing management: technology for sustaining rangeland ecosystems. *Rangeland Journal* 18: 194-215.

Hendrickson, J.R., L.E. Moser, and P.E. Reece. 2000. Tiller recruitment patterns and biennial tiller production in prairie sandreed. *Journal of Range Management* 53: 537-543.

Hodgkinson, K.C. and Baas Becking, H.G. 1977. Effect of defoliation on root growth of some arid zone perennial plants. *Australian Journal of Agricultural Research* 29:31-42.

Holechek, J.L., H. Gomez, F. Molinar, and D. Galt. 1999. Grazing studies: what we've learned. *Rangelands* 21:12-16.

Holechek, J.L., H. Gomes, F. Molinar, D. Galt, and R. Valdez. 2000. Short duration grazing: the facts in 1999. *Rangelands* 22:18-22.

Holechek, J.L., R.D. Pieper, and C.H. Herbel. 2003. *Range management: principles and practices.* 5th edition. Upper Saddle River, N.J. Prentice Hall. 624 p.

Holland, E.A., W.J. Parton, J.K. Detling, D.L. Coppock. 1992. Physiological responses of plant populations to herbivory and other consequences of ecosystem nutrient flow. *American Naturalist* 140:685-706.

Hormay, A.L. 1956. How livestock grazing habits and growth requirements of range plants determine sound grazing management. *Journal of Range Management* 9:161-164.

Hormay, A.L. 1970. Principles of rest-rotation grazing and multiple-use management. *USDA For. Serv. Training Text* 4.

Hormay, A.L. and A.B. Evanko. 1958. Rest-rotation grazing ...a management system for bunchgrass ranges. *USDA Forest Service, California Forest and Range Experiment Station Misc.* Paper 27.

Hormay, A.L. and M.W. Talbot. 1961. Rest-rotation grazing ...a new management system for perennial bunchgrass ranges. *USDA Forest Service*, Research Report 51.

Howell, J. 2006. On Twin Creek Ranch – acting on "change requests" in the sagebrush steppe. *In Practice* 109:8-9.

Hulbert, L. C. 1969. Fire and litter effects in an undisturbed bluestem prairie in Kansas. *Ecology* 50:874–877.

Hulbert, L. C. 1988. Causes of fire effects in tallgrass prairie. *Ecology* 69: 46–58.

Hunt, L.P., S. Petty, R. Crowley, A. Fisher, A.J. Ash, and N. MacDonald. 2007. Factors affecting management of cattle grazing distribution in northern Australia: preliminary observations on the effect of paddock size and water points. *Rangeland Journal* 29: 169-179.

Illius, A.W. and T.G. O'Connor. 1999. On the relevance of non-equilibrium concepts to arid and semi-arid grazing systems. *Ecological Applications* 9:798-813.

Jacobo, E.J., A.M. Rodríguez, N. Bartoloni, and V.A. Deregibus. 2006. Rotational grazing effects on rangeland vegetation at a farm scale. *Rangeland Ecology and Management* 59:249-257.

Kellner, K. and Bosch, O.J.H. (1992). Influence of patch formation in determining stocking rate for southern African grasslands. *Journal of Arid Environments* 22:99-105.

Kirby, D.R., M.F. Pessin, and G.K. Clambey. 1986. Disappearance of forage under short duration and seasonlong grazing. *Journal of Range Management* 39:496-500.

Kirkman, K.P. and A. Moore. 1995. Towards improved grazing management recommendations for sourveld. *African Journal of Range and forage Science* 12:135-144.

Kothmann, M.M. 1980. Integrating livestock needs to the grazing system, p. 65-83. *In:* K.C. McDaniel and C.D. Allison (eds.), *Grazing management systems for Southwest rangelands*. Symp., 1-2 Apr. 1980, Albuquerque, N.M. Range Improvement Task Force, New Mexico State University, Las Cruces, N.M.

Kothmann, M.M. 1984. Concepts and principles underlying grazing systems: a discussant paper, p. 903-916. *In: National Research Council. Developing strategies for rangeland management*. Westview Press, Boulder, Colo.

Kuhn, T. S. 1970. *The structure of scientific revolutions*. University of Chicago Press, Chicago.

Launchbaugh, K.L. and L.D. Howery. 2005. Understanding landscape use patterns of livestock as a consequence of foraging behavior. *Rangeland Ecology and Management* 58:99-108.

Lee, T.D. and F.A. Bazzaz. 1980. Effects of defoliation and competition on growth and reproduction of the annual plant *Abutilon theophrasti*. Journal of Ecology 68:813-821.

Levin, S.A.. 1993. Concepts of scale at the local level pp 7-20. In *Scaling physiological processes leaf to globe*. Eds. J.R. Ehleringer and C.B. Field. Academic Press, San Diego.

Lewontin, R. 2000. Gene and Organism. Pages 1-38 in T*he Triple Helix: Gene, Organism, and Environment.* Harvard University Press. Cambridge, MA.

Lisa, P.J. 1986. The Great Medical Monopoly Wars, International Institute of Natural Health Sciences, Inc., Huntington Beach, California.

Louda, S.M., K.H. Keeler and R.D. Holt. 1990. Herbivore influences on plant performance and competitive interactions. *In*: *Perspectives in Plant Competition Grace J.B. and D. Tilman* (eds.). Academic Press, New York, USA. pp. 414-444.

Manley, W.A., R.H. Hart, M.J. Samuel, M.A. Smith, J.W. Waggoner Jr., and J.T. Manley. 1997. Vegetation, cattle, and economic responses to grazing strategies and pressures. *Journal of Range Management* 50:638-646.

Martin, S.C. and D.E. Ward. 1970. Rotating access to water to improve semi desert cattle range near water. *Journal of Range Management* 23:22-26.

Matthews, E.D. 1954. *Tukulu – a pioneer's story.* 2nd Edition, Grassland Society of Southern Africa, Pietermaritzburg, South Africa.

McCollum III, F.T., R.L. Gillen, B.R. Karges, and M.E. Hodges. 1999. Stocker cattle response to grazing management in tallgrass prairie. *Journal of Range Management* 52:120-126.

McCosker, T. 1994. The dichotomy between research results and practical experience with time control grazing. *Australian Rural Science Annual* 1994:26-31.

McIvor, J.G. (2007). Pasture management in semi-tropical woodlands: dynamics of perennial grasses. *Rangeland Journal* 29: 87-100.

McNaughton S.J., M. Oesterheld, D.A. Frank and K.J. Williams. 1989. Ecosystem-level patterns of primary productivity and herbivory in terrestrial habitats. *Nature* 341:142-144.

Menke, J.W. and M.J. Trlica. 1983. Effects of single and sequential defoliations on the carbohydrate reserves of four range species. *Journal of Range Management* 36: 70-74.

Merrill, L.B. 1954. A variation of deferred rotation grazing for use under southwest range conditions. *Journal of Range Management* 7: 152-154.

Milchunas, D.G., and W.K. Lauenroth. 1993. Quantitative effects of grazing on vegetation and soils over a global range of environments. *Ecological Monographs* 63:327-366.

Morris, C.D., and N.M. Tainton 1991. Long-term effects of different rotational grazing schedules on the productivity and floristic composition of Tall Grassveld in KwaZulu-Natal. *African Journal of Range and Forage Science* 13:24-28.

Motazedian, I. and Sharrow, S.H. 1987. Persistence of a *Lolium perenne-trifolium* subterranean pasture under defoliation treatments. *Journal of Range Management* 40:232-236.

Mullahey, J.J., S.S. Waller, and L.E. Moser. 1990. Defoliation effects on production and morphological development of little bluestem. *Journal of Range Management* 43: 497-500.

Mullahey, J.J., S.S. Waller, and P.E. Reece. 1991. Defoliation effects on yield and bud and tiller numbers of two sandhills grasses. *Journal of Range Management* 44: 241-245.

Müller, B., Frank, K. and Wissel, C. 2007. Relevance of rest periods in non-equilibrium rangeland systems - A modeling analysis. *Agricultural systems* 92:295-317.

Norton, B.E. 1998. The application of grazing management to increase sustainable livestock production. *Animal Production in Australia* 22: 15-26.

Norton, B.E. 2003. Spatial management of grazing to enhance both livestock production and resource condition: A scientific argument. *In:* N. Allsopp, A.R. Palmer, S.J. Milton, K.P. Kirkman, G.I.H. Kerley, C.R. Hurt and C.J. Brown (EDS.) *Proc. VII International Rangeland Congress*, Durban, South Africa. p. 810-820.

In: Grasslands: Ecology, Management and Restoration
Editor: Hans G. Schroder
ISBN 978-1-60692-023-7

Chapter 2

EFFECTS OF LONG-TERM LIVESTOCK REMOVAL ON WATER INFILTRATION RATE IN ARID GRASSLANDS

Thomas J. Valone
Department of Biology, Saint Louis University,
St. Louis, MO 63103

ABSTRACT

Desertification models predict that arid grasslands exist in one of two stable states: grassland or desertified shrubland. This prediction is derived from an assumed positive relationship between grass cover and water infiltration rate: once grass cover is reduced, water infiltration rates are insufficient to support perennial grass. These models assume that other factors known to affect water infiltration rate are unimportant. While this alternate stable state view is widely accepted and suggests that restoration of desertified grasslands will be difficult, there have been four recent reports of the reversal of desertification (significant increases in the abundance of perennial grasses) following long-term livestock removal. At one site, recovery of perennial grass inside a long-term grazing exclosure was associated with increased water infiltration rates due to release from soil compaction. To assess the generality of this finding, I examined water infiltration rates inside and outside five other long-term livestock exclosure sites in southwestern North America that differ in degree of desertification and time since livestock removal. At each site, water infiltration rates were significantly higher inside compared to outside the grazing fence. The relative difference in water infiltration rate across the sites increased with time since livestock removal at a rate of approximately 1.7% per year. These data show that increased water infiltration rate following livestock removal appears to be a general phenomenon in arid grassland systems. In addition, they suggest that the effect of livestock on water infiltration rates should be included in desertification models to better understand the dynamics of vegetation in these systems. Finally, this work suggests that restoration of desertified grasslands may be possible given sufficient time to allow changes in soil properties including release from soil compaction and the concomitant increases in water infiltration rate that follow.

INTRODUCTION

Desertification involves a shift in dominant vegetative composition from an ecosysytem dominated by perennial grasses to one dominated by shrubs and bare soil [Daily 1995; Van Auken 2000; Scheffer *et al.* 2001; Jackson *et al.* 2002]. Thus, desertification is the process in which grasslands are converted into shrublands. Desertification is widespread in arid systems throughout the world and thought to be largely irreversible over timescales relevant to management because few studies have reported natural vegetation recovery in desertified grasslands [Laycock 1991; Van Auken 2000; Rasmussen *et al.* 2001; Valone *et al.* 2002]. Mechanisms to explain desertification include changes in climate, herbivory and fire regime, but most workers agree that overgrazing by livestock has played a major role [Schlesinger *et al.* 1990; Laycock 1991; Fleischner 1994; Archer *et al.* 1995; Daily 1995; Van Auken 2000].

In attempting to explain the apparent irreversibility of desertification, theoretical models typically focus on changes in soil properties [e.g., Walker *et al.* 1981; Schlesinger *et al.* 1990; Rosenfeld *et al.* 2001; van de Koppel and Rietkerk 2004]. Many models are based on a positive feedback between grass cover and water infiltration rate [e.g., Kelly and Walker 1976; Walker *et al.* 1981; Breman and de Wit 1983; Rietkerk and van de Koppel 1997; Rietkerk *et al.* 1997; van de Koppel *et al.* 1997]. These models demonstrate that once grass cover is reduced by livestock grazing or drought beyond a threshold that results in desertification, grass recovery is inhibited because water infiltration rates are insufficient for grass re-establishment [e.g., Walker *et al.* 1981; van de Koppel and Rietkerk 2000; Scheffer *et al.* 2001; Rietkerk *et al.* 2002; van de Koppel *et al.* 2002; van de Koppel and Rietkerk 2004]. Such alternate stable state models have come to dominate the desertification literature (e.g., Briske *et al.* 2005; Peters *et al.* 2006).

However, grass cover is only one mechanism known to affect water infiltration rate. A large literature has demonstrated that soil compaction can significantly reduce water infiltration rate [e.g., Rhoades *et al.* 1964; Rauzi and Hanson 1966; Wood and Blackburn 1981; Warren *et al.* 1986; Abdel-Magid *et al.* 1987; Gamougoun 1984; Schlesinger 1990; Fleischner 1994]. Although theoreticians do acknowledge the effect of livestock trampling on soil compaction, it is typically assumed to be negligible in relation to the effects of grass canopy cover on water infiltration rate.

Recently, four studies have reported significant increases in perennial grass abundance at formerly desertified sites: i.e., reversals of desertification. Fuhlendorf *et al.* [2001] reported significant increases in perennial grasses in a 25-year livestock exclosure in Texas, U.S.A. while Valone *et al.* [2002] reported a 300% increase in perennial grass cover at a 39-year livestock exclosure in Arizona U.S.A. Similarly, a 20-year reduction in human activities, including livestock grazing, was associated with significant vegetation recovery at a desertified site in the African Sahel [Rasmussen *et al.* 2001, Herrmann *et al.* 2005] while 30 years of protection from livestock resulted in perennial grass recovery at a desertified site in China [Zhang *et al.* 2005]. These observations suggest that desertification is potentially reversible, at least in some circumstances. Existing desertification models cannot account for these changes in vegetation or their timescale.

Castellano and Valone [2007] recently suggested a novel mechanism to explain the above reversals of desertification. They postulate that long-term livestock removal permits recovery of soil compaction which leads to increased water infiltration rates and subsequent increases in perennial grass abundance. They evaluated this hypothesis by comparing soil compaction, water infiltration rates and vegetation at three nearby grazing exclosures that differed in time since livestock removal [Castellano and Valone 2007]. The three sites, all within 6 km of one another and on similar soils, had been free of livestock grazing for 10, 26 and 45 years. At each site, they measured water infiltration, soil compaction and vegetation inside and outside the long-term grazing exclusion fence. At all three sites, water infiltration was higher and soil compaction lower inside compared to outside the grazing fence. These differences were significant for the 26 and 45 year exclusion sites but not the 10 year exclusion site. Furthermore, a comparison across sites revealed that relative water infiltration (inside [ungrazed] versus outside [grazed] measurements) and relative soil compaction (outside versus inside measurements) increased significantly with time since livestock exclosure. In other words, relative water infiltration rates were highest and relative soil compaction was lowest inside the oldest livestock exclosure. In addition, while all three sites were desertified at the beginning of the 20th century, only the 45 year exclusion site had experienced a recovery in perennial grasses. Of note, that site had experienced no perennial grass recovery during the first 20 years of livestock exclosure [Valone *et al.* 2002]. Thus, these data suggest that in these sites, more than two decades of livestock removal are required for sufficient reductions in soil compaction and concomitant increases in water infiltration to support perennial grass recovery.

The Castellano and Valone [2007] mechanism suggests that perennial grasses can recover on desertified sites following long-term livestock removal via increased water infiltration rates that occur due to subsequent recovery from soil compaction. Additionally, this mechanism can also explain the multi-decadal time lags observed between the reduction of livestock grazing and perennial grass recovery [e.g. Fuhlendorf *et al.* 2001; Rasmussen *et al.* 2001; Valone *et al.* 2002; Zhang *et al.* 2005] because soils in arid systems recover slowly from compaction. Release from compaction requires freeze-thaw or wet-dry cycles and the frequency of such cycles decreases with increasing aridity and increasing temperature [Seybold *et al.* 1999].

While the results of Castellano and Valone [2007] are insightful, they come from three nearby sites in one valley in Arizona (the San Simon), a valley that experienced dramatic desertification at the end of the 19th century [Bahre 1991]. As such, one possibility is that the patterns observed are an artifact of desertification and may not be general. Clearly, their proposed mechanism requires further evaluation and testing to assess its generality. Here, I further evaluate the effect of long-term livestock removal on water infiltration rates by reporting data from five additional sites in southern Arizona. The sites differ in time since livestock removal and degree of historic desertification but are similar in soil type and elevation.

METHODS

Between 3 June and 1 July, 2005 I collected water infiltration rate data inside and outside the following long-term livestock exclosures in southern Arizona: National Audubon Society Research Ranch near Elgin, Buenos Aires National Wildlife Refuge near Sasabee, Chiricahua National Monument near Wilcox, Fort Huachuca Army Base near Sierra Vista and San

Bernardino National Wildlife Refuge near Douglas. These sites span approximately 200 km. All sites are dominated by Thermic Semiarid soils (Tubac-Sonoita-Grabe association at both Chiricauha National Monument and San Bernardino Wildlife Refuge; White House-Bernardino-Hathaway association at the other three sites [Hendricks 1985]). Livestock grazing occurs at present outside each grazing fence (personal observation) and likely has occurred in the vicinity of each site over most of the past century (Bahre 1991).

These sites exhibit great differences in vegetation ranging from dominance by perennial grasses (e.g., Audubon Research Ranch [Bock and Bock 2000]) to dominance by mesquite shrubs (*Prosopis glandulosa*) (e.g., San Bernardino Wildlife Refuge) (Table 1). *P. glandulosa* is found at all sites and is commonly associated with the desertification of arid grassland habitats in southwestern North America [Van Auken 2000]. Differences in vegetation across the sites likely result from historic differences in grazing intensity. All sites are at an elevation that historically supported arid grassland habitat [Brown and Lowe 1974; Bahre 1991].

At each of these sites, I recorded water infiltration rate over a 10 minute period using a Turf-Tech double ring infiltrometer [Castellano and Valone 2007]. The infiltrometer was placed on level patches of bare soil at paired locations on opposite sides of, and equidistant to, the grazing fence in areas that exhibited similar vegetation composition on both sides of the grazing fence. This yielded 9 – 37 grazed-ungrazed paired locations at each of the five sites (National Audubon Research Ranch: n = 37 locations across the North, East and South boundary fences; Buenos Aires National Wildlife Refuge: n = 9 locations across the West boundary fence; Chiricahua National Monument: n = 19 locations across the West boundary fence; Fort Huachuca Army Base: n = 22 locations across the West boundary fence; San Bernardino National Wildlife Refuge: n = 24 locations across the East boundary fence).

Table 1. A description of the study sites indicating the year the grazing fence was established and the dominant vegetation at present. Grass refers to perennial grass. Shrubs refers to mesquite (*P. glandulosa*).

Name	Year	Vegetation	Elevation (m)
Audubon Society Research Ranch	1968	Grass	1500
Buenos Aires National Wildlife Refuge	1985	Grass and Shrubs	1200
Chiricahua National Monument	1977	Grass and Shrubs	1400
Fort Huachuca Army Base	1950	Grass and Shrubs	1500
San Bernardino National Wildlife Refuge	1980	Shrubs	1200

All locations were more than 10 m from the fence and adjacent locations were separated by at least 10 m. At each site, all data were collected on the same day within a 3 hour period to minimize differences in abiotic conditions during data collection (e.g., soil moisture and temperature). All data were collected during the dry season when soil was dry to a depth of greater than 10 cm (personal observation). At each site, I compared water infiltration rate values inside (ungrazed) versus outside (grazed) the grazing fence using a paired t-test.

RESULTS

At all sites, mean water infiltration rates were higher inside compared to outside the grazing fence. Across sites, mean infiltration rates outside the grazing fence (grazed areas) ranged from 33.6 mm/10 min at San Bernardino National Wildlife Refuge to 55.5 mm/10 min at Chiricahua National Monument. Mean infiltration rates inside the grazing fence (ungrazed areas) ranged from 50.1 mm/10 min at San Bernardino National Wildlife Refuge to 79.0 mm/10 min at the Audubon Society Research Ranch.

At each site, mean infiltration rate was significantly higher inside (ungrazed) compared to outside (grazed) the grazing fence (Figure 1): Audubon Society Research Ranch: mean (se) grazed = 41.3(3.2), mean (se) ungrazed = 79.0(3.0), $P < 0.001$; Buenos Aires National Wildlife Refuge: mean (se) grazed = 46.7(4.7), mean (se) ungrazed = 74.4(7.1), $P < 0.01$; Chiricahua National Monument: mean (se) grazed = 55.5(5.1), mean (se) ungrazed = 73.8(3.9), $P < 0.001$; Fort Huachuca Army Base: mean (se) grazed = 37.3(4.3), mean (se) ungrazed = 61.6(5.2), $P < 0.005$; San Bernardino National Wildlife Refuge: mean(se) grazed = 33.6(3.2), mean (se) ungrazed = 50.1(3.8), $P < 0.001$. To examine the relationship between time since livestock removal and the relative difference in water infiltration rate, I first calculated the relative difference in water infiltration rate at each site by dividing the mean infiltration rate inside the fence by the mean infiltration rate outside the fence.

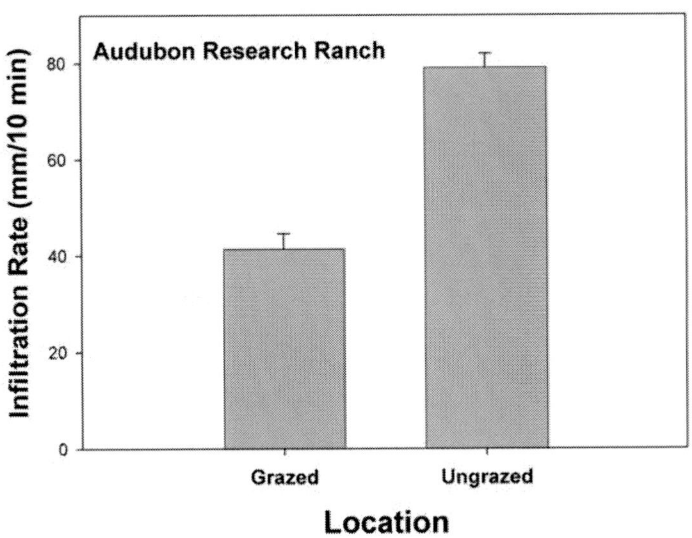

Figure 1. (Continued).

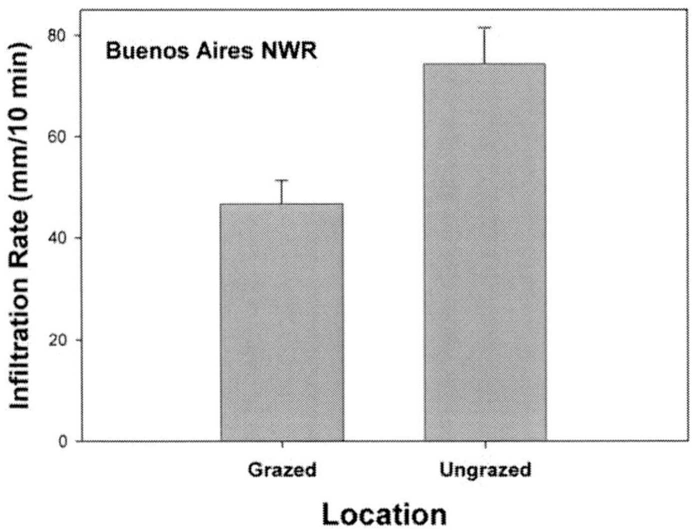
Buenos Aires NWR
Infiltration Rate (mm/10 min)
0
20
40
60
80
Grazed
Ungrazed
Location

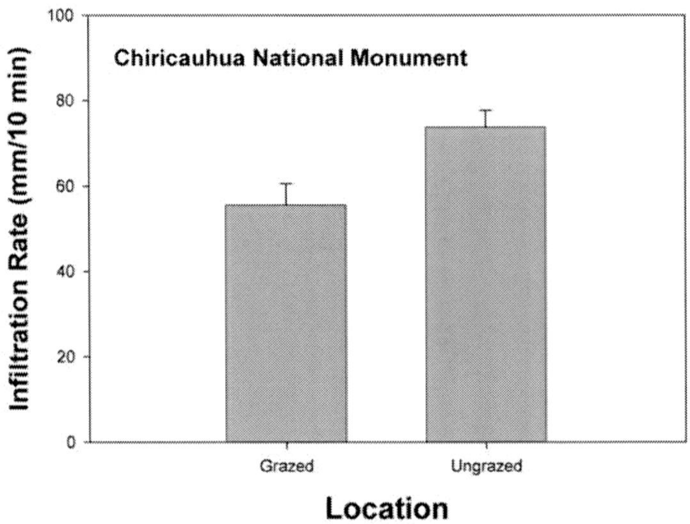
Chiricauhua National Monument
Infiltration Rate (mm/10 min)
0
20
40
60
80
100
Grazed
Ungrazed
Location

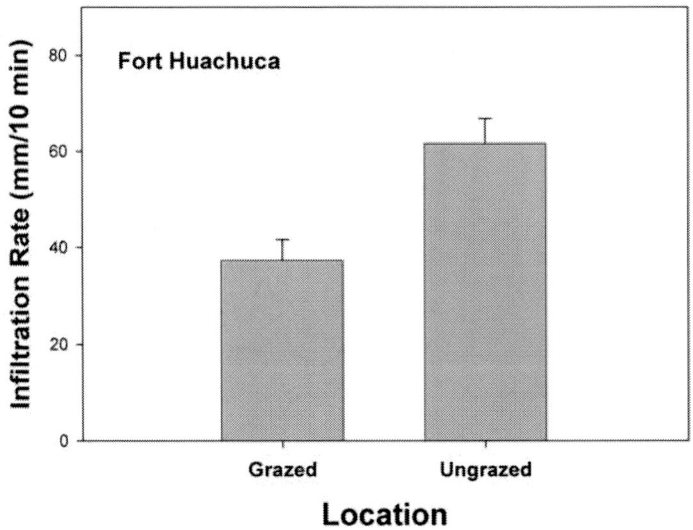
Fort Huachuca
Infiltration Rate (mm/10 min)
0
20
40
60
80
Grazed
Ungrazed
Location

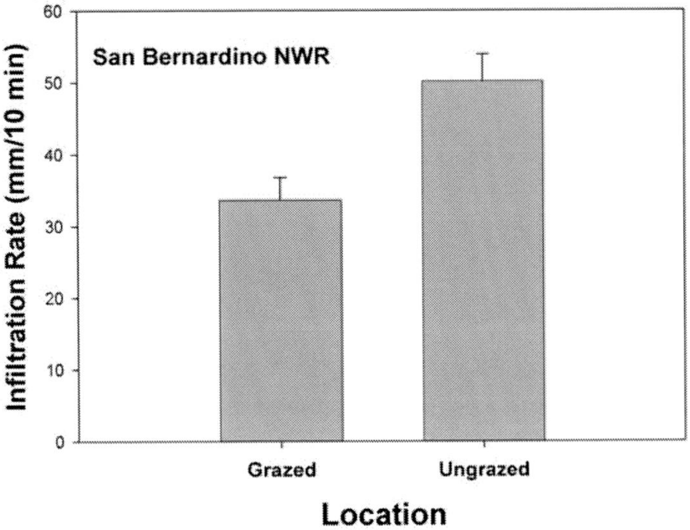

Figure 1. Mean (+ se) water infiltration rate (measured at 10 minutes) outside (grazed) and inside (ungrazed) the long-term grazing fence at the five sites studied. At each site, the mean rates differed significantly between grazed and ungrazed locations (see text).

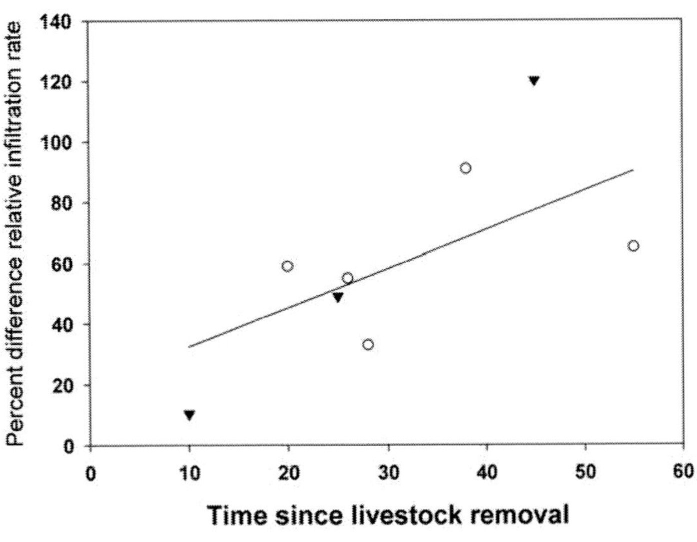

Figure 2. Relationship between time since livestock removal (number of years of grazing exclosure) and percent difference in relative water infiltration rate. Circles are for sites in this chapter while triangles are sites from Castellano and Valone (2007). Values above zero indicate higher infiltration rates inside compared to outside a grazing fence. A 100% difference means that infiltration rate is twice as high inside compared to outside the fence.

The relative difference in infiltration rate ranged from 33% higher inside the fence at Chiricahua National Monument to 91% higher inside the fence at the Audubon Society Research Ranch. Next, I plotted the relative difference in water infiltration rate values for these five sites along with the three sites reported in Castellano and Valone [2007] as a function of time since livestock removal. Figure 2 shows that there is a significant positive relationship between years since livestock removal and relative difference in water infiltration rate (Speraman Rank correlation $R_{sp} = 0.71$, $P < 0.05$). The slope of the relationship suggests that there is a 1.7% change (increase) in water infiltration rate per year following livestock removal across these sites (Linear regression: slope = 1.7, $r^2 = 0.51$, $P < 0.05$).

DISCUSSION

Despite differences in degree of desertification, current vegetation and time since livestock removal, the results presented here indicate a pervasive effect of livestock removal on arid grassland soil properties across a set of sites in Arizona: water infiltration rate increases with time since livestock removal. There is roughly a 1.7% increase in relative water infiltration rate across a grazing fence each year following livestock removal over the span of several decades.

The data presented in this chapter both reinforce and extend the findings of Castellano and Valone [2007]. They reinforce those findings because the pattern reported here, increased relative water infiltration rate with time since livestock removal, is the same. More importantly, these findings extend Castellano and Valone [2007] in two ways. First, the sites here exhibited a wide range of current vegetation (degree of desertification) as evidenced by dramatic differences in the dominance of perennial grasses and shrubs (Table 1). Thus, the severe desertification of the San Simon valley *per se* is not an artifact that produced the patterns reported by Castellano and Valone [2007]. In addition, all data reported here were collected during the summer dry season, prior to the onset of monsoon rains and so soils were dry when infiltration rate data were collected. Castellano and Valone [2007] collected data when soils were wet. Soil moisture can strongly affect water infiltration rate [Gifford *et al.* 1977; Mapfumo *et al.* 1999]. To minimize this potential concern, all cross-fence data were collected on the same day within a narrow window of time. While measurements of water infiltration rate likely will differ at different times of the year as soil moisture varies, I do not expect the general pattern to change: infiltration rates should be higher inside compared to outside the grazing fences. Additional work, conducted at the same site but at different times of the year when soil is both wet and dry is necessary to test this assertion.

While I did not measure soil compaction at these five sites, I assume that differences in soil compaction is the mechanism that explains the differences in water infiltration rate observed across the five grazing fences studied. Castellano and Valone [2007] demonstrated that differences in soil compaction related directly to water infiltration rates (the higher the compaction, the lower the water infiltration rate). In addition, a large literature has demonstrated that the presence of livestock increases soil compaction and that soil compaction strongly and negatively affects water infiltration rate [e.g., McGinty *et al.* 1979; McCalla *et al.* 1984; Warren *et al.* 1986; Abdel-Magid *et al.* 1987].

Water infiltration rates can be measured in many ways [Brady and Weil 1999]. One advantage of the double ring infiltrometer method used here is that it eliminates the potential influence of canopy cover. Previous work has shown that canopy cover, via the interception of precipitation, can strongly affect water infiltration [e.g., Thurow *et al.* 1988]. Such results have had a strong influence on the development of desertification models [e.g., Rietkerk *et al.* 2002; van de Koppel *et al.* 2002; van de Koppel and Rietkerk 2004] that focus on the effects of vegetation canopy cover on water infiltration rate. The double ring infiltrometer method eliminates this effect of vegetation canopies and thus more directly examines how soil properties affect infiltration. In addition, while not a concern at these sites, the double ring infiltrometer method is particularly useful if vegetation differs strongly across a grazing fence.

Current models of desertification assume that the effect of livestock on soil compaction and water infiltration rate are minor relative to the importance of vegetation canopies in affecting water infiltration rate. While additional work is required to better understand the relative importance of these two factors, the data presented here suggest that the presence or absence of livestock can strongly affect water infiltration rate. It seems reasonable that to more fully understand the dynamics of vegetation in arid grassland systems, we will need to include effects of livestock on water infiltration rates via their effects on soil compaction.

Finally, the work here suggests that desertified systems are not necessarily stably desertified because water infiltration rate can increase substantially following the removal of livestock. If water infiltration rate plays a key role in perennial grass establishment, restoration of even highly desertified sites may be possible, given sufficient time for the recovery of soil properties such as release from compaction. I hope that at the very least, the work presented in this chapter will help to stimulate additional theoretical and empirical work to provide a better understanding of desertification, its reversibility, and the restoration of arid grasslands.

REFERENCES

Abdel-Magid, A. H., Trlica, M. J. and Hart, R. H. (1987). Soil and vegetation responses to simulated trampling. *Journal of Range Management*, *40*, 303-306.

Archer, S., Schimel, D. S.and Holland, E. A. (1995). Mechanisms of shrubland expansion: land use, climate or CO_2? *Climatic Change*, *29*, 91-99.

Bahre, C. J. (1991). *A Legacy of Change*. Tucson, AZ: University of Arizona Press.

Bock, C. E. and Bock, J. H. (2000). The *View from Bald Hill.* Berkeley, CA: University of California Press.

Brady, N. C. and Weil, C. T. (1999). The *Nature and Properties of Soils.* Upper Saddle River, NJ: Prentice-Hall.

Breman, H. and de Wit, C. T. (1983). Rangeland productivity and exploitation in the Sahel. *Science*, *221*, 1341-1347.

Briske, D. D., Fuhlendorf, S. D., and Smeins, F. E. (2005). State-and-Transition models, thresholds, and rangeland health: a synthesis of ecological concepts and perspectives. *Rangeland Ecology and Management, 58,* 1-10.

Brown, D. E. and Lowe, C. H. (1974). A digitized computer compatible classification for natural and potential vegetation in the Southwest with particular reference to Arizona. *Journal of the Arizona Academy of Sciences. Supplement 2*, 11.

Castellano, M. J. and Valone, T. J. (2007). Livestock, soil compaction and water infiltration rate: evaluating a potential desertification recovery mechanism. *Journal of Arid Environments 71*, 97-108.

Daily, G. C. (1995). Restoring value to the worlds degraded lands. *Science*, *269*, 350-354.

Fleischner, T. L. (1994). Ecological costs of livestock grazing in western North America. *Conservation Biology*, *8*, 629-644.

Fuhlendorf, S. D., Briske, D. D. and Smeins, F. E. (2001). Herbaceous vegetation change in variable rangeland environments: The relative contribution of grazing and climatic variability. *Applied Vegetation Science*, *4*, 177-188.

Gamougoun, N. D., Smith, R. P., Wood, M. K. and Pieper, R. D. (1984). Soil, vegetation, and hydrologic responses to grazing management at Fort Stanton, New Mexico. *Journal of Range Management*, *37*. 538-541.

Gifford, G. F., Faust, R. H. and Coltharp, G. B. (1977). Measuring soil compaction on rangeland. *Journal of Range Management*, *30*, 457-460.

Hendricks, D. M. (1985). *Arizona Soils*. Tucson, AZ: University of Arizona Press.

Herrmann, S. M., Anyamba A. and Tucker C. J. (2005). Recent trends in vegetation dynamics in the African Sahel and their relationship to climate. *Global Environmental Change 15*, 94-404.

Jackson, R. B., Banner, J. L., Jobbagy, E. G., Pockman, W. T. and Wall, D. H. (2002). Ecosystem carbon loss with woody plant invasion of grasslands. *Nature*, *418*, 623-626.

Kelly, R. D. and Walker, B. H. (1976). The effects of different forms of land use on the ecology of a semi-arid region in southeastern Rhodesia. *Journal of Ecology*, *64*, 553-576.

Laycock, W. A. (1991). Stable states and thresholds of range condition on North American rangelands: a viewpoint. *Journal of Range Managemant*, *44*, 427-433.

Mapfumo, E., Chanasyk, D. S., Naeth, M. A. and Baron, V. S. (1999). Soil compaction under grazing of annual and perennial forages. *Canadian Journal of Soil Science*, *79*, 191-199.

McCalla, G. R., Blackburn, W. H. and Merrill, L. B. (1984). Effects of livestock grazing on infiltration rates, Edwards plateau of Texas. *Journal of Range Management 37*, 265-269.

McGinty, W. A., Smeins, F. E. and Merrill, L. B. (1979). Influence of soil, vegetation, and grazing management on infiltration rate and sediment production of Edwards plateau rangeland. *Journal of Range Management 32*, 33-37.

Peters, D. P. C., Bestelmeyer, B. T., Herrick, J. E., Fredrickson, E. L., Monger, H. C. and Havstad, K. M. (2006). Disentangling complex landscapes: new insights into arid and semiarid system dynamics. *BioScience, 56,* 491-501.

Rasmussen, K., Fog, B. and Madsen, J. E. (2001). Desertification in reverse? Observations from northern Burkina Faso. *Global Environmental Change*, *11*, 271-282.

Rauzi, F. and Hanson, C. L. (1966). Water intake and runoff as affected by intensity of grazing. *Journal of Range Management 19*, 351-355.

Rhoades, E. D., Locke, L. F., Taylor, H. M. and McIlvain, E. H. (1964). Water intake on a sandy range as affected by 20 years of differential cattle stocking rates. *Journal of Range Management 17*, 185-190.

Rietkerk, M., van den Bosch, F. and van de Koppel, J. (1997). Site-specific properties and irreversible vegetation changes in semi-arid grazing systems. *Oikos*, *80*, 241-252.

Rietkerk, M. and van de Koppel, J. (1997). Alternate stable states and threshold effects in semi-arid grazing systems. *Oikos*, *79*, 69-76.

Rietkerk, M., Boerlijst, M. C., van Langevelde, F., HilleRisLambers, R., van de Koppel, J., Kumar, L., Prins, H. H. T. and de Roos, A. M. (2002). Self-organization of vegetation in arid ecosystems. *American Naturalist*, *160*, 524-530.

Rosenfeld, D., Rudich, Y. and Lahav, R. (2001). Desert dust suppressing precipitation: A possible desertification feedback loop. *Proceedings of the National Academy of Sciences U.S.A.*, *98*, 5975-5980.

Scheffer, M., Carpenter, S., Foley, J. A., Folke, C. and Walker, B. (2001). Catastrophic shifts in ecosystems. *Nature*, *413*, 591-596.

Schlesinger, W. H., Reynolds, J. F., Cunningham, G. L., Huenneke, L. F., Jarrell, W. M., Virginia, R. A. and Whitford, W. G. (1990). Biological feedbacks in global desertification. *Science*, *247*, 1043-1048.

Seybold, C. A., Herrick, J. E. and Brejda, J. J. (1999). Soil resilience: A fundamental component of soil quality. *Soil Science*, *164*, 224-234.

Thurow, T. L., Blackburn, W. H. and Taylor C. A. Jr. (1988). Infiltration and interrill erosion responses to selected livestock grazing strategies, Edwards plateau, Texas. *Journal of Range Management 41*, 296-302.

Valone, T. J., Meyer, M., Brown, J. H. and Chew, R. M. (2002). Timescale of perennial grass recovery in desertified arid grasslands following livestock removal. *Conservation Biology*, *16*, 995-1002.

Van Auken, O. W. (2000). Shrub invasions of North American semiarid grasslands. *Annual Review of Ecology and Systematics*, *31*, 197-215.

van de Koppel, J., Rietkerk, M. and Weissing, F. J. (1997). Catastrophic vegetation shifts and soil degradation in terrestrial grazing systems. *Trends in Ecology and Evolution*, *12*, 352-356.

van de Koppel, J. and Rietkerk, M. (2000). Herbivore regulation and irreversible vegetation change in semi-arid grazing systems. *Oikos*, *90*, 253-260.

van de Koppel, J., Rietkerk, M., van Langevelde, F., Kumar, L., Klausmeier, C. A., Fryxell, J. M., Hearne, J. W., van Andel, J., de Ridder, N., Skidmore, A., Stroosnijder, L. and Prins, H. H. T. (2002). Spatial heterogeneity and irreversible vegetation change in semiarid grazing systems. *American Naturalist*, *159*, 209-218.

van de Koppel, J. and Rietkerk, M. (2004). Spatial interactions and resilience in arid ecosystems. *American Naturalist*, *163*, 113-121.

Walker, B. H., Ludwig, D., Holling, C. S. and Peterman, R. M. (1981). Stability of semi-arid savanna grazing systems. *Journal of Ecology*, *69*, 473-498.

Warren, S. D., Thurow, T. L., Blackburn, W. H. and Garza N. E. (1986). The influence of livestock trampling under intensive rotation grazing on soil hydrologic characteristics. *Journal of Range Management, 39*, 491-495.

Wood, M. K. and Blackburn, W. H. (1981). Grazing systems: their influence on infiltration rates in the rolling plains of Texas. *Journal of Range Management*, *34*, 331-335.

Zhang, J. Y., Wang, Y., Zhao, X., Xie, G. and Zhang, T. (2005). Grassland recovery by protection from grazing in a semi-arid sandy region of northern China. *New Zealand Journal of Agricultural Research 48*, 277-284.

In: Grasslands: Ecology, Management and Restoration
Editor: Hans G. Schröder
ISBN 978-1-60692-023-7

Chapter 3

Floods Recover the Deterioration of Grasslands and Soils Caused by Cattle Grazing in the Flooding Pampas of Argentina

Pedro Insausti, Miguel A. Taboada and Raúl S. Lavado
Facultad de Agronomía, Universidad de Buenos Aires. Av. San Martín 4453
C1417DSE Buenos Aires, Argentina

Introduction

At the times of America's discovery in 1492 the temperate part of South America was covered by grasslands. This area includes today Central and Northeastern parts of Argentina, a great deal of Uruguay and Southern Brazil (Figure 1) (Soriano et al., 1991). Among specific features, those grasslands had a unique characteristic: they were developed without high grazing pressures because the region lacked of large herbivores. This differs from the situation found in equivalent African and North American grasslands (Mc Naughton et al., 1988; Sala et al., 1986). The floristic composition of these grasslands varies in accordance with the regional climatic gradients and the local substrate heterogeneity (León et al., 1979; Burkart et al., 1990; Perelman et al., 2001; Soriano et al., 1991). Since European colonization, this vegetation has been progressively modified by the introduction of domestic cattle and the appearance of annual crops, pastures and forestation. These disturbances have led to the invasion and naturalization of several exotic vegetal species, provoking changes in the distribution of the native species (Ghersa and León, 1999) and alterations in the structure and functioning of the vegetal communities (Sala et al., 1986; Rusch and Oesterheld, 1997). Nowadays, only semi-natural grasslands remain in some locations within the region- They are located where the agricultural possibilities of the lands are seriously constrained (Ghersa and León, 1999). This typically occurs in the so called Flooding Pampa. Two main reasons affect agricultural development in this subregion: the occurrence of frequent ponding and/or floods alternating with summer drought cycles, and the prevalence of halo-hydromorphic soils (León et al., 1975; Batista et al., 1988; Lavado and Taboada, 1988).

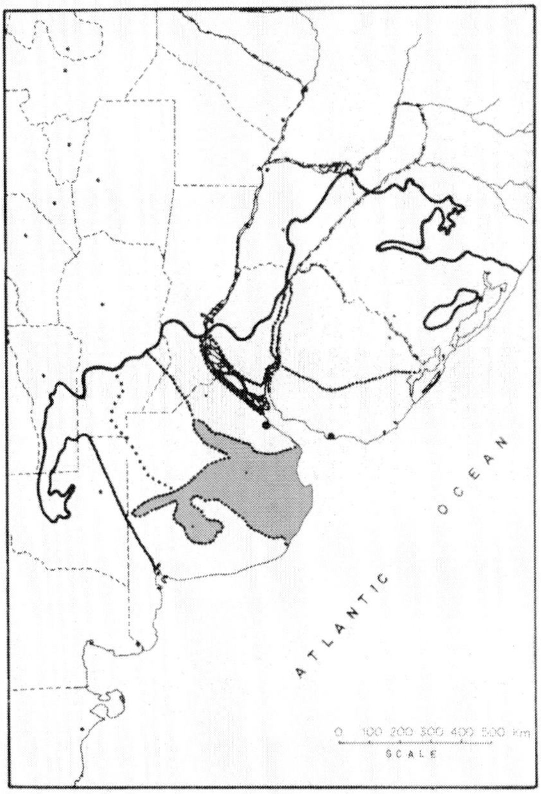

Figure 1. Rio de la Plata grasslands with Flooding Pampa in green.

The vegetation growth there owes many important aspects of its current dynamics and structure to the grazing. As a result of this intervention, the differences among the grasslands communities associated with varying types of soil is reduced and a significant part of the resident flora is formed by exotic species whose living habits are quite different from that of the native flora (Sala et al., 1986; Perelman et al., 2001). These effects are modified by floods, a natural disturbance whose influence on the conservation of native grasslands and their productive potential has not been sufficiently studied (Insausti et al., 1999).

THE FLOODING PAMPA GRASSLANDS

The Flooding Pampa grasslands cover a plain of about 6 Mha in the Buenos Aires province, in Argentina, including the lowlands known as the Rio Salado Basin (Figure 1) (Soriano et al., 1991). Natural grasslands are the main forage resource that supports the cattle activity of the region and continuous grazing is one of the main determinants of the structure and functioning of the vegetation (León et al., 1984; Sala et al., 1986; Rusch and Oesterheld, 1997).

As previously mentioned, periodical floods are a typical environmental event in the region. Floods are determined by a climatic, soil properties, and topographical factors, and their duration and intensity vary from year to year. Floods are partially associated to the seasonal cyclic behavior of the water table and to the occurrence of intense precipitation. Two kind of floods can be distinguished. The ones are marked by a lower duration and intensity, take place at the end of the winter and the onset of spring. The others are more intense and prolongued, and do not occur in a precise period of the year (Paruelo and Sala, 1990; Taboada and Lavado, 2003).

As a natural disturbance, flood consequences vary according to the system upon which they impact and the way this impact is evaluated (Chaneton and Facelli, 1991). Traditionally, floods have been considered disturbances that alter the vegetation structure (White, 1979) and, owing to the fact that flood tolerance varies significantly among the different species, great changes in the floristic composition of the community can be expected. Thus, the disturbed ecosystems may experience major changes in the dominant species, an important part of their dynamics (Odum, 1985). Similarly, floods are often believed to have degradative effects on soils (Taboada y Lavado. 2003)

Flooding generates gaps in the vegetal canopy and modifies the physical and biotic environment explored by the different species, thus determining changes in the quantity and time in which resources are available (Tilman, 1982; Insausti et al., 1995). Natural grasslands are systems that generally show remarkable flood tolerance (Holberg and Bischoff, 1980). This differs from many other cultures, including sown grasslands, which may be significantly damaged by flooding (Jackson and Drew, 1984; Pezeshki, 1994; 2001). Responses to poor soil aeration depend upon the species. Many experience severe limitations for root growth, with soil oxygen diffusion rates below 20 $\mu g cm^{-2} min^{-1}$. However, other species such as *Agrostis palustris* and *Eleusine indica* experience solid radical growth in soils with an oxygen diffusion rate below the aforementioned value (Meek and Stolzy, 1978). Depending on the species, the ecological, physiological, morphological and anatomical flood responses may vary, but with only two biological results: tolerance or susceptibility to flooding.

At the individual level and depending on the behavior of each species, the restriction of the soil oxygen caused by flooding reduces the respiration rate and water and nutrient absorption through plant roots. This occurs when plants are not adapted for internal oxygen transfer from the aerial part (e.g. aerenchyma) (Ponnamperuma, 1984; Voesenek et al., 2006). In the short term, physiological stress induced by soil flooding reduces the productivity of species sensitive to flooding, whereas in the long term, flooding may increase the mortality rate in the population, thus decreasing the presence of those species (Wehier and Keddy, 1995). Nevertheless, in the Flooding Pampa grasslands, species from flooded sites like *Paspalum dilatatum* grow during spring floods and present a higher photosynthesis rate and water potential during flooding than at field capacity (Insausti et al., 1999; 2001).

The Flooding Pampa is one of the areas in the world with extensive areas covered by Solonetz type soils (Bui et al., 1998). These soils were developed in a large scale by pedogenetic factors that are well known (Lavado and Taboada, 1988; Moscatelli and Scoppa 1984; Tricart 1973). Most soils (60%) in the region under study are included in the aquic regime (Salazar Lea Plaza and Moscatelli, 1989); most of them show a natric horizon and excess of soluble salts. The main Great Groups of Soil Taxonomy in this region are Natraquolls (28 000 km2) and Natraqualfs (11 000 km2). These soils do not occupy patches of different size, as in other regions in the world, but exist in large areas: the same soil Series may cover a continuous area of a hundred thousands hectares (Moscatelli and Salazar Lea Plaza, 1989). Natralbolls, Argialbolls, etc. are also present, but in a smaller extent. Most soils belong to loamy or loamy-fine texture families. They show a strong textural contrast between horizons, with a tough Bt (natric) horizon limiting soil water movements through the profile (Lavado and Taboada, 1988; Taboada et al., 2001).

In the region cattle grazing was found to be a major factor causing topsoil salinization. This process was associated to the decovering action of grazing that favors upward salt movements from deep soil layers (Lavado and Taboada, 1987; 1988).

Grazing and Grasslands

Grazing dramatically alters the vegetation structure and, among the different vegetation types, grasslands are the most affected by this activity (Mc Naughton, 1985). Grazing produces changes in the architecture of the canopy, the composition of species, spatial heterogeneity and diversity (Sala et al., 1986; Chaneton and Facelli, 1991). In addition, herbivory may alter the functional properties of grasslands in terms of the energy flow and nutrient circulation in the ecosystem (McNaughton, 1985; Detling, 1987; McNaughton et al., 1988). The disturbance of grassland species by herbivory is caused directly by the removal of leaves and other photosynthetic organs; indirectly, grazing provokes environmental alterations (e.g. alterations in the water balance and soil thermal regime, in the ligth environment, and in nutrient availability) (Deregibus et al., 1985; Detling, 1987; Lavado et al., 1996). Grazing also alters the competition among neighboring plants (Harper, 1977; Crawley, 1986). Selective herbivory determines the defoliation of some individuals and not others; the selective impact of grazing on dominant species may provoke the release of resources that can be exploited by the neighboring species which are less preferred by herbivores. Thus, grazing may help less competitive species and contribute to the preservation of diversity among the vegetal community (Harper, 1977). Nevertheless, when preferential herbivory affects the less competitive species, grazing may cause a decrease in the species diversity (Crawley, 1986).

In the Flooding Pampa grasslands, the relatively recent introduction of large domestic herbivores has provoked an increase in the abundance and diversity of the native dicots species, as well as the invasion of exotic dicots (Sala et al., 1986; Rusch and Oesterheld, 1997).

Grazing represents a major constraint in grasslands, primarily through its effect on biomass loss rates (McNaughton 1985) and plant regeneration patterns (Bullock et al. 1995). In the Flooding Pampa, continuous grazing by cattle maintains the grassland standing biomass at low levels and may also limit above-ground primary productivity (Rusch and Oesterheld 1997). These effects are related to changes in life form and species composition promoted by selective herbivory and gap disturbance (Sala et al.1986, Rusch and Oesterheld 1997).Typically, in the argentine pampas grazed grasslands become dominated by low-growing perennial forbs at the expense of more palatable and productive caespitose grasses and sedges (Sala et al., 1986; Facelli et al. 1989).

The species that form the Flooding Pampa grasslands may be divided in two functional groups (Wilson and Roxburgh, 1994; Insausti et al., 1999): dicots and gramineous. The later species, along with other families (Cyperaceous and Juncaceous, etc.) of similar structure, are called graminoids (Insausti et al., 1999). There are differential responses to the disturbances that may be attributed to those two functional groups and there are clear evidences that both grazing and floods affect the grassland dicots and graminoids in different ways (Sala et al., 1986; Insausti and Soriano, 1987; Chaneton et al., 1988; Rusch and Oesterheld, 1997; Insausti et al., 1999). Nevertheless, the fact that certain species have similar responses to a particular disturbance does not mean that other important functional features are homogeneous (Lavorel et al., 1997). For example, grazing-tolerant species may differ in flooding response patterns and, in turn, flooding-tolerant species may differ according to their seasonal pattern and phenology (Elmquist et al., 2003). The capacity of a plant species to recover from a disturbance depends on this "response diversity" within the functional groups (Walker et al., 1999; Luck et al., 2003).

Grazing and Soils

Cattle grazing are usually considered a severe disturbance on grassland soil properties. In the case of the Flooding Pampa soils, unlike other grasslands, grazing caused no effect on soil organic matter content, exchangeable cations and nitrogen contents (Chaneton et al., 1996; Lavado and Taboada, 1986, Lavado et al., 1996). Only slight losses of soil phosphorus accredited to cattle production were observed (Chaneton et al., 1996; Lavado et al., 1996). Instead, grazing affected soil physical properties through soil decovering and through the stress caused by animal hooves (i.e. treading and trampling). Soil decovering results in higher evaporation rates, lower water contents, reduced infiltration rates (Dreccer and Lavado, 1993; Lavado and Alconada, 1994; Taboada and Lavado, 1988) and –in salt-affected areas- also higher salt contents in topsoil (Lavada and Taboada, 1987, 1988).

Trampling often causes a decrease in porosity in grassland topsoils (Taboada and Lavado 1988; Warren *et al.* 1986). The magnitude of the stress caused by animal hoof impact mainly depends on soil water content (Mulholland and Fullen 1991; Warren *et al.* 1986). When soil of low to medium water content is trampled, the main process is compression beneath the hoof (Scholefield *et al.* 1985). This collapses the larger soil pores by the mechanical disruption of aggregates causing shallow soil compaction (Warren *et al.*, 1986). When wetter soil is trampled there is plastic flow around the hoof (Scholefield *et al.* 1985). Repeated treading in these conditions produces deep hoof-prints, which damage the sward (Scholefield and Hall 1986). This effect (poaching) generally coincides with the presence of free surface water (Mulholland and Fullen 1991), and creates dense, unstable surface clods (Warren *et al.* 1986). Previously damaged soil pores can be regenerated, however, during wetting-drying cycles. This depends on soil shrink-swell (Dexter, 1988), provided the externally applied stress is decreased or removed. Grazing often causes decreases in structural stability associated with detrimental effects on soil macroporosity (Warren et al., 1986) and water infiltration (Gifford and Hawkins, 1978).

Floodings and Grasslands in the Flooding Pampa Grasslands

The greatest difficulty when studying the effects of floods on the vegetation is that floods constitute stochastic phenomena, difficult to predict and control. For this reason, when a major flood occurs, generally all coetaneous control species have been affected by the disturbance.

In this chapter, we present the results of a manipulative experiment in which a natural flood was studied on a field using non-flooded control plots. This experiment corrected the aforementioned problems of studying the effects of flooding because coetaneous control plots not affected by the natural flood were available in the field.

The general working hypothesis is based on dynamic system in which flooding is a directional force opposed to grazing. Both disturbances affect the composition and structure of the vegetal community, and the intensity and duration of each of these disturbances determines which functional group (dicots and graminoids) prevails. Thus, grasslands might switch back and forth between two alternative states in response to both disturbing agents considered (Figure2).

To analyze the effect of natural flooding under field conditions, two treatments were carried out in established plots within the grazing-free area throughout the experiment. One group of plots was totally flooded during a great part of winter and the onset of spring, while another group of plots was extracted as grasslands blocks of 1.5 x 0.8 x 0.45 m each and elevated "in situ" above soil level to keep them from flooding. In this manner, a series of flood-free control plots was established. The naturally flooded plots in the field during winter-spring had a higher total aerial biomass than the plots elevated above flooding level (Figure 3). The total graminoid biomass was higher in the flooded plots than in the controls ones; nevertheless, the most abundant species was the winter-spring gramineous growth *Danthonia montevidensis*, which increased 2.5 times in the flooded plots with respect to the elevated plots (Table 1). The differences have been more evident on the upper layer of the canopy, where several native graminoids (*Paspalidium paludivagum, Paspalum vaginatum, Panicum gouinii* and *Leersia hexandra*) recorded higher biomass increases with respect to the non-flooded controls.

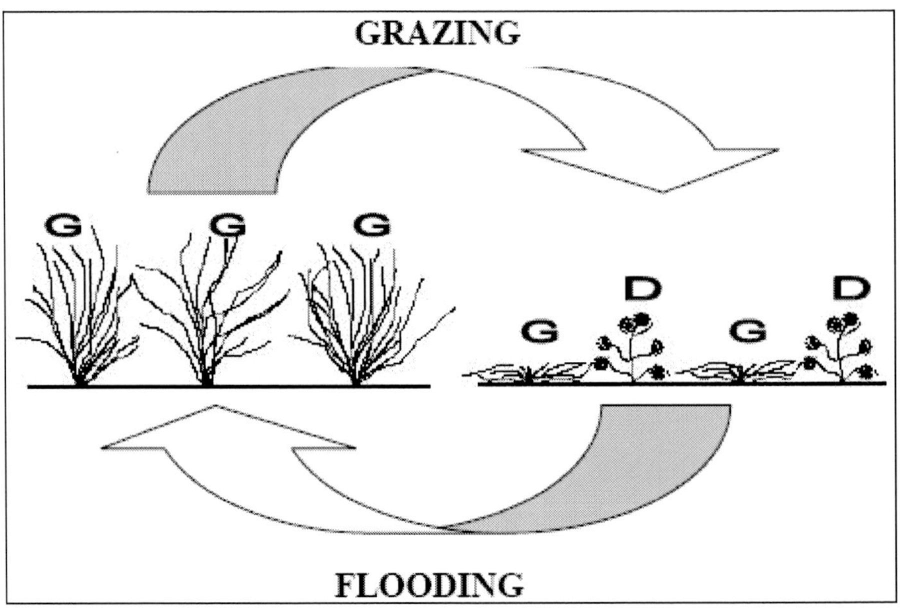

Figure 2. Alternative states of grassland in response to grazing or flooding (G: graminoids, D: dicots).

Table 1. Above-ground biomass of grassland plant species, after flooding period in flooded and non-flooded conditions. Values are means ± SE of five replications. * Significant differences ($P < 0.05$) among treatments within each species.

Species	Control	Flooding
Warn-season Dicots		
Ambrosia tenuifolia	3.67 (1.05)	0.07 (0.03)*
Aster squamatus	3.49 (0.21)	1.95 (0.91)
Cool-season Dicots		
Eryngium echinatum	2.48 (0.58)	2.25 (1.89)
Leontodon taraxacoides	15.84 (1.62)	14.77 (4.58)
Mentha pulegium	1.45 (1.03)	1.23 (0.72)
Plantago lanceolata	31.36 (0.78)	10.51 (4.52)*
Warn-season Graminoids		
Leersia hexandra	0.88 (0.25)	5.20 (0.65)*
Panicum berguii	0.74 (0.34)	1.14 (0.86)
Panicum gouinii	2.97 (0.65)	16.69 (5.55)*
Panicum milioides	3.52 (1.94)	3.83 (1.99)
Paspalidium paludivagum	1.85 (0.67)	6.24 (1.70)*
Paspalum vaginatum	1.14 (0.89)	17.06 (1.6)*
Setaria parviflora	0.78 (0.38)	0.84 (0.80)
Cool-season Graminoids		
Danthonia montevidensis	12.80 (4.66)	26.23 (4.8) *
Chaetotropis elongata	3.39 (1.34)	15.95 (5.95)*
Sisyrinchium platense	6.19 (1.54)	18.18 (2.48)*

The two latter species boasted almost six times as much biomass in the flooded plots than in the non-flooded controls (Table 1). Dicots biomass decreased drastically due to flooding under field conditions (Figure 3). The biomass of *Plantago lanceolata* is very abundant during winter-spring, however it decreased by 65% as the result of flooding (Table 1). *Leontodon taraxacoides* behaved as a co-dominant species in the non-flooded controls. Changes related to the relative contribution of both functional groups to the community biomass coincided with the results obtained by Insausti et al. (1999) in an experiment in mesocosms, outside the natural grasslands, with a controlled spring- summer flooding. Nevertheless, when regrouping the species according to their growth season, the specific responses depended on the flooding period considered.

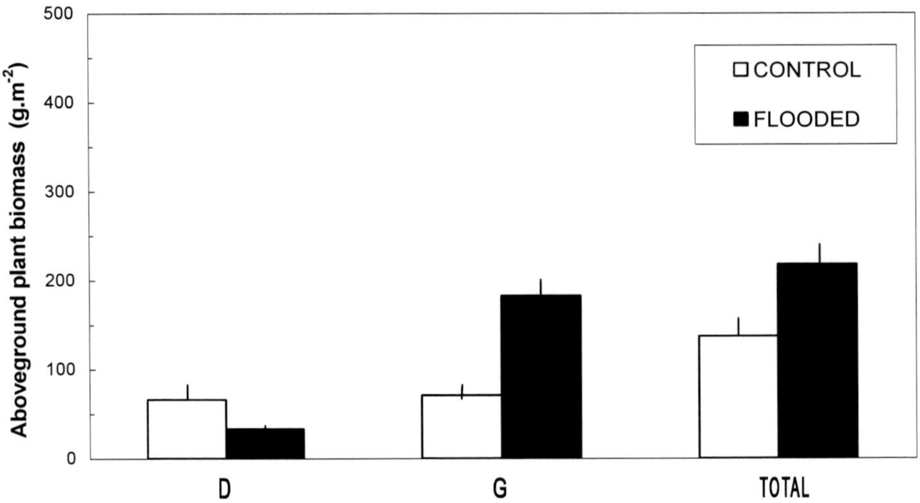

Figure 3. Aboveground plant biomass (g dry weight) of dicots (D) and graminoids (G) after flooding period in flooded and non-flooded conditions in the field. Vertical bars denote 1 SE.

Besides, the most significant responses have been observed in the canopy layer where the biomass of each species is usually located, according to its type of growth. Dicots grew more in the lower layer when they had not been affected by floods, whereas graminoids performed better in the upper layer, especially during flooding conditions.

Flooding increased the mean in field canopy height and modified the vertical biomass distribution. While gramineous positioned almost 50% of their biomass above water level (6 cm) during flooding, only 25% of the biomass from this group appeared above this height when flooding did not occur. Eighty-five percent of the dicots were under 6 cm high in the control plots owing to the prostrated growing habit exhibited by the majority of dicots at that time of the year. The decrease of dicots during flooding was very remarkable in the lower canopy layer. Nevertheless, some species like *Leontodon taraxacoides, Phyla canescens* and *Eryngium ebracteatum*, modified the position of their leaves, thus showing remarkable morphological adaptability, by localizing a larger fraction of the foliar area in higher layers, above the flooded level (Grimoldi et al., 1999; Jackson, 2008).

In the Flooding Pampa, major floods are a regular occurrence. According to our results, within a relative short period of time, these floods slip back the cumulative effect of grazing on the structure and composition of the grasslands.

The alteration of the community structure produced by prolonged flooding depends on the existence of functional groups with different responses to that disturbance factor. Dicots, the majority of which are not native, find favorable conditions to expand in the grazed fields, but decrease their abundance (or almost disappear, depending on the species) during floods. At the same time, the aerial biomass of grasslands increases with flooding. This biomass increase is caused by the positive response of native graminoids to flooding. This functional group decreases its biomass as a result of continuous grazing, because it represents the main foraging resource for domestic herbivores.

Graminoids increase during flooding as a result of the decrease of dicots in terms of the total aerial biomass of the community. This allows us to infer competitive interference between those functional groups in terms of resource usage. The significant growth of graminoids under flooding conditions, then, could be indirectly owed to the effects of flooding on resource availability. According to this hypothesis, flooding reduces the competition that is favorable for the species (in this case, graminoids) that are not significantly affected by flooding. Graminoids then occupy greater canopy and soil volumes due to the disappearance of dicots; more specifically, grass adaptations to flooding would allow them to use the nutrients available in the soil during flooding (Rubio and Lavado, 1999). These soil resources would, to some extent, be shared with the dicots in the grazed grasslands during flood-free periods.

Effect of Floods and Grazing on Flooding Pampa Soils

The combined effect of flooding and grazing was investigated in a field 2 ½ year study, in which the variation of soil physical properties was compared in nearby grazing enclosured and a continuously grazed (about 0.5 cow ha^{-1} $year^{-1}$) areas (Taboada and Lavado, 1993; Taboada et al., 1999). It was found that soil water content and not grazing was the primary cause of changes in soil total porosity. Because of the continuous trampling by cattle, the applied stress never ceased during the study period. Despite this, soil pores damaged in summer were regenerated during the subsequent period of surface ponding, when the soil swelled and both total porosity and macroporosity > 30 μm increased significantly in the grazed area. The regeneration of pores could be related to the appearance of microcracks (Dexter 1988), which are induced by air entrapment when the surface is flooded (Taboada et al., 2001). Parker *et al.* (1977) suggested that the magnitude of volume change on swelling can be increased by compaction and disruption of aggregates. In Flooding Pampas soils pore damage by trampling during the summer could have led to the increased soil expansion when it rewetted in winter.

Soil water content was the primary cause of changes in total porosity, because of shrink-swell processes (Taboada and Lavado, 1993). Contrary to what was expected, trampling caused macropores larger than 60 μm to collapse and decreased the size of water-stable aggregates in dry soil. Because of this collapse and the greater shrinkage, the soil under grazing tended to have significantly less total porosity during summer than during winter. The damaged soil pores and the aggregates both increased in size some months later, during the period of surface ponding. Trampling accentuated soil swelling during these periods, and no poaching damage was observed.

Grazing effects on soil structural stability were superimposed on the environmental effects. Aggregate mean weight diameter (MWD) was often lower in the soil under grazing (Taboada and Lavado, 1993; Taboada et al., 1999). This aggregate size reduction is attributable to the fracturing and pulverizing of dry soil caused by the mechanical action of trampling, as shown by Warren *et al.* (1986). This increased the percentage of water-stable microaggregates (< 0.3 mm) both in dry summers in the soil under grazing. A highly significant decrease in wet MWD was also found in both grazed and old exclosure areas at the end of the dry summer. The low soil matric potential during that period (Lavado and Taboada 1988) brought about significant shrinkage. It was demonstrated that planes of weakness are generated within the dried and shrunken mass of soils with a certain proportion of active clay, as a consequence of water stress (Utomo and Dexter 1982). Likewise, in smectitic clay soil, most disaggregation could be caused by macroscopic and microscopic shrinkage planes developed when the soil was dry. When the soil was rewetted during the fall, the aggregate size increased again in both treatments.

Aggregate stability under grazing increased when the soil was wet and decreased when the soil was dry (Taboada and Lavado, 1993; Taboada et al., 1999). Results on aggregate MWD and on soil porosity agree well. Those results showed detrimental effects by trampling during the summer when the soil was dry. Figure 4 shows a conceptual model that postulates decreases in structural stability resulting from crushing air-filled pores by cattle hooves. This yields smaller water-stable aggregates, as shown by the higher proportion of aggregates < 0.3 mm usually found in the soil of the grazed area compared to the soil in the enclosure area. Only at low water contents was the structure of the topsoil destabilized by grazing. The recovery of structural stability began in the fall and was completed in the winter, when the soil was ponded (Figure 4). The structural recovery results from swelling, when the smaller aggregates created by trampling of dry soil are bound again into larger structural units.

The ultimate cause of this unusual behaviour was studied by Taboada et al. (2001). Topsoil structural improvement found during flooding appears to be a consequence of soil inflation. In a laboratory experiment, Gäth and Frede (1995) found a process of soil "inflation", under quick saturation of a rigid material. Our field results showed the same process. A conceptual model describing the changes taking place in the soil profile was proposed (Figure 5). The unusually high entrapped air volumes at the maximum swollen condition can be regarded as the result of two coincident wetting fronts during winter – spring periods (Lavado and Taboada, 1988). Air bubbles entrapped between the ponded soil surface and the rising water table could not escape. This explains why air entrapment exerted so great influence on the swelling of our soils.

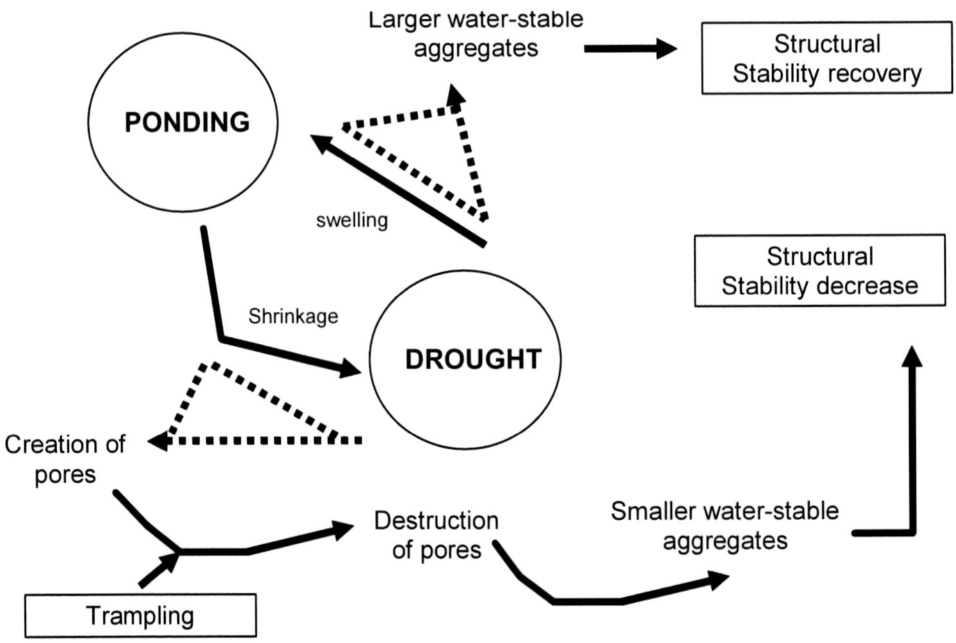

Figure 4. Conceptual model showing the process of soil structural destabilization when the soil dries and the process of structural recovery when soil wets.

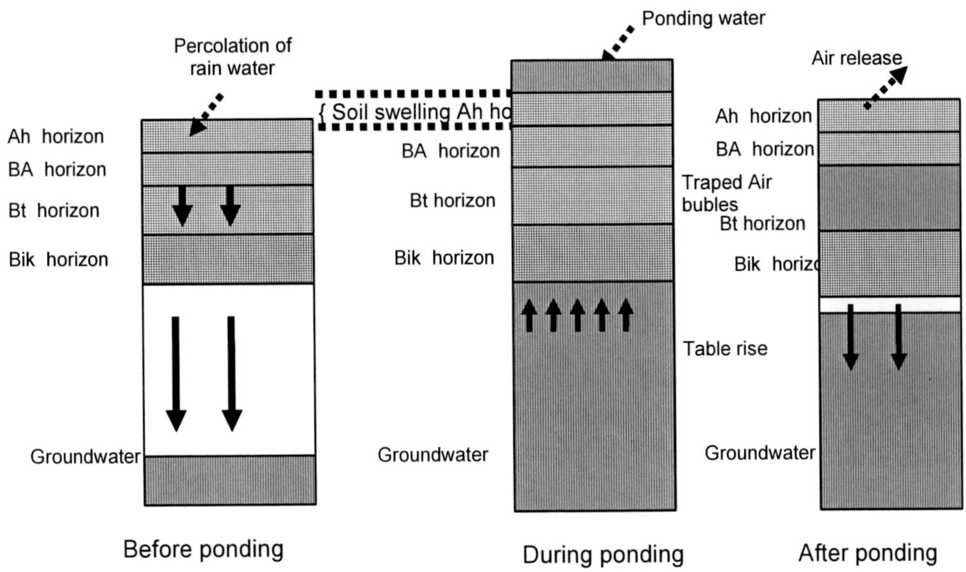

Figure 5. Conceptual model describing the process of soil swelling because of air entrapment. a) before ponding: free water and air movements throughout the profile (bio-opened system); b) during ponding: soil swelling by air trapped within water table perched over the impervious Btk horizon (air-occluded system); c) after ponding: soil shrinkage after perched water and water table decreases, and rapid air escape to the atmosphere.

We can then conclude that air entrapment due to water table rises from depth and surface ponding was the main factor determining the swelling of soils. These results explain the structural behaviour of seasonally flooded soils and their recovery from soil deterioration cause by grazing in general and trampling in particular.

During flooding the tough natric horizon separates two kind of water qualities: the salty groundwater below it and the rainwater above it (Lavado and Taboada, 1988; Taboada et al., 2001). The process of ponding can be then regarded as a huge perched water table, which accumulates over the impervious natric horizon during water excess periods. This actually impedes the arrival of salty groundwater to soil surface. So, rather than being a disturbance, ponding with rainwater is a key factor preserving soil physical properties in this lowland.

Conclusions

Prolonged flooding constitutes a directional force in grasslands, opposed to that of grazing. The advance of one functional group over another depends upon the intensity and duration of each flooding event. The fact that flooding may act in this ecosystem as a subsidy for graminoids and a stress factor for dicots might allow us to consider flooding a “service of nature" (Ehrlich and Mooney, 1983) that is, an event capable of reverting the decay of vegetation induced by grazing, at least partially and transitorily. From the point of view of the use of natural grasslands, this “service” improves the quantity and quality of the available forage. Nevertheless, it should be mentioned that the duration of the beneficial effects of flooding on grasslands depends, to a larger extent, on grazing management (charge and frequency) and on the prevailing climatic conditions after flooding.

Under the study conditions, the stocking rates were low and, according with the scarce results obtained in the area, the removal of nutrients by grazing would operate in the long term. Changes in nutrient supply at root level can not be rule out, but it was not determined. Instead, it was definitively found that soil physical properties like porosity and soil structural stability are markedly affected by trampling and that the flooding revert those negative effects.

The effect of grazing on grassland species and soil properties occurs in a parallel way. Far from being an environmental disturbance, the floods recovered the graminoid importance in the grassland and the impaired soil physical properties. The degree of linking between these two processes is still unknown.

REFERENCES

Bui, E., L. Krogh, R.S. Lavado, F. Nachtergaele, T. Tóth and R.W. Fitzpatrick. 1998. Distribution of sodic soils: The world Scene. en M.E. Sumner y R. Naidu (Eds.). *Sodic Soils: Distribution, Processes, Management and Environmental Consequences.* Oxford University Press. Pp. 19-34.

Batista, W. B., R. J. C. Leon y S. B. Perelman. 1988. Las comunidades vegetales de un pastizal natural de la Region de Laprida, Prov. de Buenos Aires, Argentina. *Phytocoenologia* 16:465-480.

Bullock, J.M., Clear Hill, B., Silvertown, J. and Sutton, M. 1995. Gap colonization as a source of grassland community change: effects of gap size and grazing on the rate and mode of colonization by different species. *Oikos* 72: 273-282.

Burkart, S. E., Leon, R. J. C. and Movia, C. P. 1990. Inventario fitosociologico del pastizal de la Depresion del Salado (Prov. Bs. As.) en un area representativa de sus principales ambientes. - *Darwiniana* 30: 27-69.

Crawley, M. 1986. The structure of plant communities. Plant Ecology, M. Crawley, ed.: 1-50.

Chaneton E.J., Facelli J.M. and León R.J.C. 1988. Floristic changes induced by flooding on grazed and ungrazed lowland grasslands in Argentina. *J. Range Manage.* 41: 497-501.

Chaneton E.J. and Facelli J.M. 1991. Disturbance effects on plant community diversity: Spatial scales and dominance hierarchies. *Vegetatio* 93: 143-155.

Chaneton, E.J. and R.S. Lavado. 1996. "Soil nutrients and salinity after long-term exclusion in a Flooding Pampa grassland". *Journal of Range Management* 49: 182-187.

Chaneton, E. J., Lemcoff, J. H. and Lavado, R. S. 1996. Nitrogen and phosphorus cycling in grazed and ungrazed plots in a temperate subhumid grassland in Argentina. *J. Appl. Ecol.* 33: 291-302.

Deregibus V.A., Sánchez R.A., Casal J.J. and Trlica M.J. 1985. Tillering responses to enrichment of red light beneath the canopy in a humid natural grassland. *J. Appl. Ecol.* 72: 199-206.

Detling J.K. 1987. Grass response to herbivory. In *A Shortgrass Prairie Perspective*: 56-68. J.L. Capinera (ed.) Westview Press, Colorado.

Dreccer, M.F. y Lavado, R.S. 1993. Influence of cattle trampling on the preferential flow paths in alkaline soils. *Soil Use and Management* 9: 143-148.

Ehrlich P.R. and Mooney H.A. 1983. Extinction, substitution, and ecosystem services. *BioScience* 33: 248-254.

Elmquist, T., Folke, C., Nystrom, M., Peterson, G., Bengtsson, J., Walker, B. and Norberg, J. 2003. Response diversity, ecosystem change, and resilience. *Frontiers Ecol. Environ.* 1: 488-494.

Facelli, J.M., León R.J.C. and Deregibus V.A. 1989. Community structure in grazed and ungrazed grasland sites in the Flooding Pampa, Argentina. *Am. Midl. Nat.* 121: 125-133.

Ghersa, C. M. y R. Leon. 1999. Successional changes in agroecosystems of the Rolling Pampa. en L. Walker, editor. *Ecosystems of the World 16.* Elsevier, Amsterdam.

Grimoldi A.A., Insausti P., Roitman G.G and Soriano A. 1999. Responses to flooding intensity in *Leontodon taraxacoides. New Phytol.* 141:110-119.

Harper, J.L. 1977. *Population biology of plants*. Academic Press. London.

Holberg H. and Bischoff H.M. 1980. Flooding tolerance yield and feed quality of some grasses in a model experimental with simulated flood. Archiv. *Für Acker-Und Planz. Und. Bodenkunde* 244: 513-521.

Insausti P. y Soriano A. 1987. Efecto del anegamiento prolongado en un pastizal de la Depresión del Salado (Provincia de Buenos Aires-Argentina): dinámica del pastizal en conjunto y de *Ambrosia tenuifolia. Darwiniana*, 28: 397-403.

Insausti, P., Soriano, A. and Sanchez, R. A. 1995. Effect of flood-induced factors on seed germination of *Ambrosia tenuifolia. Oecologia* 103: 127-132.

Insausti, P., Chaneton, E.J. and Soriano, A. 1999. Flooding reverted grazing effects on plant community structure in mesocosms of lowland grassland. *Oikos* 84: 266-276.

Insausti, P., Grimoldi, A., Chaneton, E.J. andVasellati, V.A. 2001. Flooding induces a suite of adaptive plastic responses in the grass *Paspalum dilatatum*. *New Phyt*ol. 152: 291-300.

Jackson M. B. and Drew M. 1984. Effects of flooding on growth and metabolism of herbaceous plants. In Flooding and plant growth (edited by T.T. Kozlowski). Academic Press, Florida, 10-42.

Jackson M.B. 2008. Ethylene-promoted Elongation: an Adaptation to Submergence Stress. *Annals of Botany* 101(2): 229-248.

Lavado, R.S. and Alconada, M. 1994. "Soil properties behavior on grazed and ungrazed plots of a grassland sodic soil". *Soil Technology* 7(1): 75-81.

Lavado, R.S. and Taboada, M.A. 1985. Influencia del pastoreo sobre algunas propiedades químicas de un Natracuol de la Pampa Deprimida. Ciencia del Suelo 3: 102-108.

Lavado, R.S. and Taboada, M.A. 1987. Soil salinization fluxes as an effect of grazing in a native grassland soil in the Flooding Pampa in Argentina. *Soil Use and Management* 4(3): 143-148.

Lavado, R.S. and Taboada, M.A. 1988. Water, salt and sodium dynamics in a Natraquoll in Argentina. *Catena* 15(4): 577- 594.

Lavado, R.S.; J.O. Sierra and P.N. Hashimoto. 1996. Impact of grazing on soil nutrients in a pampean grassland. *Journal of Range Management* 49: 452-457.

Lavorel, S., McIntyre, S., Landsberg, J. and Forbes, T. D. A. 1997. Plant functional classifications: from general groups to specific groups based on response to disturbance. *Trends Ecol. Evol.* 12: 474-478.

León, R.J.C. 1975. Las comunidades herbáceas de la región Castelli-Pila. En Productividad Primaria Neta de Sistemas Herbáceos. *Monografías* 5: 75-101. CIC. La Plata.

Leon, R.J.C., Burkart, S.E. and Movia, C.P. (1979): Relevamiento fitosociologico del pastizal del norte de la Depresion del Salado (Partidos de Magdalena y Brandsen). INTA, *Serie Fitogeografica* 17.

León, R.J.C., Rusch, G.M. and Oesterheld, M. 1984. Pastizales pampeanos - impacto agropecuario. *Phytocoenologia* 12: 201-218.

Luck, G.W., Daily, G.C. and Ehrlich, P.R. 2003. Population diversity and ecosystem services. *Trends Ecol. Evol.* 18: 331-336.

Mc Naughton S.J. 1985. Ecology of a grazing system: The Serengeti. *Ecol. Monogr*. 55: 259-294.

Mc Naughton S.J., Ruess R.W. and Seagle S.W. 1988. Large mammals and process dynamics in African ecosystems. *BioScience* 38: 794-800.

Meek J. and Stolzy L.H. 1978. Short-term flooding. In Plant life in anaerobic environments. Ann. *Arbor Sci. Publ.* (O.H. Hook and R.M.M. Crawford eds.) pp 331-373. Michigan.

Odum E.P. (1985). Trends expected in stressed ecosystems. *Bio Science*, 35, 419-422.

Paruelo J.M. and Sala O.E. 1990. Caracterización de las inundaciones en la Depresión del Salado (Buenos Aires, Argentina): Dinámica de la capa freática.. *Turrialba* 140: 5-11.

Perelman SB, León RJC and Oesterheld M (2001) Cross-scale vegetation patterns of Flooding Pampa grasslands. Journal of Ecology 89: 562-577.

Pezeshki S.R. 1994. Plant response to flooding. In: Wilkinson RE, ed. *Plant-Environment Interactions*. New York, USA: Marcel Dekker Inc., 289-321.

Pezeshki S.R. 2001. Wetland plant responses to soil flooding. *Environmental and Experimental Botany* 46: 299–312.

Ponnamperuma F.N. 1984. Effects of flooding on soils. In Flooding and plant growth (edited by T.T. Kozlowski). Academic Press, pp 10-42, Florida.

Rubio G. and Lavado RS. 1999. Acquisition and allocation of resources in two waterlogging-tolerant grasses. *New Phyto.* 143: 539-546.

Rusch, G. M. and Oesterheld, M. 1997. Relationship between productivity, and species and functional group diversity in grazed and non-grazed pampas grassland. *Oikos* 78: 519-526.

Sala O.E., Oesterheld M., León R.J.C. and Soriano A. 1986. Grazing effects upon plant community structure in subhumid grasslands of Argentina. *Vegetatio* 67: 27-32.

Soriano A., León R.J.C., Lavado R.S., Deregibus V.A., Cahuepé M.A., Scaglia O.A., Velázquez C.A. and Lemcoff J.H. (1991). *Río de La Plata grassland. In Ecosystem of the World-Natural grasslands* (ed. R.T. Coupland) Chapter 19, Volumen 8A. Elsevier Science Publishers, Ansterdam, 367-407.

Taboada, M.A. and Lavado, R.S. 1988. Grazing effects on soil bulk density in the Flooding Pampa of Argentina. *J. of Range Management* 41: 502-505.

Taboada, M.A., R.S. Lavado, G. Rubio and D.J. Cosentino. 2001. Soil volumetric changes in natric soils caused by air entrapment following seasonal ponding and water table rises. *Geoderma* 101: 49-64.

Taboada, M.A. and Lavado R.S. 1993. Influence of cattle trampling on soil porosity under alternate dry and ponded conditions. *Soil Use and Management* 9: 139-143.

Taboada, M.A.; R.S. Lavado; H. Svartz and A.M.I. Segat. 1999. Structural stability changes in a grazed grassland Natraquoll of the Flooding Pampa (Argentina). Wetlands 19: 50-55.

Taboada, M.A.and R.S. Lavado. 2003. Cuando el agua se retira. Inundaciones, consecuencias sobre los suelos. *Encrucijadas* 21: 28-37.

Tilman, D. 1982. *Resource competition and community structure*. Princeton University Press. Princeton, New Jersey.

Voesenek L. A. C. J., T. D. Colmer, R. Pierik, F. F. Millenaar and A. J. M. Peeters. 2006. How plants cope with complete submergence. *New Phytologist* 170: 213–226.

Walker, B., Kinzig, A. and Langridge, J. 1999. Plant attribute diversity, resilience, and ecosystem function: the nature and significance of dominant and minor species. *Ecosystems* 2: 95-113.

Weiher, E. and Keddy, P. A. 1995. The assembly of experimental wetland plant communities. - *Oikos* 73: 323-335.

White, P.S. 1979. Pattern, process and natural disturbances in vegetation. *Bot. Rev*. 45: 229-299.

Wilson J.B. and S.H. Roxburgh. 1994. A demonstration of guild-based assembly rules for a plant community, and determination of intrinsic guilds. *Oikos* 69: 267-276.

In: Grasslands: Ecology, Management and Restoration
Editor: Hans G. Schröder
ISBN 978-1-60692-023-7

Chapter 4

ON THE ROLE OF SOIL SEED BANK OF RICH PASTURE COMMUNITIES IN A FRAGMENTED SUBMEDITERRANEAN LANDSCAPE

Josep M. Ninot[a]*, Albert Petit[a] and Carme Casas[b]
[a]Grup de Geobotànica i Cartografia de la Vegetació
Departament de Biologia Vegetal, Universitat de Barcelona
Av. Diagonal 645, E-08028 Barcelona, Spain
[b]Grup de Geobotànica i Cartografia de la Vegetació
Departament d'Indústries Agroalimentàries i Ciències Ambientals
Universitat de Vic. C. de la Laura 13, E-08500 Vic, Spain

ABSTRACT

Here we report the main trends observed in the soil seed bank of some species-rich pastures located near Vic (NE Spain), evaluating the relationship between soil seed bank and standing vegetation in two contrasting communities (xerophilous pasture and mesoxerophilous grassland), and the potential role of the soil bank in community regeneration. We analyzed 140 soil samples via germination tests and direct observation. The total seed pool (including soil seed bank and standing vegetation) was characterized in terms of morpho-functional traits.

The soil seed content was high and showed strong seasonal variation. We found 12,178.8 to 785.7 seeds m^{-2} in the germination tests, and 6,909.1 to 4,882.5 seeds m^{-2} through direct observation. The shallow soil horizon contained much higher densities and species richness than the deep horizon, and higher relative frequencies of elongated or appended seeds. In both communities, 35-45% of the species richness in the soil seed bank corresponded to opportunists, which were rarely found in the established vegetation. In contrast, half of the taxa in the aboveground xerophilous pasture and two-thirds in the mesoxerophilous grassland were not found in the soil bank.

The persistent seeds in the soil bank belonged to a few taxa, mainly not found or rare in the established vegetation. Most of the main species of this vegetation, such as dominant grasses and abundant sub-shrubs, were poorly represented in the soil seed bank.

* Corresponding author. Fax: +34 934112842. E-mail address: jninot@ub.edu

These trends indicate the limited role of this bank in the regeneration of the communities studied, particularly in the mesoxerophilous grassland. The changes in land use occurred during the last half century have strongly increased the fragmentation of semi-natural plant communities, and impair the conservation of plant diversity of this species-rich landscape.

Keywords: biodiversity decline, community ecology, habitat fragmentation, grasslands, landscape ecology, seed persistence, seed traits, xerophilous pastures

1. INTRODUCTION

Soil seed banks are important components of terrestrial ecosystems, being crucial vegetation drivers after perturbation events and in annual regeneration. Their composition and dynamics are highly evident in therophytic communities, found in dry areas or subject to frequent disturbance (Recasens et al., 1991; Lyaruu and Backéus, 1999; Caballero et al., 2003; Mayor et al., 2003; Reiné et al. 2004). On the contrary, the relationship between soil seed bank and standing vegetation is more obscure where resprouting perennials are dominant (Thompson and Grime, 1979; Leck et al., 1989). Thus, the crucial role played by soil seed banks in maintaining the biodiversity of vegetation over space and time is related to the particular constrictions and perturbation dynamics of each habitat (Fenner and Thompson, 2005). Besides the theoretical interest in this relationship, there are also implications for ecological restoration (Bekker et al., 1997; Willems and Bik, 1998; Lyaruu and Backéus, 1999).

Grime and co-workers have developed a promising approach based on seed features, concerning vegetation dynamics (Grime et al., 1981; Thompson, 1993). According to their methods and focus, the analysis of a few 'soft' morpho-functional seed traits allows one to predict for each taxa the chances of dispersal, persistence and germination. Various studies have shown that these predictions work well enough at least in temperate and Mediterranean floras (Thompson et al., 1993; Cerabolini et al., 2003; Peco et al., 2003). Thus, this approach may assist in understanding the role the soil seed bank plays in shaping standing vegetation, according to the capacities and limitations of the different morphotypes.

A good deal of information on species composition, distribution and the main habitat descriptors in Mediterranean and submediterranean vegetation has been obtained (Rivas-Martínez et al., 2002, and references therein). However, only a few studies have targeted functional aspects, while those devoted to regeneration functions are scarce. Moreover, most of these studies deal with sclerophyllous forests, whereas scrubs and pastures remain much less investigated (but see Russi et al., 1992; Peco et al., 1998a; Guàrdia et al., 2000; etc.), despite the diversification and species richness of these communities. In the Plana de Vic area, a wide range of species-rich pasture communities stand out against a farmed landscape, forming heterogeneous mosaics on the slopes of a number of small hills (Casas and Ninot, 1994, 1995). These hilly surfaces represent noticeable biodiversity headquarters, with very low connectiveness among them (Figure 1). Traditionally, sheep flocks maintained these pastures exploited, also acting as diaspore vectors among the pastures and between pasture areas, fallows and other marginal lands (Poschold and Bonn, 1998).

Figure 1. General view of Plana de Vic, with scattered hills (strongly eroded on South facing slopes) over the farmed plateau.

However, the abandonment of these activities, in tandem with farming intensification during the past half century has increased the actual fragmentation of the pasture units, furthering their encroachment and impoverishment (Casas, 2001).

The aim of this paper is to provide an analysis of the soil seed bank of these pastures, which on the whole are a good example of a seminatural, species-rich landscape. We have selected the two most extensive communities, which moreover show contrasting site conditions, species composition, and functional structure. The analysis included comparisons between standing vegetation and soil seed bank, between communities, and between seasons, on the basis of species composition, functional and ecological plant groups, and seed traits. Specifically, we wanted to evaluate (i) the size and the dynamics of the soil seed bank in these transitional pastures, (ii) the appropriateness of this ecological compartment in community regeneration, and (iii) the functional differences between the two communities studied in terms of soil seed bank. Scientific interest in these aspects has increased along with awareness of the ongoing biodiversity impoverishment in most seminatural landscapes, since knowledge of the structure and dynamics of soil seed banks is a key foundation for restoration management (Bekker et al., 1997; Willems and Bik, 1998).

2. Materials and Methods

2.1. The Area and the Communities Studied

This study deals with the pastures situated on several small hills located in the Plana de Vic basin (Catalonia, Spain; 41° 53' N, 2° 15' E; Figure 2), lying between the Mediterranean, Catalanidic mountains south- and eastwards, and the pre-Pyrenean ranges northwards. The main outcrops are flat, lime rich, Tertiary rocks, chiefly Eocene marls. The macroclimate is submediterranean axeromeric (Bolòs and Vigo, 1984), a transition type between Mediterranean and montane (Pyrenean). The mean annual temperature in Vic lies between 12 and 13 °C and the annual rainfall normally ranges from 650 to 750 mm (Figure 2).

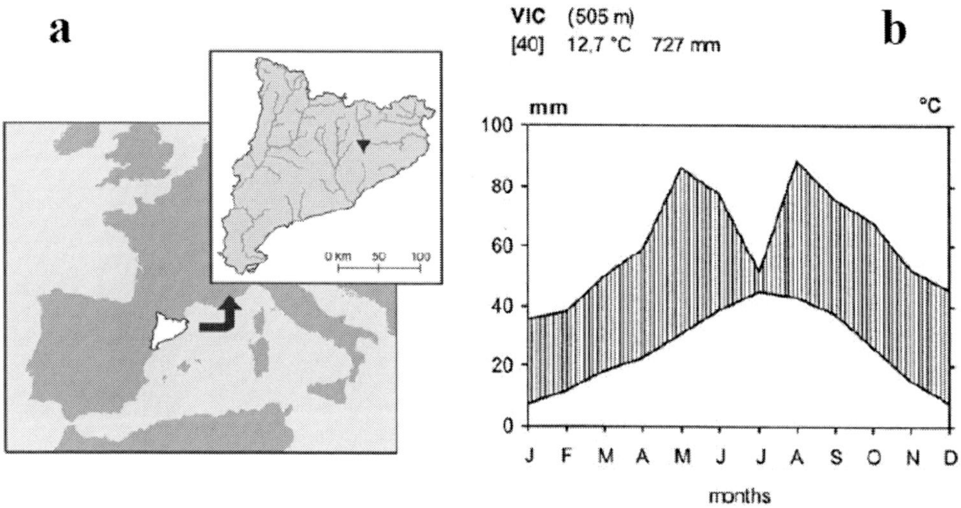

Figure 2. Location of the area studied (a), and climatic diagram of Vic, corresponding to a submediterranean climate with a continental tendency (b).

The vegetation is mainly mesoxerophilous, with xerophilous Mediterranean communities being restricted to drier habitats, such as the south facing slopes and rocky shelves. The hills and the peripheral slopes around Vic stand out from the intensively farmed surrounding plains, and are the main location of varied mosaics of semi-natural communities (Figure 3).

Figure 3. Hill slope bearing a complex mosaic of semi-natural vegetation: *Quercus pubescens* stands, diverse xerophilous pastures and eroded surfaces with sparse vegetation (badlands).

The climax vegetation is a deciduous, mesoxerophilous oak forest (*Buxo sempervirentis-Quercetum pubescentis* Br.-Bl. 1932), which remains on a few gently sloping sites. Deforested slopes, on any aspect and with soils ranging from well preserved to highly eroded, are prime locations for a wide variety of pasture communities (Figure 4).

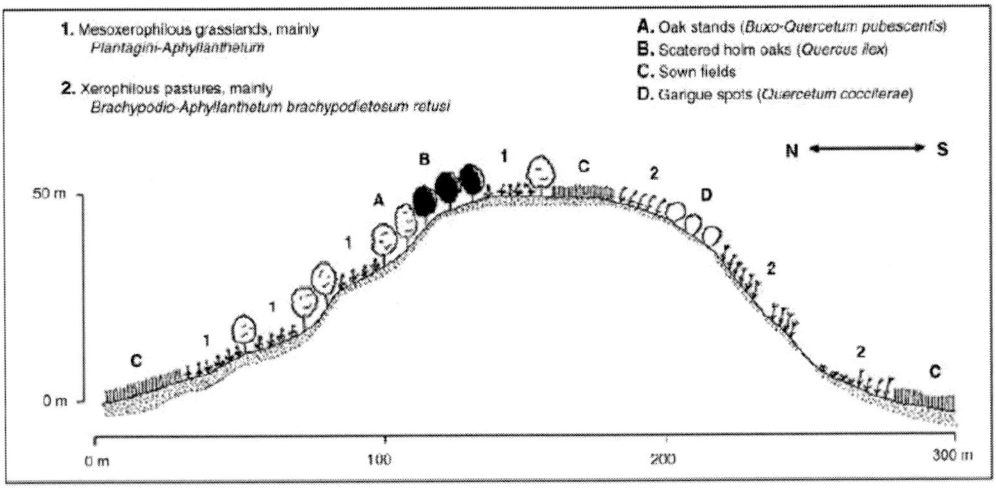

Figure 4. Cross-section of a hill in the area studied, with semi-natural plant communities surrounded by farmland.

Depending on each precise habitat, plant communities include varying percentages of Mediterranean species (xerophilous plants, more abundant in drier environments), Medio-European species (mesophilous plants, more abundant in milder environments), and pluriregional species (ubiquituous plants, widespread).

Most of the pasture surfaces correspond to calcicole perennial communities, from xerophilous to mesophilous. Excluding the extremes of this range, we selected two pasture communities showing contrasting ecological conditions, structure and phenology (Table 1), which are rather extensive in the current landscape (Casas and Ninot, 1994). These communities are:

- *Brachypodio-Aphyllanthetum* O. Bolòs (1956) 1967 *brachypodietosum retusi* O. Bolòs (1959) 1960. Irregular pastures dominated by the chamaephytic, xerophilous grass *Brachypodium retusum*, bearing some dwarf shrubs, therophytes and small patches of bare soil (Figure 5). They develop on south facing slopes with stony, eroded soil.
- *Plantagini-Aphyllanthetum* O. Bolòs (1948) 1956. Dense grasslands dominated by *Aphyllanthes monspeliensis* (Liliaceae) and various mesoxerophilous graminoids, such as *Brachypodium phoenicoides* (Figure 6). They cover deep, stable soils occurring on north facing or intermediate slopes.

2.2. Field Sampling

The two communities were sampled at two sites (Malla and Montrodon) representative of the relief discussed above. In the resulting 4 plots (2 communities x 2 sites) we recorded the standing vegetation in terms of phytosociological relevés (complete lists of the plant taxa found in 20 m^2 areas, with evaluation of cover/abundance), and completed them with additional taxa occurring in pasture adjacent to the plots. Nomenclature and taxonomy of taxa follows Bolòs et al. (2005) from sampling to here onwards.

Table 1. Main aspects of the communities studied (summarised from Casas, Ninot 1994). Percentages of life forms and phytogeographic types refer both to species number and to relative cover, respectively.

	Plantagini-Aphyllanthetum	*Brachypodio-Aphyllanthetum brachypodietosum retusi*
Life forms (%)	Hemicryptophytes: 60 / 78 Chamaephytes: 20 / 20 Geophytes: 5 / 2 Therophytes: 13 / <1	Hemicryptophytes: 47 / 50 Chamaephytes: 31 / 47 Geophytes: 3 / <1 Therophytes: 19 / 3
Dominant species (mean cover)	Brachypodium phoenicoides: *64* *Avenula pratensis*: 35 *Aphyllanthes monspeliensis*: 32 *Carex caryophyllea*: 11 *Plantago media*: 11	Brachypodium retusum: *68* *Thymus vulgaris*: 30 *Koeleria vallesiana*: 17 *Teucrium polium*: 13 *Aphyllanthes monspeliensis*: 12
Aspect, slope (°)	N (any), 4	S (SW-SE), 17
Topographic situation	Toeslopes, gentle slopes	Abrupt slopes
Cover (%)	99	89
Phytogeographic types (%)	Mediterranean: 28 / 53 Medio-European: 44 / 34 Pluriregional: 28 / 13	Mediterranean: 56 / 78 Medio-European: 21 / 16 Pluriregional: 23 / 6

Figure 5. Stand of the xerophilous pasture (*Brachypodio-Aphyllanthetum*) on a dry slope, showing typical patchy structure and dominance of *Brachypodium retusum* and dwarf shrubs (*Santolina chamaecyparissus, Teucrium polium,* etc.).

Figure 6. Aspect of the mesoxerophilous grassland (*Plantagini-Aphyllanthetum*) in fulll bloom of *Aphyllanthes monspeliensis*.

We collected 10 soil samples from each plot using a steel cylinder, 5.5 mm in diameter, knocked 10-12 cm into the soil. Each cylindrical sample (core) was cut into two samples (0 to 5 cm and 5 to 10-12 cm deep) representing shallow and deep soil horizons, respectively, as in other studies addressing soil seed banks (Thompson, 1993). Thus, for each sampling date we obtained 80 samples, which were laid out to dry for 15 days or more in the laboratory, to prevent both germination and loss of viability.

Soil sampling took place three times during the year, with the aim of obtaining data related to the three basic seasonal states of the soil seed bank in temperate regions, following Thompson and Grime (1979): at the end of spring, when the seeds in the soil correspond to the persistent bank; at the end of summer, when the currently produced seeds have been incorporated; and in winter, after the autumn germinating species may have left the soil bank. Samples were collected in February, at the end of June and in September (October) of 2000. Moreover, we took an extra set of samples in June of 1999 (5 cores per plot, total 40 samples), to provide some estimate of the inter-annual variability of the persistent bank.

2.3. Laboratory Analyses

Given the difficulties and constraints involved when analysing the composition of soil seed banks (Gross, 1990; Recasens et al., 1991; Russi et al., 1992), we used two complementary methods. From each set of samples, one-half (40 samples: i.e., 5 samples from each plot and depth) was laid out to germinate in the greenhouse, and the other half was hand sorted to separate and identify seeds. It is known that direct observation of soil samples involves the inventorying of non-viable seeds, and may neglect a number of cryptic seeds (small, irregular, brown). On the other hand, germination tests can result in under-estimates, because this procedure may miss a number of species with special germination requirements, or with dormancy mechanisms (Grime et al., 1981; Gross, 1990).

In this paper, what we call seed corresponds more precisely to the diaspore or dispersule: i.e., the seed of a legume, the entire fruit of a compositae, or the fruit with the concealing lemma and palea of a grass.

2.3.1. Hand Sorting and Visual Observation

The soil samples were washed and hand screened through a column of sieves decreasing in mesh width (1, 0.5 and 0.2 mm) resulting in three sub-samples per sample. The material retained in each sieve contained diverse plant materials (debris, seeds) and mineral particles (aggregates, sand); these were particularly abundant in the medium and finest sieves. The three sub-samples obtained from each sample were maintained separately to facilitate their handling; this gave 120 sub-samples per sampling date.

The sorting and identification of the seeds was done under a stereoscopic microscope, at 10-40X magnification. Seeds showing evidence of damage (broken cover, anomalous appearance) were rejected. Identification was supported by illustrations or descriptions of seeds (Hanf 1982; Jávorka-Csapody, 1991; Bolòs and Vigo, 1984-2001; Castroviejo et al., 1996-2004; etc.) and by comparison with herbarium samples or by seeds collected *ad hoc*.

Sorting the sub-samples retained in the 0.2 mm sieve proved a very hard task, and yielded almost no seeds. Thus, we laid a selection of these samples in a culture chamber with standard spring conditions (14 light hours at 25 °C and 10 dark hours at 15 °C), watered with a 10^{-3}M giberelic solution. Seedling emergence obtained after one month was less than 1 per sub-sample, and none coincided with the few species found in the pastures with diaspores smaller than 0.2 mm (*Blackstonia perfoliata*, *Hypericum perforatum*, *Helichrysum stoechas* and *Sedum sediforme*, and Orchidaceae). Consequently, given the time needed and the almost sterile results expected, we abandoned the exhaustive sorting of these samples, basing this part of the study on the remaining 80 samples retained in the 1 mm and 0.5 mm sieves.

2.3.2. Germination

For this treatment the 5 samples from each plot, depth and date were pooled, and then used as a single soil sample. Each of these samples was spread out in a 40 x 60 cm container, forming a layer a few mm deep, on a standard substrate made of peat, mica and perlite. The containers were laid on greenhouse tables, sprayed with water daily, and protected with a fine net to avoid contamination by new seeds.

The samples obtained were cultivated for 6 months, from November, 2000, to April, 2001. During this period the seedlings were identified as soon as possible, recorded, and removed to avoid interference with other possible germinations. The identification of seedlings was done by means of comparison with herbarium material, when available, or with the help of some specific studies (Hanf, 1982; Jávorka-Csapody, 1991). Seedlings that were not easily identified were grown on in small containers until identification was possible.

2.3.3. Morpho-Functional Analysis

This analysis addressed the seeds of the entire taxa pool of the pastures; i.e., taxa found in the standing vegetation of the plots studied and taxa found only in the soil seed bank, which produced a list of 95 taxa. The analysis is based on the classification of diaspores given by Grime et al. (1981), and refers to their general morphology, surface and appendages. We introduced small modifications to the given categories (Table 2), consisting of the union of certain traits not clearly differentiated, and also in the omission of a few not occurring in our local Flora. In addition, the attribute 'teeth and hairs' was moved into the group of appendages. We also measured length, width and thickness of the seeds, averaged from 5 samples, following Thompson (1993).

Table 2. Classification of diaspores into categories relating to general morphology, appendages and surface (slightly modified after Grime et al. 1981)

Morphology	Appendages	Surface
1. Spherical, or nearly so	1. Absent	1. Smooth
2. Ovoid, rhomboidal, turbinate	2. Straight awn(s) or spine(s)	2. Rugose, muricate, tuberculate or reticulate
3. Trigonous or triquetrous	3. Hygroscopic awn or spine	3. Striate
4. Lenticular, reniform or subulate	4. Persistent pappus or calyx	4. Hairy
5. Cylindrical or ligulate	5. Large hook(s) or barbed spine(s)	5. Striate and hairy
6. Clavate	6. Elaiosome	6. Mucilaginous
7. Winged	7. Wing	
8. Tadpole-shaped	8. Antrorse hairs or teeth	
9. Conical		

In the classification of the diaspores, we took into account previous studies dealing with the morphology and function of seeds (Grime et al., 1981; Hanf, 1982; Jávorka-Csapody, 1991) and also standard floras (Bolòs and Vigo, 1984-2001; Castroviejo et al., 1996-2004). For taxa not included in those studies, or for data not recorded previously, we examined a number of seeds from samples available in the Herbarium of the University of Barcelona (BCN), or obtained from the field. We omitted from our analysis a few species, namely the Orchidaceae since, due to the very small size and specific germination requirements or their seeds, they would not be detected in germination tests, or by hand sorting.

3. Results

3.1. Seed Density and Species Richness in the Soil Bank

The soil seed content varied noticeably throughout the year (Table 3), this being more evident in the germination study (values ranging from 12,178.8 to 785.7 seeds m^{-2}) than via direct observation (values between 6,909.1 and 4,882.5 seeds m^{-2}). In seasonal terms, the densities obtained in 2000 followed the expected dynamics, with the lowest values occurring in spring, when the soil would contain only the persistent seeds. But the densities found in the spring of 1999 were surprisingly high, according to germination tests. The differences in seed density between soil horizons revealed by visual observation also followed the expected pattern (Thompson, 1993); i.e., the topsoil was much richer (Table 4).

Table 3. Seed densities (number of seeds m-2 and standard deviation) found in the sites Malla (Ma) and Montrodon (Mo) for the two pasture types, Plantagini-Aphyllanthetum (P-A) and Brachypodio-Aphyllanthetum (B-A), and the four sample dates, using direct observation (obs.) and germination tests (germ.)

	Ma, P-A	Ma, B-A	Mo, P-A	Mo, B-A	Means
June 1999					
obs.	11,867	2,267	-	6,720	6,951 ± 4,404
germ.	13,400	18,000	-	9,360	13,587 ± 4,323
total	12,633	10,133	-	8,040	10,269 ± 5,469
February 2000					
obs.	6,482	8,081	6,061	2,441	5,766 ± 2,382
germ.	2,525	2,240	3,283	1,179	2,306 ± 871
total	4,504	5,161	4,672	1,810	4,037 ± 2,485
June 2000					
obs.	4,293	3,788	5,556	5,809	4,862 ± 975
germ.	-	926	337	1,094	786 ± 398
total	4,293	2,357	2,946	3,451	3,115 ± 2,297
September 2000					
obs.	5,303	2,525	4,883	8,418	5,282 ± 2,422
germ.	1,599	589	926	10,439	3,388 ± 4,719
total	3,451	1,557	2,904	9,428	4,335 ± 3,617

Table 4. Mean seed densities (seeds m^{-2} and standard deviation) found for the four sample dates, the two soil horizons and the two communities, using direct observation (obs.) and germination tests (germ.). P-A, *Plantagini-Aphyllanthetum*; B-A, *Brachypodio-Aphyllanthetum*

	shallow	deep	P-A	B-A
obs.	4,799 ± 200	782 ± 78	6,081 ± 445	5,149 ± 399
germ.	1,972 ± 584	1,639 ± 655	2,469 ± 1 362	4,438 ± 2,228
total	3,385 ± 1,999	1,210 ± 606	4,275 ± 2,554	4,794 ± 503

However, differences based on germination tests were smaller, although this was mainly due to the high germination rates attained in deep samples collected in the spring of 1999.

The total species number found visually in the soil samples was slightly lower than that from germination tests (41 and 45, respectively), although the amount of seeds was clearly higher in the former method (946 compared to 574, across all samples). The established vegetation of *Plantagini-Aphyllanthetum* at the two sites included 70 species of vascular plants, whereas in the corresponding soil samples we found only 55 species (of which 7 could not be identified); 26 species were found in both the soil bank and the established vegetation. As for *Brachypodio-Aphyllanthetum,* the vegetation consisted of 52 species while soil sampling yielded 46 species (of which 4 could not be identified); in this case, 25 species were found in both. It is interesting to point out that although the relevés of *Plantagini-*

Aphyllanthetum included almost 40% more species than that of *Brachypodio-Aphyllanthetum*, the species richness found in each soil seed bank was similar.

With regards species frequencies, we found noticeable differences between the observation and the germination methods (Table 5). Whereas the 10 species recorded most often by visual observation were frequently found, or even dominant, in the established vegetation (e.g. *Aphyllanthes mospeliensis*, *Brachypodium retusum*, *Thymus vulgaris*, *Carex caryophyllea* and *Dichanthium ischaeumum*), the germination method yielded more uneven results. Among the taxa most recorded by this method, some were only sparsely found in the established vegetation (*Blackstonia perfoliata* or *Centaurium pulchellum*) or were even absent (*Capsella bursa-pastoris, Sorghum halepense* and *Conyza sumatrensis*), while a few were either frequently found or abundant in the relevés (*Thymus vulgaris* or *Brachypodium retusum*). Some of the most abundant species in the vegetation, mainly in *Plantagini-Aphyllanthetum,* appeared very sparsely in the germination tests (such as *Aphyllanthes mospeliensis* or *Brachypodium phoenicoides*).

Table 5. Taxa most frequently found in the soil seed bank of the two associations, based on germination and observation, expressed as the mean of seeds per sample and the percentage vs. the numbers of seeds found

Plantagini-Aphyllanthetum	Seeds/ sample	%	Brachypodio-Aphyllanthetum	Seeds/ Sample	%
Aphyllanthes monspeliensis	2.79	20.78	*Brachypodium phoenicoides*	1.36	11.81
Linum tenuifolium subsp. *milletii*	1.90	14.11	*Plantago lanceolata*	1.34	11.70
Helianthemum sp.	0.88	6.51	*Blackstonia perfoliata*	0.91	7.91
Centaurium pulchellum	0.81	6.05	*Euphorbia flavicoma* subsp. *flavicoma*	0.83	7.22
Carex caryophyllea	0.56	4.19	*Linum tenuifolium* subsp. *milletii*	0.82	7.11
Ononis pusilla	0.46	3.41	*Dichanthium ischaemum*	0.82	7.11
Coronilla minima subsp. *minima*	0.44	3.26	*Capsella bursa-pastoris*	0.55	4.82
Potentilla neumaniana	0.44	3.26	*Thymus vulgaris*	0.49	4.24
Thymus vulgaris	0.44	3.26	*Coris monspeliensis*	0.42	3.67
Helianthemum oleandicum subsp. *italicum*	0.44	3.15	*Helianthemum oleandicum* subsp. *italicum*	0.36	3.00
sp. 1	0.42	3.10	*Convolvulus cantabrica*	0.36	3.10
Blackstonia perfoliata	0.38	2.79	*Aphyllanthes monspeliensis*	0.34	2.98
Ranunculus bulbosus	0.31	2.33	*Teucrium* sp.	0.33	2.87
sp. 2	0.23	1.71	*Lavandula* sp.	0.32	2.75
Teucrium sp.	0.23	1.71	*sp.* 1	0.30	2.64
TOTAL	13.14		TOTAL	11.47	

3.2. Life Forms, Families and Ecological Groups

In the visual observation of soil samples, the relative frequencies of life forms showed small variability among communities, depths and sampling dates, in terms of the numbers of taxa identified. Hemicryptophytes and chamaephytes were the main groups (25-35% and 45-60%, respectively), whereas therophytes accounted for about 15%. In contrast, when these frequencies were based on seed numbers, the soil seed bank reflected better the distinct life-form spectra of each community. That is, chamaephytes were more abundant than hemicryptophytes in the xerophilous *Brachypodio-Aphyllanthetum* (more than 45%, versus 35%); and hemicryptophytes surpassed chamaephytes in the mesoxerophilous *Plantagini-Aphyllanthetum* (65%, versus 50%). Other life forms were represented by less than 5% of the seeds. If soil depth is taken into account, chamaephytes were more frequent in the shallow horizon of the samples, decreasing to around 30% in the deep horizon, compared to the other main group, hemicryptophytes. As to sampling dates, variations in percentages did not show clear patterns, except for the frequency of therophytes in spring of 1999, which exceeded 10% of the seeds, following the expected trend whereby short-lived species show a relative increase in the persistent soil bank.

The germination analysis yielded generally higher percentages of therophytes (45% in spring 2000, and in all the deep soil samples), as well as greater differences between sampling dates (40-50% in the spring versus less than 25% in the winter and autumn). Therophytes also stood out in the deep soil horizon, accounting for 45% of the taxa identified, versus less than 35% of chamaephytes.

The main plant families (compositae, grasses, legumes) were found in the seed bank in proportions similar to those occurring in the vegetation, although the germination analysis yielded some irregularities. Grasses were more frequent in the xerophilous pasture than in the mesoxerophilous grassland (15-20% and 5-10%, respectively), and were mainly concentrated in the shallow soil horizon, which would corroborate the feeble contribution of this group to seed regeneration in dense grasslands, as well as their generally low presence in persistent banks. Legumes were more frequent in the observation analysis than after cultivation, probably due in part to generalised requirements for dormancy breaking in this family (Russi et al., 1992). On the other hand, compositae germinated easily, their frequency being higher in the germination test than via observation.

Overall the main phytosociological group represented was the order *Rosmarinetalia* s.l. (2,034.44 seeds m^{-2}, average of plots and sampling dates), a syntaxon characterized by mesoxerophilous, slow-growing perennials. Caespitose herbs like *Aphyllanthes monspeliensis, Carex humilis* and *Potentilla neumanniana*, and sub-shrubs like *Coris monspeliensis, Helianthemum oelandicum* and *Linum tenuifolium* were the core of this ecological group (Figure 7). The second most dominant group (1,120.91 seeds m^{-2}) was the class *Thero-Brachypodietea*, characterized by therophytes (*Linum strictum, Odontites lutea*) and xerophilous perennials (*Botriochloa ischaemum, Brachypodium retusum, Thymus vulgaris*) that are more adapted to Mediterranean conditions and partly to perturbation. Such adaptation implies a more 'ruderal' character, since greater soil moisture irregularities cause more seasonal variation in standing biomass (Grime 2001). The other groups were less well represented (average densities lower than 400 seeds m^{-2}), that of low shrubs (*Dorycnium pentaphyllum, Genista scorpius*) being the lowest (27.1 seeds m^{-2}).

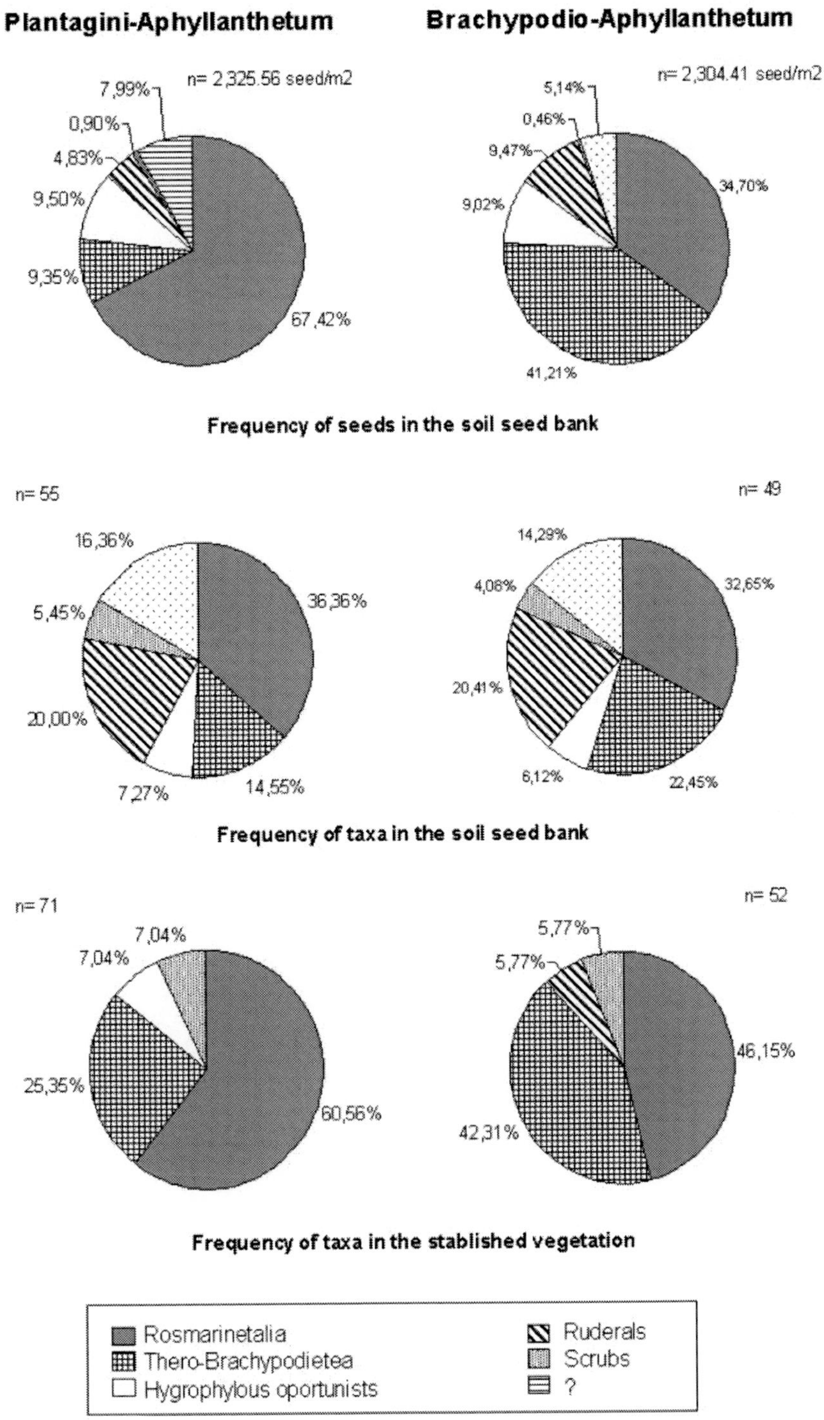

Figure 7. Contribution of ecological groups to the species pool in the pastures studied, the mesoxerophilous grassland *Plantagini-Aphyllanthetum* (left column) and the xerophilous pasture *Brachypodio-Aphyllanthetum* (right column), by means of the percentage of seeds found in the soil samples (top), the percentage of taxa in the soil samples (middle), and the percentage of taxa in the standing vegetation (bottom).

In the mesoxerophilous grassland (*Plantagini-Aphyllanthetum*), the dominance of *Rosmarinetalia* seeds was even higher, accounting for one-third of the total species, but nearly 70% of seeds. Here, the hygrophilous opportunists were the second most dominant group (220.98 seeds m^{-2}), though consisting of only 4 species. Samples from the xerophilous pasture (*Brachypodio-Aphyllanthetum*) contained the most seeds of *Thero-Brachypodietea*: 22.45% of the species found in the soil bank and more than 40% of the total amount of seeds. The *Rosmarinetalia* taxa were also well represented (16 taxa, 32.65%). The hygrophilous opportunists and the ruderals showed moderate seed densities (more than 200 m^{-2}), the latter ecological group (ruderals) being more frequent here than in the mesoxerophilous grassland.

3.3. Morpho-Functional Trends in the Seed Pool

The established vegetation of the two communities showed very similar spectra of seed types. Percentages of the different morphologies varied by less than 10% (the maximum corresponded to trigonous diaspores, close to 25% in *Brachypodio-Aphyllanthetum* and nearly 35% in *Plantagini-Aphyllanthetum*). The absence of appendages was dominant, slightly higher than 60% in both communities, whereas in the other categories only small differences appeared. Awns, spines and pappus were slightly more abundant in *Brachypodio-Aphyllanthetum*, while diaspores with teeth, antrorse hairs or hygroscopic spines were more numerous in *Plantagini-Aphyllanthetum*. With regards to seed surface, the two communities were similar, with 50% of smooth seeds in both (Figure 8).

Most of the species identified visually from the soil samples had more or less compact seeds (from spherical to reniform) with smooth surfaces; only a few had pappus or appendages (Figure 4). The same trends occurred across the sampling periods, the two communities, and the two soil depths. When taking into account the number of seeds, the spring samplings yielded higher percentages of compact seeds, mainly ovoid, and also of seeds without appendages; these seeds were also more abundant in the deep soil horizon than in the shallow horizon. In general, cylindrical or ligulate seeds (mainly of grasses) showed higher percentages in terms of number of seeds rather than number of species.

Figure 8. Some representative seeds from the pastures (from 1.8 to 5.6 mm in length).

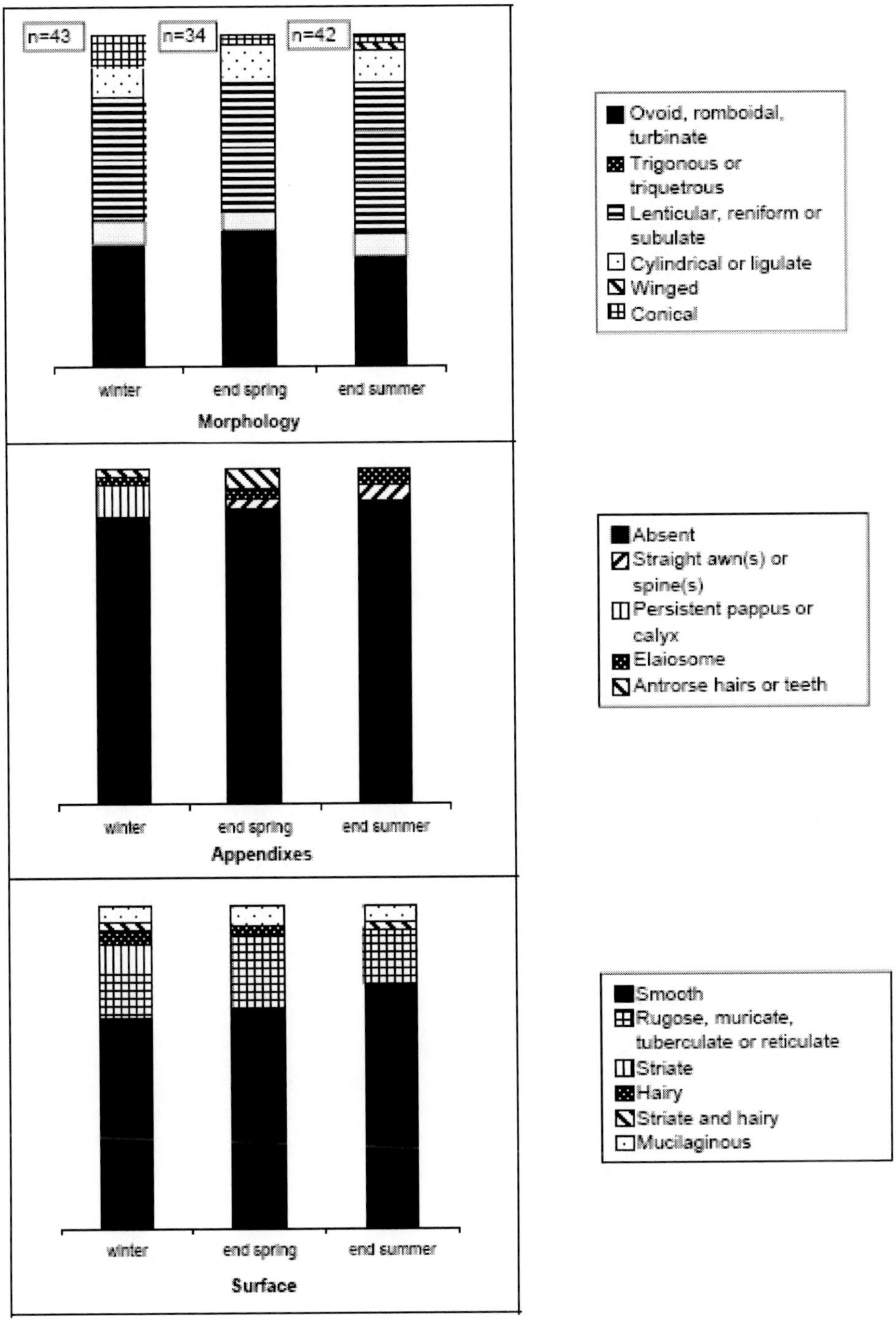

Figure 9. Morpho-functional spectra (percents on each vertical bar) of the seeds found in the soil bank, in terms of general morphology, appendages and surface. Seasonal dynamics is evident as changing percentages (*n* stands for number of taxa).

The morphological spectrum of the germinated seeds was slightly different from that of seeds found visually (Figure 9). There was a noticeable decrease in lenticular or reniform seeds (almost 20% lower) and an increase in conical diaspores (including achenes with pappus) and, less pronounced, of ovoid seeds.

As for differences between depths, species with lenticular or reniform seeds were more abundant in the deep horizon, which would result in better persistence of these types. On the contrary, seeds with appendages were clearly more frequent near the soil surface, as previously hypothesised (Grime et al., 1981). Species with conical seeds (including seeds with pappus) occurred in both horizons with similar percentages, but the number of these seeds was clearly lower in the deep horizon.

Differences between established flora and soil seed bank were more noticeable than that between the two communities. Isometric types (rounded, ovoid, rhomboidal, turbinate, lenticular, reniform or subulate) were more dominant in the soil bank (almost 75% of the total species found) than in the vegetation (less than 65%). The absence of appendages was slightly more frequent in the soil bank (about 70%) than in the vegetation (65%), but the different types of appendages showed similar percentages in soil and vegetation. As for surface types, smooth diaspores were more abundant in the soil seed bank (almost 70% of species) than in the vegetation (less than 55%), where striate or hairy seeds were more frequent.

4. Discussion

4.1. Soil Seed Bank and Standing Vegetation Face to Face

The size of the soil seed bank of the pastures studied is similar to those of other temperate pasture communities (Leck et. al., 1989; Reiné et al., 2004; Bisteau and Mahy, 2005), but clearly lower than those corresponding to dryer open communities, such as gypsophilous scrubs of central Spain (Caballero et al., 2003) or semiarid Syrian pastures (Russi et al., 1992). This would corroborate the contention that the role played by the soil seed bank is clearly less crucial in the perennial, mesoxerophilous pastures of Vic than in therophytic, xerophilous pastures and low scrubs, which are subject to strong seasonal dynamics (Guàrdia et al., 1998).

Following a general trend in perennial pastures (Thompson, 1989), some of the most abundant species in the communities studied appeared very sparsely in the soil bank. Such impoverishment could have been enhanced by the low grazing levels in the pastures studied (Milberg, 1995; Handlová and Münzbergová, 2006). This fact, and the occurrence in the soil seed bank of opportunists rarely found in the established vegetation, resulted in relatively low similarity between the two ecosystem compartments. The Sørensen index (Sørensen, 1948) was 0.4407 in *Plantagini-Aphyllanthetum* and to 0.5319 in *Brachypodio-Aphyllanthetum*. These figures should actually be even lower, since the seeds not identified in the visual observation were not included in the ratio; in fact they probably do not correspond to any of the established species. Higher similarities have been found in communities subjected to severe constraints, such as badlands vegetation in the Pyrenees (Guàrdia et al., 2000) or salt catchments (Gul and Weber, 2001). However, these values are higher than those obtained by Lyaruu and Backéus (1999) in semiarid pastures, and similar to the autumn values calculated by Peco et al. (1998b) for Mediterranean pastures in the Iberian Sistema Central. Such values reveal the impaired potential of the soil seed bank in perennial, dense pastures for community regeneration after severe disturbance.

In both communities, species richness was lower in the soil seed bank than in the standing vegetation. In general, the higher numbers of seeds without appendages and of more or less compact seeds did not compensate for the lack of or weak representation of other seed types. Therophytes showed higher frequencies in the soil bank than in the standing vegetation, a trend more clearly apparent in the germination analysis. Moreover, they reached the highest frequencies in those samples more representative of the persistent bank (spring, and deep soil horizon), a tendency already found in previous studies (Milberg and Hanson, 1993). This fact was less obvious in our visual observation, reflecting the difficulty in detecting the small seeds of most annuals (*Blackstonia perfoliata*, *Centaurium pulchellum*, *Capsella bursa-pastoris*, *Cardamine hirsuta*, etc.). On the other hand, the fairly high percentage of therophytes in the mesoxerophilous *Plantagini-Aphyllanthetum* may be evidence of the greater role taken in the soil bank by the persistent taxa, compared with that of the transient taxa. This is consistent with the observations made by Peco et al. (1998b), who pointed out the prevalence of persistent seeds in the soil bank of ungrazed pastures.

4.2. Space and Time Heterogeneity in the Soil Seed Bank

The seasonal dynamics in the soil seed bank of the pastures studied was apparent from several points of view. Spring samples included lower taxa richness, lower seed densities, higher frequency of therophytes, and higher proportion of compact morphotypes (rounded, ovoid, lenticular) without appendages. This is consistent with the expected seasonal dynamics, in which the soil seed bank becomes impoverished in spring, while the resulting persistent species become more proportionally represented (Thompson and Grime, 1979).

However, inter-annual heterogeneity seems to be higher than seasonal variations, at least with regards to seed densities. This trend seems to become common towards Mediterranean and arid pastures (Coffin and Lauenroth, 1989; Russi et al., 1992; Peco et al., 1998a). Occasional high spring densities, like that obtained from germination of the 1999 samples, might be due to good emergence rates in persistent seeds accumulated in the soil over several years (Grime et al., 1981).

The differences between shallow and deep soil horizons proved to be the highest found in our study, since both density and species richness were clearly higher in the superficial samples. Moreover, ornamented seeds (with appendages, hairs, awns, etc.) and heterometric seeds (conical, clavate, winged) were clearly more frequent near the surface as well. These results are consistent with the hypothesis of Grime et al. (1981), in the sense that compact seeds without appendages become more easily buried, and thus are dominant in the persistent soil banks. Hairy or striate diaspores showed no differences between the two horizons, which suggests these traits have little effect on the chances of the seeds reaching deep soil horizons.

A number of lenticular or reniform seeds were from papilionaceae, which are known to produce very persistent seeds with strong testae and, consequently, are more prone to forming part of the persistent soil seed bank (Grime et al., 1981). On the other hand, the impoverishment of conical seeds (mainly of compositae) in the deep horizon is consistent with this group being mainly transient (Leck et al., 1989). A similar trend may be deduced for the dominant grasses of the pastures studied, based on their diaspore morphology (including awns and lemmas). Thus, persistent seeds in the soil bank are confined to a few taxa in the pastures, as found in various studies (Fenner and Thompson, 2005, and references herein; but see Bossuyt et al., 2007). This demonstrates the limited role of these soil banks in regeneration at the community level, compared to other processes (resprouting, immigration).

4.3. Differences between the Two Communities Studied

The higher seed content shown by germination tests in *Brachypodio-Aphyllanthetum* is consistent with the Mediterranean character of this community, made of high proportions of annuals and dwarf shrubs producing large numbers of seeds (Casas and Ninot, 1994). This is partly because the small gaps and the low intensity disturbance typical of this xerophilous pasture create more opportunities for annuals, which may occur sparsely here and there. Also, the more stressful conditions of this community would slow down plant succession (Casas, 2001; Figure 10), and allow most non-dominant species (small chamaephytes, sparse forbs, therophytes) to persist and contribute to the soil seed bank. This may explain the higher similarity between soil bank and standing vegetation in *Brachypodio-Aphyllanthetum*.

The ongoing pasture abandonment had stronger effects in *Plantagini-Aphyllanthetum*. There, the increasingly dominant grasses (especially *Avenula pratensis* and *Brachypodium phoenicoides*) and some encroachment of shrubs would inhibit the flowering and fruiting of the typical pasture species (Figure 11), as found in various studies (Poschold et al., 1998; Willems and Bik, 1998). The higher seed content in *Plantagini-Aphyllanthetum* revealed by some visual analyses could be partially due to the higher proportion of non-viable seeds included, since seed ageing (without germination) would be enhanced under these conditions.

Figure 10. Even under low grazing pressure, the xerophilous pasture keeps most of its typical irregular structure and species richness.

Figure 11. Pasture abandonment in *Plantagini-Aphyllanthetum* involves densification of dominant grasses (top) and encroachment of shrubs and *Quercus pubescens* (bottom), which lead to species loss via declining of typical pasture plants.

Seeds without appendages and more or less compact were more frequent in the soil bank of this grassland. These and other features indicate a more persistent bank, i.e. *Plantagini-Aphyllanthetum* is more sensitivite to the effects of grazing abandonment. Thus, the potential for regeneration from the soil seed bank in this grassland is lower, compared with that of *Brachypodio-Aphyllanthetum.*

4.4. Declining Biodiversity

The pastures studied include high species richness both within each community (areas of a few square meters harbour 30-40 species) and between communities (a mosaic of 3 different communities may include 50-75 species). The number of species in a whole hill may reach two hundreds, most of them from pastures and others from forests, thickets, rocky areas and ruderal sites (Casas and Ninot, 1994, 1995).

Differences between hills are noticeable, mainly where they include different rock outcrops. These rich systems are semi-natural, small landscape patches standing out from a matrix of farmed land, where vegetation consists of relatively poor flora, very distinct than that secluded in the hills (Figure 12). In landscapes like that of Plana de Vic, pastures include high species percentages from local floras (25-33 %), and together with neighbouring habitats (shrubs, forest clearings, rocky places) may account for 48-53 % of species richness (Ninot and Ferré, 2008) at the landscape level.

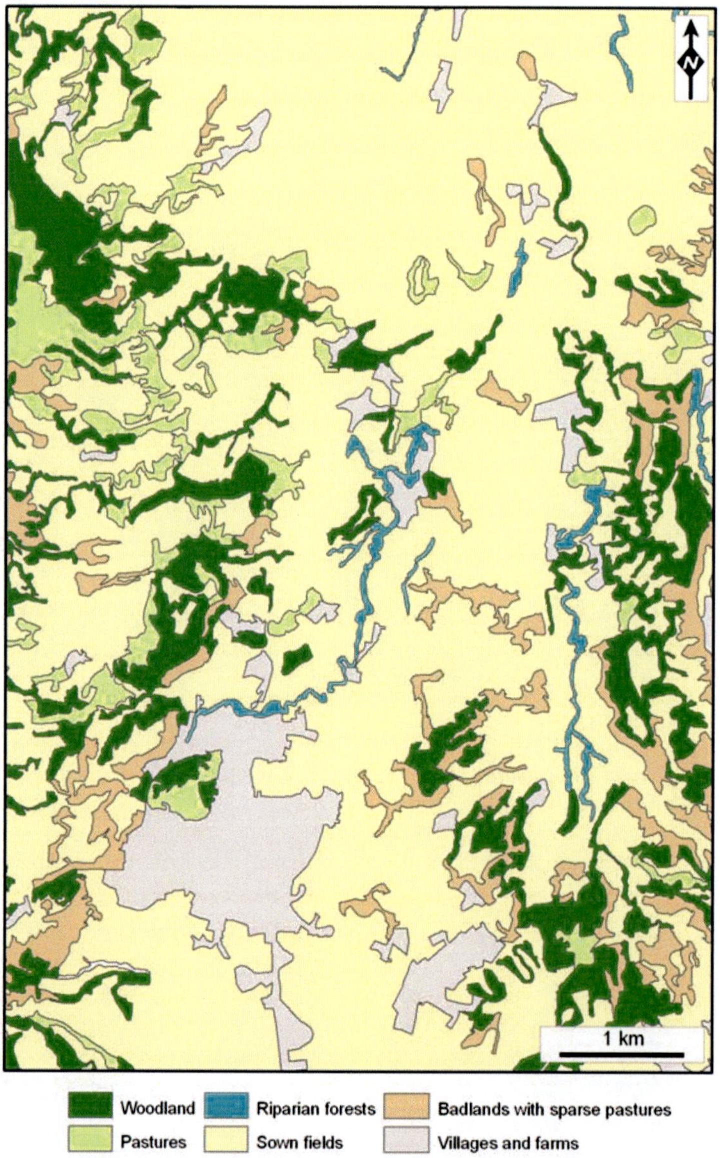

Figure 12. Simplified vegetation map in one example from Plana de Vic, which shows the fragmented structure of semi-natural hills (rangeland = pastures + badlands) on the agricultural matrix.

The traditional land use of Plana de Vic and similar areas involved rotation of crops and fallows and livestock ranging extensively. Small sheep and goat flocks took some profit from fallows, marginal areas and hills (pastures, open woodland, eroded slopes), and manured fields. In parallel, they acted as diaspore vectors between all these habitats, both via endo- and epizoochory, together with other domestic animals. Given the poor dissemination ability of most of the pasture species, these processes carried out most of the seed flow between different hills (Poschold and Bonn, 1998). During the last half century, farming intensification has reduced crop rotation and almost avoided fallow areas. The use of chemical fertilizers and the specialization in farming practices has almost eliminated sheep ranging; the few remaining flocks are much more sedentary than decades ago. These changes in land use have strongly increased the fragmentation of the semi-natural plant communities of the hills (Figure 13).

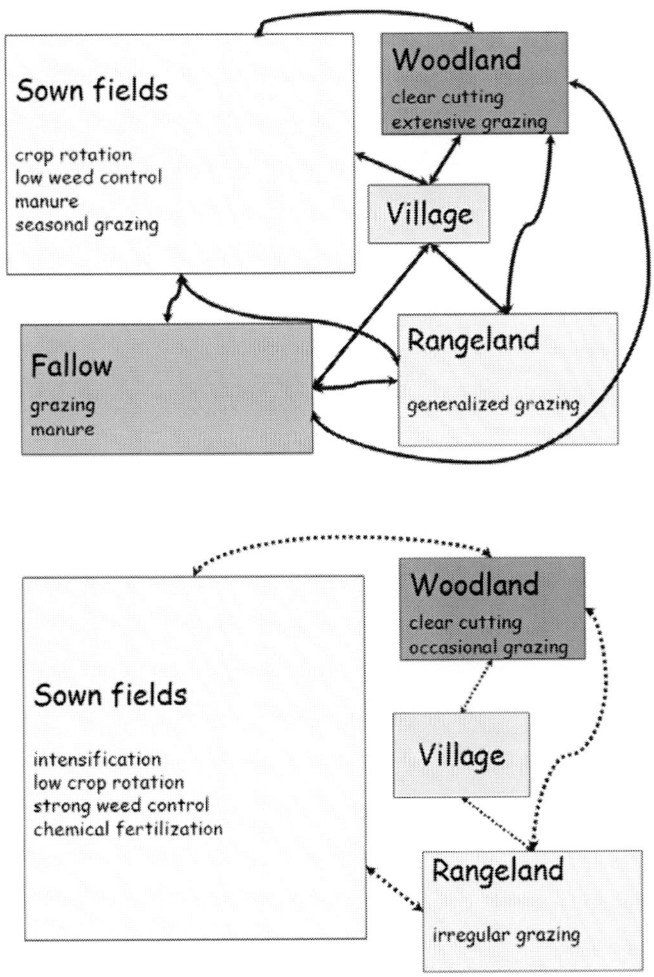

Figure 13. Scheme of landscape structure and seeds flow in traditional (top) and present (bottom) land use. Besides field intensification and vanishing of fallow, seed transfer between distinct habitats via livestock and other processes has changed from diverse and active to few and scarce.

Wind maintains seed flow between populations for anemochorous plants, such as a number of ruderals and opportunists. Other ruderals and therophytes with persistent seeds keep good regeneration potential in the soil seed banks following occasional colonization and seed production. But most of perennial pasture species remain strongly secluded as small populations in some of the hills. Low seed production due to ongoing pasture densification and encroachment, and poor dispersal ability make the less common of these species prone to local extinction. Therefore, present landscape functioning impairs the conservation of plant diversity of this species-rich landscape, and biodiversity restoring is not very promising if soil bank is the main seed source, given the low potential of this ecosystem compartment (Poschold et al., 1998; Willems and Bik, 1998; Bisteau and Mahy, 2005; Rosef, 2008)

CONCLUSION

In semi-natural pastures of Plana de Vic, most of the main species in the stablished vegetation are poorly represented in the soil seed bank, but are able to thrive or even dominate via vegetative persistence and expansion. The changes in land use occurred during the last half century have promoted species impoverishment at stand scale, and fragmentation of semi-natural plant communities at landscape scale. This, together with the low dissemination ability of most pasture species and the limited role of the soil seed bank, impair the conservation or recovery of plant diversity of this species-rich landscape.

REFERENCES

Bekker R.M., Verweij G.L., Smith R.E.N., Reiné R., Bakker J.P. and Schneider S. (1997). Soil seed banks in European grasslands: does land use affect regeneration perspectives? *J. Appl. Ecol.* 34: 1293-1310.

Bisteau E. and Mahy G. (2005). Vegetation and seed bank in a calcareous grassland restored from a Pinus forest. *Appl. Veg. Sci.* 8: 167-174.

Bolòs O. de and Vigo J. (1984-2001). *Flora dels Països Catalans*. Vols. I-IV. Barcino, Barcelona.

Bolòs O. de, Vigo J., Masalles R.M., Ninot J.M. (2005). *Flora manual dels Països Catalans* (3d. ed.). Pòrtic, Barcelona.

Bossuyt B., Cosyns E. and Joffmann M. (2007). The role of soil seed banks in the restoration of dry acidic dune grassland after burning of Ulex europaeus scrub. *Appl. Veg. Sci.* 10: 131-138.

Caballero I., Olano J.M., Loidi J. and Escudero A. (2003). Seed bank structure along a semi-arid gypsum gradient in Central Spain. *J. Arid Environ.* 55: 287-299.

Casas C. (2001). Estudi tipològic, ecològic i funcional de les pastures de la Plana de Vic. Ph. Thesis, University of Barcelona.

Casas C. and Ninot J.M. (1994). Estudi fitocenològic de les pastures de la Plana de Vic. I: comunitats vivaces (Mesobromion i Aphyllanthion). *Butll. Inst. Catalana Hist. Nat.* 62: 25-52.

Casas C. and Ninot J.M. (1995). Estudi fitocenològic de les pastures de la Plana de Vic. II: comunitats terofítiques (Thero-Brachypodietea) i síntesi. *Butll. Inst. Catalana Hist. Nat.* 63: 27-50.

Castroviejo S. et al. (eds.) (1996-2004). *Flora iberica.* Vols. 1-9, 10, 14. CSIC, Madrid.

Cerabolini B., Ceriani R.M., Caccianiga M., De Andreis R. and Raimond B. (2003). Seed size, shape and persistence in soil: a test on Italian flora from Alps to Mediterranean coasts. *Seed Sci. Res.* 13: 75-85.

Coffin D.P. and Lauenroth W.K. (1989). Spatial and temporal variation in the seed bank of a semiarid grassland. *Amer. J. Bot.* 76: 53-58.

Fenner M. and Thompson K. (2005). *The ecology of seeds.* Cambridge University Press.

Grime J.P. (2001). *Plant Strategies, Vegetation Processes and Ecosystem Properties.* John Wiley and Sons, Chichester.

Grime J.P., Mason G., Curtis A.V., Rodman J., Band S.R. Mowfrth M.A.G., Neal A.M. and Shaw S. (1981). A comparative study of germination characteristics in a local flora. *J. Ecol.* 68: 1017-1058.

Gross K.L. (1990). A comparison of methods for estimating seed numbers in the soil. *J. Ecol.* 78: 1079-1093.

Guàrdia R., Casas C. and Ninot J.M. (1998). Phenological patterns in Mediterranean pastures and scrubs of Catalonia. *Acta Bot. Barc.* 45: 557-576.

Guàrdia R., Gallart F. and Ninot J.M. (2000). Soil seed bank and seedling dynamics in badlands of the Upper Llobregat basin (Pyrenees). *Catena* 40: 189-202.

Gul B. and Weber D.J. (2001). Seed bank dynamics in a Great Basin salt playa. *J. Arid Environ.* 49: 785-794.

Handlová V. and Münzbergová Z. (2006). Seed banks of managed and degraded grasslands in the Krkonose Mts., Czech Republic. *Folia Geobot.* 41: 275-288.

Hanf M. (1982). Ackerunkräuter Europas mit ihren Keimlingen und Samen. DLG-Verlag, Frankfurt (Main).

Jávorka S. and Csapody V. (1991). Iconographia florae partis austro-orientalis Europae centralis. Akadémiai Kiadó, Budapest.

Leck M.A., Parker V.T. and Simpson R.L. (eds.) (1989). Ecology of soil seed banks. Academic Press, London.

Lyaruu H.V.M. and Backéus I. (1999). Soil seed bank and regeneration potencial on eroded hill slopes in the Kondoa Irangi Hills, central Tanzania. *Appl. Veg. Sci.* 2: 209-214.

Mayor M.D., Bóo R.M., Peláez D.V. and Elía O.R. (2003). Seasonal variation of the soil seed bank of grasses in central Argentina as related to grazing and shrub cover. *J. Arid Environ.* 53: 467-477.

Milberg P. (1995). Soil seed bank after eighteen years of succession from grassland to forest. *Oikos* 72: 3-13.

Milberg P. and Hanson M.L. (1993). Soil seed bank and species turnover in a limestone grassland. *J. Veg. Sci.* 4: 35-42.

Ninot J.M. and Ferré A. (2008). Plant diversity across five vegetation belts in the Pyrenees (Catalonia, Spain). *Collect. Bot.* (Barcelona) 27: 67-76.

Peco B., Espigares T. and Levassor C. (1998a). Trends and fluctuations in species abundance and richness in annual Mediterranean pastures. *Appl. Veg. Sci.* 1: 21-28.

Peco B., Ortega M. and Levassor C. (1998b). Similarity between seed bank and vegetation in Mediterranean grassland: a predictive model. J. *Veg. Sci.* 9: 815-828.

Peco B., Traba J., Levassor C., Sánchez, A.M. and Azcárate, F.M. (2003). Seed size, shape and persistence in dry Mediterranean grass and scrublands. *Seed Sci. Res*. 13: 87-95.

Poschold P. and Bonn S. (1998). Changing dispersal processes in the central European landscape since the last ice age: an explanation for the actual decrease of plant species richness in different habitats? *Acta Bot. Neer.* 47: 27-44.

Poschold P., Kiefer S., Tränkle U., Fischer S. and Bonn S. (1998). Plant species richness in calcareous grasslands as affected by dispersability in space and time. *Appl. Veg. Sci.* 1: 75-90.

Recasens J., Riba F., Izquierdo J. and Taberner A. (1991). L'anàlisi del banc de llavors de males herbes de sòls agrícoles. Aspectes metodològics. *Ilerda "Ciències"* 49: 83-102.

Reiné, R. Chocarro C. and Fillat F. (2004). Soil seed bank and management regimes of semi-natural mountain meadow communities. *Agr. Ecosyst. Environ.* 104(3): 567-575.

Rivas-Martínez S., Díaz T.E., Fernández-González F., Izco J., Loidi J., Lousa M. and Penas A. (2002). Vascular plant communities of Spain and Portugal. Addenda to the syntaxonomical checklist of 2001). *Itinera Geobot.* 15(1-2): 5-922.

Rosef, L. (2008). Germinable soil seed banks in abandoned grasslands in central and western Norway and their significance for restoration. *Appl. Veg. Sci.* 11: 223-230.

Russi L., Cocks P.S. and Roberts E.H. (1992). Seed bank dynamics in a Mediterranean grassland. *J. Appl. Ecol.* 29: 763-771.

Sørensen T.A. (1948). A method for establishing groups of equal amplitude in plant sociology based on similarity of species content, and its application to analyses of the vegetation on Danish commons. K. Dan. Videnskab. *Selsk. Biol. Skr.* 5: 1-34.

Thompson K. (1989). Small-scale heterogeneity in the seed bank of an acidic grassland. *J. Ecol.* 74: 733-738.

Thompson K. (1993). Persistence in soil. In: Hendry G.A.F. and Grime J.P. (Eds.), *Methods in comparative ecology. A laboratory manual*, Chapman and Hall, London, pp. 199-202.

Thompson K., Band S. and Hodgson, J.C. (1993). Seed size and shape predict persistence in soil. *Funct. Ecol.* 7: 236-241.

Thompson K. and Grime J.P. (1979). Seasonal variation in the seed banks of herbaceous species in ten contrasting habitats. *J. Ecol*. 67: 893-921.

Willems J.H. and Bik L.P.M. (1998). Restoration of high species density in calcareous grassland: the role of seed rain and soil seed bank. *Appl. Veg. Sci.* 1: 91-100.

In: Grasslands: Ecology, Management and Restoration
Editor: Hans G. Schroder
ISBN 978-1-60692-023-7

Chapter 5

RE-CREATING PLANT AND BEETLE ASSEMBLAGES OF SPECIES-RICH CHALK GRASSLANDS ON EX-ARABLE LAND

Ben A. Woodcock*, Simon R. Mortimer[2], Andrew R. Edwards[2], Clare S. Lawson[2], Duncan B. Westbury[2], Alex J. Brook[3], Stephanie J. Harris [2] and Valerie K. Brown[2]

[1]Biodiversity and Conservation Management Group,
Centre for Ecology and Hydrology, Maclean building,
Crowmarsh Gifford,
Wallingford, OX10 8BB, UK
[2]Centre for Agri-Environmental Research,
Department of Agriculture,
Policy and Development, University of Reading,
Reading, RG6 6AR, UK
[3]CABI Europe - UK, Bakeham Lane, Egham,
Surrey, TW20 9TY, UK

ABSTRACT

During the last century changing patterns of land use in the UK have led to the large scale loss and degradation of species-rich chalk grasslands, a habitat of conservation importance across Europe. In particular, the conversion of chalk grasslands into arable land has represented a major threat to this habitat type. The re-creation of chalk grasslands on ex-arable land therefore has the potential to both increase the area of this threatened habitat, and to provide links between existing remnant fragments. We present a four year study investigating the effects of three different methods of introducing plant species during grassland re-creation: 1) a simple grass seed mix, intended to suppress the

* Corresponding author: Ben Woodcock. E-mail: BAWood@ceh.ac.uk Tel.: +44(0)1491692415; Fax: +44(0)1491692424. Contact address: Biodiversity and Conservation Management Group, Centre for Ecology and Hydrology, Maclean Building, Crowmarsh Gifford, Wallingford, OX10 8BB, UK.

establishment of pernicious weeds; 2) spreading of hay; and 3) sowing of brush harvested seeds. The latter two methods introduce local provenance seeds collected from species-rich chalk grassland. Using a replicated block design we investigate the efficacy of these management practices in promoting the re-creation of both plant and phytophagous beetle assemblages. We compare re-creation success relative to a target chalk grassland and contrast differences in the responses between these two trophic levels. The introduction of local provenance seeds by hay spreading and brush harvesting is shown to be beneficial during re-creation. However, if used in combination with simple grass-only seed mixtures the establishment of some chalk grassland species may be reduced. We discuss the relevance of these findings for the re-creation of plant and phytophagous beetle assemblages of species-rich grasslands on ex-arable land. We also consider the implications of differences in establishment rates between these two taxa for the long term success of habitat creation.

Keywords: brush harvesting, chalk grassland, hay spreading, hay strewing, leaf beetles, re-creation, seed mixes, weevils

INTRODUCTION

Agriculturally unimproved chalk grasslands represent an important semi-natural habitat of high conservation importance, both in the UK and north-western Europe (Willems 2001; WallisDeVries et al. 2002; Woodcock et al. 2005). While there are many types of chalk grassland, these lowland habitats are known to support a particularly diverse flora, rich in forbs, including many rare or declining species (Gibson and Brown 1991; WallisDeVries et al. 2002). The invertebrate community associated with chalk grasslands is diverse (van Swaay 2002; WallisDeVries *et al.* 2002; Woodcock *et al.* 2005), and similarly includes many threatened species, perhaps the most widely recognised of which are the butterflies (van Swaay 2002). Traditional management of chalk grasslands is typically characterized by low intensity grazing, normally by sheep, although hay cutting is also used in some situations (Gibson and Brown 1991; Willems 2001; WallisDeVries et al. 2002). Without the removal of a relatively large proportion of the annual primary production, succession of the grassland will occur to either scrub / woodland or coarse species-poor grassland types (Gibson and Brown 1991).

Agricultural intensification by the application of inorganic fertilisers or excessively high grazing rates combined with the conversion of chalk grasslands to arable cultivation have resulted in the large scale loss and degradation of this species-rich habitat type across north-western Europe (Blackstock *et al.* 1999; Willems 2001). This decline in both the quality and quantity of calcareous grassland has meant that of *c.* 400 plant species associated with this habitat, approximately 42 % are now considered to be threatened (reviewed in WallisDeVries *et al.* 2002). Similarly, although less well documented, the invertebrates of chalk grasslands have also undergone massive population declines in response to these changing patterns of land use and management (Steffan-Dewenter and Tscharntke 2002; van Swaay 2002). Reversing this trend requires multiple approaches, including the protection of remnant patches of chalk grassland, as well as the re-establishment of species-rich chalk grassland on either agriculturally-improved grassland or ex-arable land (Willems 2001; WallisDeVries *et al.* 2002; Walker *et al.* 2004). In this paper we will focus on the last of these, the re-creation of chalk grassland on land that has been converted to arable agriculture within the last 60 years. Here, re-creation is used to refer to the re-establishment of a grassland community on land that was no longer covered by a permanent grass sward of any type. The re-creation of chalk grassland will serve a number of potential purposes. For example, it will increase the overall area of this habitat type and so reduce the effects of fragmentation and patch isolation. Similarly if used strategically it may also provide buffer zones reducing the impacts of edge effects in remnant grassland patches (Bakker and Berendse 1999; Steffan-Dewenter and Tscharntke 2002). In the UK such re-creation schemes are now incorporated as part of agri-environmental policy, and as such are associated with financial reimbursement to support farmers for income forgone (DEFRA 2005).

The re-creation of chalk grasslands has been achieved by a wide variety of approaches, all of which can be broadly classified as promoting either the introduction of target plant species or promoting their subsequent establishment and persistence (Bakker and Berendse 1999; Willems 2001; Walker et al. 2004). The introduction of target plant species, i.e. those considered typical of chalk grasslands, has been achieved by direct sowing of commercially available seeds or the introduction of local provenance seeds collected using a variety of methods, including vacuum harvesting (e.g. Stevenson et al. 1997), brush harvesting (Morgan and Collicutt 1994; Edwards et al. 2007) or hay spread directly onto the site (Jones et al. 1999; Edwards et al. 2007). The sowing of a seed mix can have agronomic advantages over natural regeneration by reducing the establishment of undesirable pernicious weeds (Willems 2001; Bakker *et al.* 2003). Promoting establishment and persistence of species once they have been introduced using these methods falls into two main phases. The first is preperation prior to seed introduciton. This include practices such as the reduction of soil nutrients by soil stripping or growing unfertilised crops (Walker *et al.* 2004; Edwards *et al.* 2007) or the creation of germination niches by low levels of soil disturbance (e.g. Woodcock *et al.* 2008). Secondly, once seeds have been introduced, extensive grazing or mowing regimes have been used to both reduce the dominance of some plant species while continuing to reduce soil fertility by the removal of this above ground biomass (Willems 2001; Walker *et al.* 2004).

This study investigates the effects of management practices used to introduce plant species during the re-creation of chalk grassland on ex-arable land. We compare two fundamentally different approaches that while promoting the establishment of plants, have fundamentally different goals. The first approach is the sowing of a simple commercially available seed mix of grasses that are intended to control the pernicious weeds by the rapid establishment of grass cover. The second approach considers two methods that introduce local provenance seeds typical of chalk grasslands, namely hay spreading and the sowing of brush harvested seeds. The impact of these management practices on the re-creation of chalk grassland was assessed for both plant and phytophagous beetle assemblages. Phytophagous beetles were chosen in addition to the plants as they are a functionally diverse component of grassland fauna that represent a major proportion of the total abundance and species composition of grassland invertebrates (Woodcock *et al.* 2005; Woodcock *et al.* 2006; Woodcock *et al.* 2007).

METHODS

Study Site an Experimental Design

The re-creation experiment was undertaken at Aston Rowant, Oxfordshire, UK (Lat: 51:41:23N; Lon 0:56:35W) on land previously managed for cereal production under conventional arable agricultural practices. In 2001, this field was taken out of arable production and a replicated block design experiment was set up to investigate the effect of introducing a grass-only seed mixture (2 factor levels) vs. seeds of local provenance (5 factor levels). Forty experimental plots, each of 10 × 10 m, were established in four blocks of 10 plots, each separated from any adjacent plots by 5 m. Within each block the 2 × 5 replicated block design of treatments was established.

Treatments one had two factor levels represented by a control and a grass-only seed mix sown at a rate of 25 kg/ha. This was a commercially available grass seed mix (Emorsgate Seeds, Ltd., UK) typical of those used in agri-environment scheme options for chalk grassland creation in the UK (DEFRA 2005). These seeds were not of local provenance but originated from populations with southern England. By weight the seed mix contained the grass species *Agrostis capillaris* L. (5%), *Cynosurus cristatus* L. (20%), *Dactylis glomerata* L. (10%), *Festuca ovina* L. (15%), *Festuca rubra* L. (20%), *Phleum pratense* L. (10%), *Poa pratensis* L. (20%). The seed mix was applied only in August 2001, prior to any of the subsequent management practices applied to the experimental plots.

The second treatment was the introduction of local provenance seeds harvested from an agriculturally unimproved chalk grassland (Aston Rowant National Nature Reserve, Oxfordshire, UK) adjoining the experimental field. This grassland was classified as a CG2 *Festuca ovina - Avenula pratensis* grassland under the UK's National Vegetation Classification (Rodwell 1992), and had a forb rich sward, that included the species *Campanula rotundifolia* L. (Campanulaceae), *Sanguisorba minor* Scop. (Rosaceae) and *Scabiosa columbaria* L. (Dipsacaceae). To ensure full phenological development of the plants present within this donor site the grassland remained ungrazed in 2001 to allow for the effective harvesting of seeds.

Local provenance seeds were then harvested from the donor site using two methods, hay spreading and brush harvesting. The local provenance seed treatment had five factor levels: 1) a control; 2-3) hay spreading at high and low rates; 4-5) brush harvesting at high and low rates. Hay was harvesting using conventional grass cutting machinery in August and October of 2001 and applied to the experimental plots within 1 week of each harvest (Jones *et al.* 1999). Brush harvested seeds were collected again in August and October using specialised machinery that strips seeds from the sward using rotating brushes, therefore avoiding the need to harvest the bulk of the foliage (Morgan and Collicutt 1994). The seeds were then air-dried, cleaned and broadcast onto the plots. The low rate of application for both hay and brush harvesting comprised the application of material from one unit area of the donor site to three times the area on the experimental site, whilst the high rate comprised a 1:1 ratio. The August and October applications of hay or brush harvested seeds were applied respectively to only one half of each experimental plot. However, as all subsequent analyses are based on the whole plot this split plot effect has been ignored.

Throughout the duration of the study all experimental plots were grazed by sheep at rates of between 12 – 15 ewes ha^{-1}. Grazing occurred in blocks of approximately 4-5 weeks, commencing in May of each year. Once the sward had been grazed to a height of 4-7 cm grazing was ended for a 3-4 month period until the re-growth of the sward had reached a height > 15-20 cm. No grazing was permitted over the winter period and the site was not cut for hay. Some additional horse grazing on the site occurred, although on an *ad hoc* basis and at a low intensity. This grazing rate was approximately equivalent to stocking densities of 0.25 livestock units ha^{-1} yr^{-1} that have been recommended for UK lowland calcareous grassland (Crofts and Jefferson 1999).

Botanical Assessments

Plant community composition was recorded for all experimental plots in August from 2002 to 2004. In each split-plot, ten 0.5 x 0.5 m quadrats were randomly positioned at least 1 m from the edge of the split-plot and the presence of all vascular plant species recorded. Plant species were scored on a scale of zero to ten based on their frequency in the ten quadrats. The botanical composition of the donor site was assessed in 2004 using the same methodology in 15 randomly positioned 10 × 10 m 'dummy' plots in each of the two blocks of vegetation used. Plant nomenclature follows Stace (1997).

Beetle Sampling

Phytophagous beetle assemblages were sampled from the experimental plots using a Vortis suction sampler (Burkhard Ltd, Rickmansworth, UK). As with the plants, beetle sampling was carried out annually from 2002 to 2004. During each year, experimental plots were sampled on three occasions (May, July and September) to provide a comprehensive measure of the full seasonal variation in the beetle fauna. On each sampling occasion, the suction sampler was placed in 15 positions, representing a total sampled area of $0.3m^2$. For each of these individual positions the sampler was held in location for 10 seconds. Suction sampling is a quantitative method suitable for the collection of adult invertebrates inhabiting short grassland swards (Mortimer *et al.* 2002; Woodcock *et al.* 2005; Woodcock *et al.* 2006; Woodcock *et al.* 2008). All beetle counts were summed for individual years. Samples were also taken from the Aston Rowant NNR donor site based on a similar set of fifteen samples (each representing 15 individual suction samples) taken in May, July and September 2001. Each sample was separated by at least 15 m. The phytophagous beetles from the families Apionidae, Curculionidae, Chrysomelidae, and Bruchidae were identified to species following the nomenclature of Strejcek (1993) and Morris (2003).

Measuring Re-Creation Success

Re-creation of the plant community and associated phytophagous beetle assemblages would be considered to be successful if the experimental plots were to contain the same species with the same frequencies as those found at the donor site. To provide a measure of the success of re-creation the species composition of the plants and beetles were compared to those of the donor site (Aston Rowant NNR) using Jaccard's similarity coefficient. Jaccard's similarity index (S_j) is a binary coefficient that considers only the presence or absence of species, and is defined as: $S_j = a / (a + b + c)$ where: a = number of species shared by the donor site and the experimental plot; b = number of species in the donor site only; c = number of species in the experimental plot only (Krebs 1999).

Data Analyses

For the beetles all analyses are based on summed values of either abundance or species richness for individual experimental plots for each year. For the plants mean frequency values were derived from the 10 quadrats taken within each experimental plot. The response of plant and beetle species richness to the treatments of 'grass-only seed mix' (GRASS-ONLY) and 'seeds of local provenance' (LOCAL SEEDS) were assessed using a temporal split-plot ANOVA. At the whole plot level the ANOVA tested the effects of block (4 levels), GRASS-ONLY (2 levels), LOCAL SEEDS (5 levels) and the interaction between GRASS-ONLY × LOCAL SEEDS. These whole plot factors were tested against the error term of block × GRASS-ONLY × LOCAL SEEDS. The temporal split-plot explanatory variables were year (3 levels), year × GRASS-ONLY, year × LOCAL SEEDS and year × GRASS-ONLY × LOCAL SEEDS. All species richness values were Log_e N+1 transformed to normalise the data. These analyses were then repeated for the measures of Jaccard's similarity between the experimental plots and the donor site at Aston Rowant NNR for the plants and beetles. No transformations of Jaccard's similarity were used. Analyses were carried out in SAS 9.01. The *post hoc* comparisons of means presented on graphs were performed using the CONTRAST statement within SAS.

Using the linear ordination method, partial redundancy analysis (pRDA), changes in the structure of both the plant and phytophagous beetle assemblages were assessed in response to the re-creation management treatments. For the phytophagous beetles, species represented by only one individual were excluded from the analysis. For both beetle abundances and plant cover $\log_{10}$ transformations of the data set were used. Following ter Braak and Šmilauer (2002), temporal change in beetle assemblage structure was assessed based on interactions between management treatment variables and sample years (e.g. Env.Var.*2002, Env.Var.*2003 and Env.Var.*2004). Included as covariables within these analyses were sample year (2002, 2003 and 2004) and replicate block, with the later of these designated as a blocking factor. Monte Carlo permutation tests (999 permutations) were then used to assess for differences in assemblage structure in response to the fixed treatment effect interactions with sample year. To account for the repeated measures taken from the same experimental plots on subsequent years Monte Carlo permutations were restricted so that these repeated measures were treated as split-plots in time, in a similar fashion to the ANOVA model described above. Monte Carlo permutations were permitted only between whole plot factors. The two management treatments of GRASS-ONLY and LOCAL SEEDS were tested individually. For the GRASS-ONLY treatment, with its two factor levels, this required only one test on the interaction between year and the presence of the grass-only seed mix. For the LOCAL SEEDS treatments, which had five factor levels, separate tests for the interaction between 'control × year', 'low rate hay spreading × year', 'high rate hay spreading × year', 'low rate brush harvesting × year' and 'high rate brush harvesting × year' were required. To provide insight into the interaction between the two re-creation management treatments the ten separate interactions terms from the 2 × 5 factorial design of the experiment were coded for by dummy environmental variables. The interaction of each of these dummy environmental variables with year was then tested following the methodology described above. The analysis was carried out in CANOCO 4.5.

RESULTS

In total 3,242 phytophagous beetles were identified to one of 56 species over the three year period from 2002 – 2004 at the Aston Rowant re-creation site. Of these, the beetle fauna was dominated by the leaf beetle *Longitarsus pratensis* (Panzer) (Chrysomelidae), although both *Sitona lineatus* L. (Curculionidae) and *Altica palustris* (Weise) (Chrysomelidae) were represented by summed abundances in excess of 100 individuals. Only six species of beetle found at the donor site were not found in the experimental plots, however, in terms of relative abundances the beetle fauna of the donor site differed considerable from those seen within the experimental plots. A total of 115 species of vascular plant were found within the experimental plots, with an additional 7 species of forbs being found at the donor site only.

Plant and Beetle Species Richness

For the plants there were significant effects of both the GRASS-ONLY and LOCAL SEED treatments, although in the case of the LOCAL SEED treatment a significant interaction with year was also found (Table 1). Plant species richness was highest on those experimental plots that had not had been sown with the grass-only seed mix in 2001, independent of the effect of the LOCAL SEEDS treatment (Figure 1). The interaction between LOCAL SEEDS and year showed that the control treatment in 2001 had a much lower plant species richness than where either brush harvested seeds or hay had been applied. The species richness of the plants in the control did, however, increase to a level similar to that seen for the hay and brush harvested seed plots. In contrast, there was little change in the species richness of those plots receiving either brush harvested seed or hay (either at high or low application rates) over the same three year period. The plant species richness where seeds had been introduced by brush harvesting or hay was in general higher than that seen for the control, although typically did not change to any great extent between 2002 and 2004 (Figure 2). A significant overall effect of year did indicate a general increase in plant species richness over this time.

For the beetles, significant effects of GRASS-ONLY, LOCAL SEEDS and the interaction between both of these treatments were found for species richness (Table 1). This interaction suggested that there was a general trend of lower species richness being found in those levels of the LOCAL SEEDS treatment when found in combination with the grass-only seed mix (Figure 3). This trend was, however, only significant for the LOCAL SEEDS control, which had significantly lower species richness when in combination with a grass-only seed mix. The highest overall beetle species richness was found for high rates of hay applied without the presence of a grass-only seed mix. On the whole, however, the remaining factor levels of the GRASS-ONLY × LOCAL SEEDS interaction did not differ significantly. A significant year effect was found for the beetles, although this showed no interaction with either the GRASS-ONLY or LOCAL SEEDS treatment (Table 1).

Table 1. Response of plant and phytophagous beetle species richness (Ln N+1) to the management practices of applying seeds from a local provenance seed source (LOCAL SEEDS) or the sowing of a simple grass-only seed mix (GRASS-ONLY).

	Plants species richness	Beetles species richness
Whole plot effects		
LOCAL SEEDS	$F_{4,27}$ = 19.1***	$F_{4,27}$ = 3.51*
GRASS-ONLY	$F_{1,27}$ = 12.4**	$F_{1,27}$ = 5.34*
LOCAL SEEDS × GRASS-ONLY	$F_{4,27}$ = 0.35 ns	$F_{4,27}$ = 0.75*
Block	$F_{3,27}$ = 1.15 ns	$F_{3,27}$ = 0.53 ns
Error = Prov. × Grass × Block		
Split plot effects		
Year	$F_{2,60}$ =26.5***	$F_{2,60}$ = 15.1***
Year × LOCAL SEEDS	$F_{8,60}$ = 2.84**	$F_{8,60}$ = 0.97 ns
Year × GRASS-ONLY	$F_{2,60}$ = 0.87 ns	$F_{2,60}$ = 0.05 ns
Year × LOCAL SEEDS × GRASS-ONLY	$F_{8,60}$ = 0.79 ns	$F_{8,60}$ = 1.16 ns

ns = non-significant (p>0.05); *** = p<0.001.

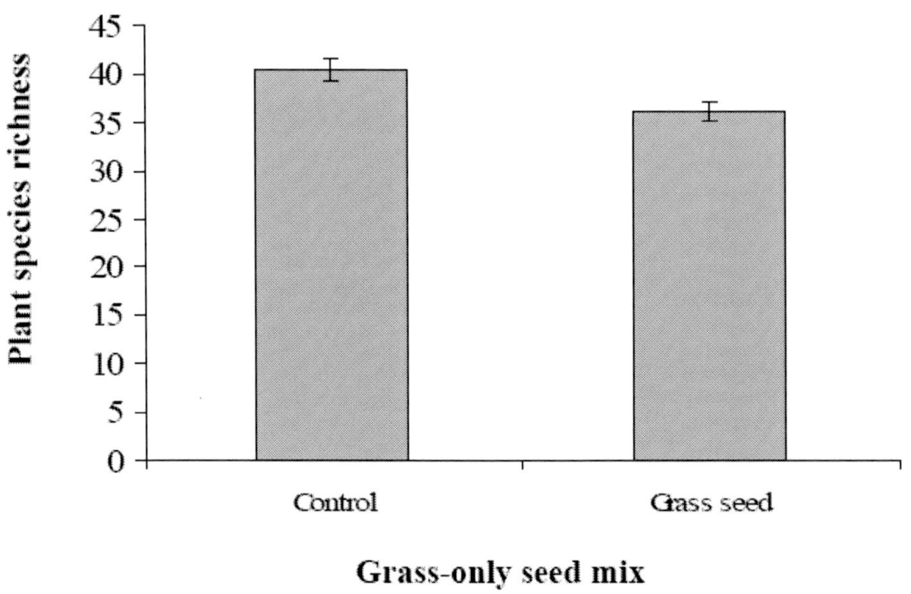

Figure 1. The effect of sowing a simple grass seed mixture on plant species richness during the re-creation of chalk grassland.

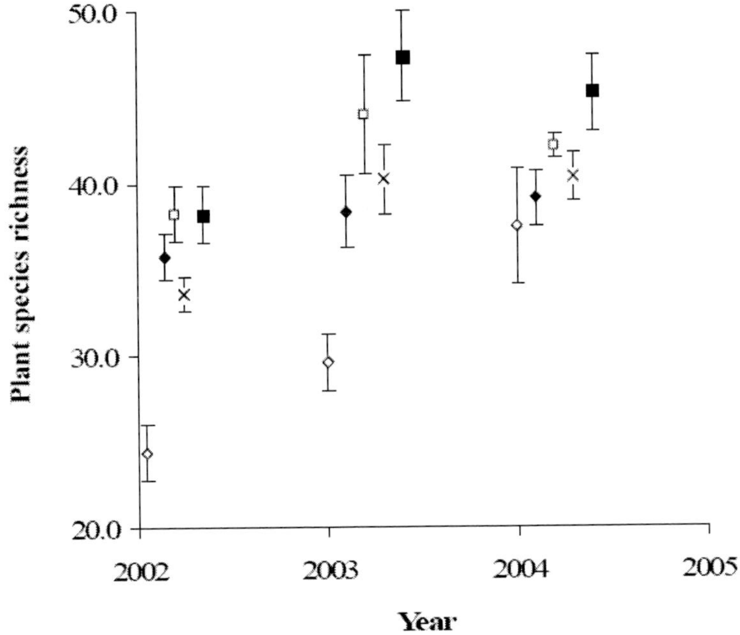

Figure 2. Plant species richness from 2002-2004 in response to the addition of local provenance seeds by either hay spreading or brush harvested seeds. Where the levels of the seed addition treatment were: ◊ = control; ♦ = low rate of hay spreading; □ = high rate of hay spreading; × = brush harvested seeds at a low rate; ■ = brush harvested seeds at a high rate. Note that for clarity an offset 0.05 years has been used for consecutive seed addition methods from 2002 –2004.

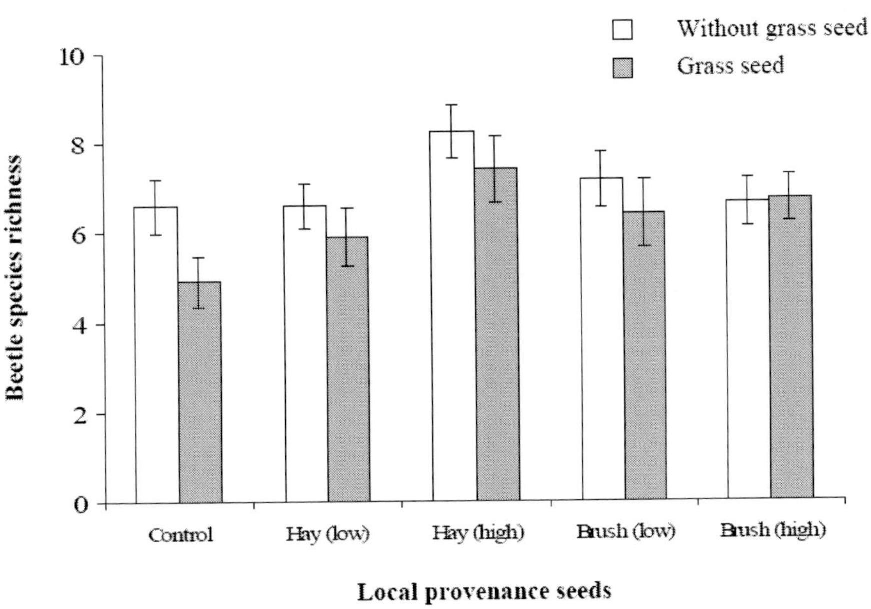

Figure 3. The effect on beetle species richness of introducing local provenance seeds by either hay spreading (low or high application rates) or the application of brush harvested seeds (low or high application rates) relative to a control. The interaction of this effect with the presence or absence of a sown simple grass-only seed mixture is also shown.

This effect of year demonstrated an increase in beetle species richness from 2002 to 2004, independent of management practice (Figure 4).

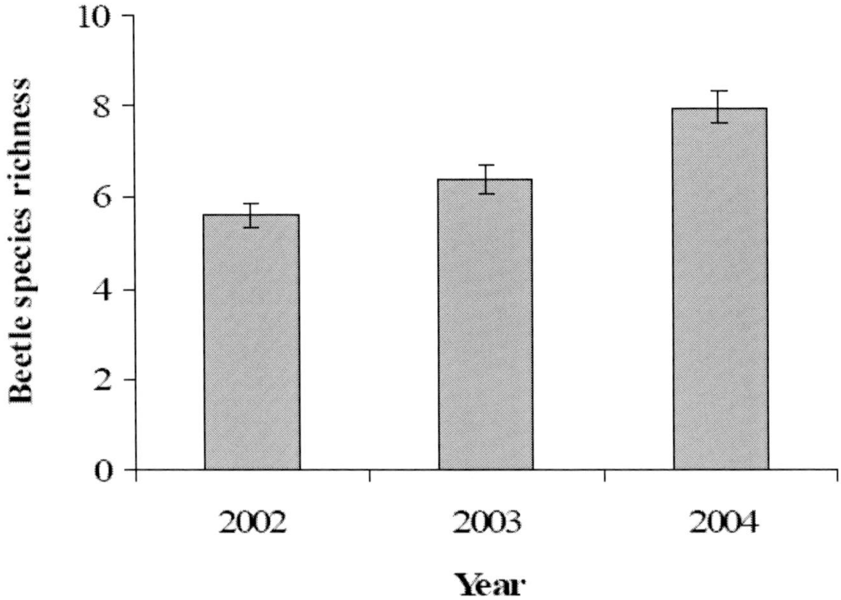

Figure 4. Changing beetle species richness across all treatments with year.

Plant Responses to Re-Creation Management

The effect of the two management treatments on the community structure of the plants was assessed using RDA. Significant overall effects were found for the interactions of year with the treatments GRASS-ONLY and LOCAL SEEDS. There was also a significant GRASS-ONLY × LOCAL SEEDS × Year interaction which explained 52.0 % of the variance in the plant community structure. Considering each of the ten treatment levels of the GRASS-ONLY × LOCAL SEEDS interaction only the factor levels of 'low rate brush harvesting × year' and 'low rate hay spreading with grass-only seed mix × year' did not individually have a significant effect on the structure of the plant communities (Table 2). The RDA biplots suggest that the changes in plant community structure over the three year period were relatively small for all treatment level interactions (Figure 5). Notably, however, there was a split along axis 1 of the RDA biplot between treatment combinations which included the grass-only seed mix, which tended to have negative axis 1 scores, and those without the seed grass-only seed mix which had positive axis 1 scores.

Beetle Responses to Re-Creation Management

Similar to the effect found for plants, the assemblage structure of the beetles changed significantly in response to the interaction between year and the treatment effects of GRASS-ONLY, LOCAL SEEDS and the GRASS-ONLY ×LOCAL SEED interaction (Table 2).

Table 2. Results for redundancy analysis of beetle assemblage responses to establishing seed mixture, margin management and the floristic composition and architectural structure of the field margins. All significances were tested using Monte Carlo permutation tests (1000 permutations) of both canonical axes, where 'F' is the F-statistic of this test. See the Methods for environmental variable abbreviations. × = Interaction effect; DF = degreed freedom; * = significant at $p < 0.05$; ** = $p < 0.01$; *** = $p < 0.001$; ns = p > 0.05. The variance in the species data explained by each fixed effect parameter is given in parentheses.

Environmental interaction	Plants	Beetles
Treatment effects		
LOCAL SEEDS × Year	F = 2.42*** (21.8 %)	F = 2.75*** (24.1 %)
GRASS-ONLY × year	F = 11.5 *** (23.3 %)	F = 3.89*** (9.3 %)
LOCAL SEEDS × GRASS-ONLY × Year	F = 3.56 *** (52.0 %)	F = 2.37*** (41.7 %)
Tests for Individual LOCAL SEEDS × GRASS-ONLY interactions		
Control × Year	F = 3.37**	F = 2.36**
Low hay × Year	F = 2.13*	ns
High hay × Year	F = 2.70**	F = 2.71**
Low brush × Year	ns	ns
High brush × Year	F = 2.29**	ns
Grass × Year	F = 2.88***	F = 4.01**
Grass and Low hay × Year	ns	ns
Grass and High hay × Year	F = 2.21**	ns
Grass and Low brush × Year	F = 1.94*	F = 1.86*
Grass and High brush × Year	F = 2.38*	ns

Table 3. Significance values for tests of how similarity (Jaccard) in the plant and phytophagous beetles to a target unimproved species-rich chalk grassland changed in response to the management practices of applying seeds from a local provenance seed source (LOCAL SEEDS) or the sowing of a simple grass-only seed mix (GRASS-ONLY). Where: ns = non-significant (p>0.05); *** = p<0.001.

	Plant similarity	Beetles similarity
Whole plot effects		
LOCAL SEEDS	$F_{4,27} = 36.4$***	$F_{4,27} = 9.02$ ***
GRASS-ONLY	$F_{1,27} = 0.04$ *	$F_{1,27} = 20.1$***
LOCAL SEEDS × GRASS-ONLY	$F_{4,27} = 0.53$ ns	$F_{4,27} = 0.74$ ns
Block	$F_{3,27} = 1.20$ ns	$F_{3,27} = 0.64$ ns
Error = Prov. × Grass × Block		
Split plot effects		
Year	$F_{2,60} = 53.8$ ***	$F_{2,60} = 27.6$***
Year × LOCAL SEEDS	$F_{8,60} = 3.97$ **	$F_{8,60} = 2.09$*
Year × GRASS-ONLY	$F_{2,60} = 0.36$ ns	$F_{2,60} = 3.40$*
Year × LOCAL SEEDS × GRASS-ONLY	$F_{8,60} = 0.39$ ns	$F_{8,60} = 0.56$ ns

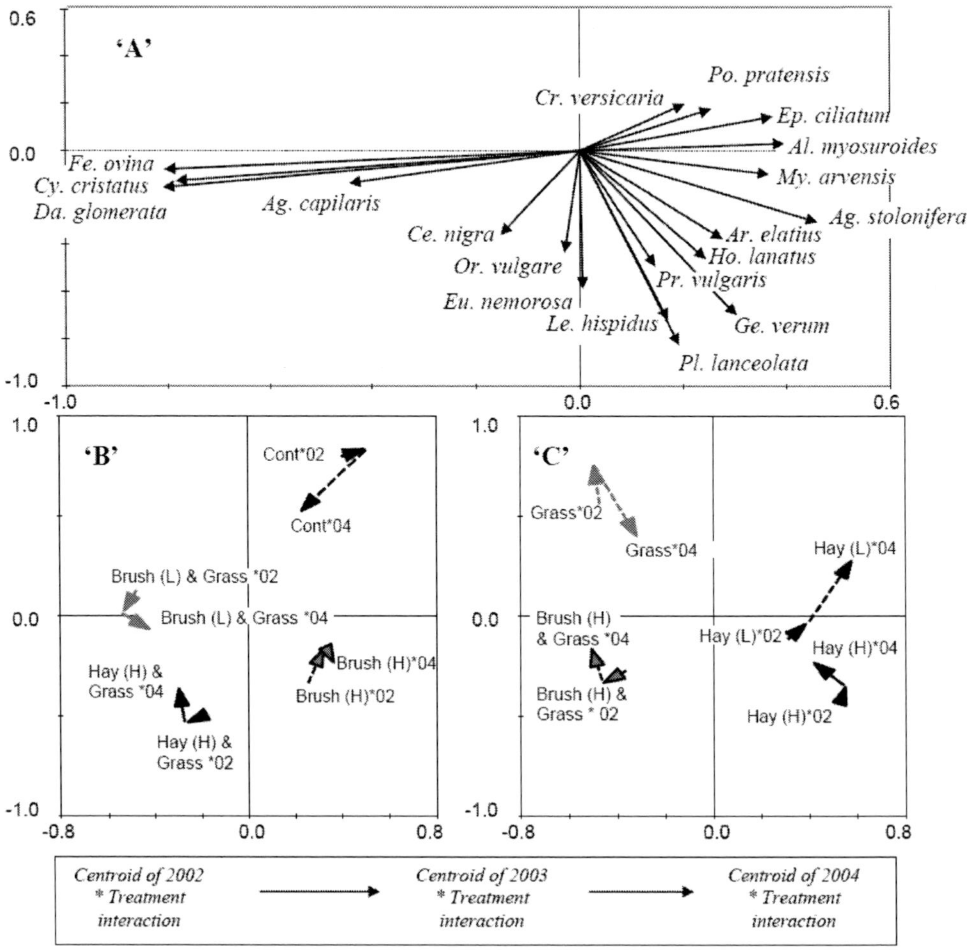

Figure 5. RDA Ordination diagrams for the sample years 2002-2004 showing the temporal response of the plant assemblages to the significant re-creation management treatments. Figure 'A' represents the companion plant species scatter plot to figure 'B' and 'C'. Due to the large number of significant treatment*year interactions two separate biplots 'B' and 'C' using the same scales are shown. The change with time of the plant assemblages in response to the management treatments in figure 'B' and 'C' is emphasised by the connection of the centroids of year*treatment interaction with arrows, from 2002*treatment (start of first arrow) to 2004*treatment (end of final arrow). Only species with high axis 1 or 2 scores are shown (thin line vectors) are shown in figure 'A'. Note year has been abbreviated to the last two digits so that 2002 and 2004 are respectively 02 and 04. The label for the interaction between treatment and year for 2003 has not been shown on the biplot figure. Treatment abbreviations are: Cont. = Control, no re-creation management; Grass = simple grass-only seed mixture; Hay = hay spreading; Brush = brush harvested seeds. Where 'H' or 'L' in parenthesis indicate high or low rates of application or either hay or brush harvested seeds. Specific names are given in full, while generic names are abbreviated as follows. Asteraceae: Ce. = Centaurea; Cr. = Crepis; Le. = Leontodon. Boraginaceae: My. = Myosotis. Lamiaceae: Or. = Origanum; Pr. = Prunella. Onagraceae: Ep. = Epilobium. Plantaginaceae: Pl. = Plantago. Rubiaceae: Ga. = Galium. Scrophulariaceae: Eu. = Euphrasia. Poaceae: Ag. = Agrostis; Al. = Alopecurus; Ar. = Arrhenatherum; Cy. = Cynosurus; Da. = Dactylis; Fe. = Festuca; Ho. = Holcus; Po. = Poa.

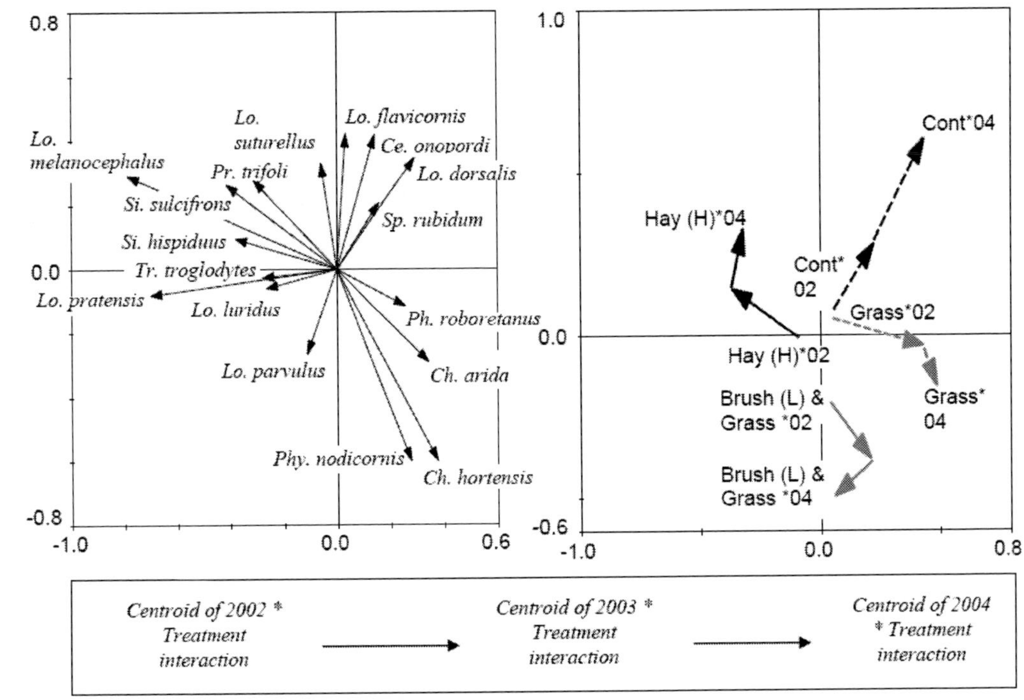

Figure 6. RDA Ordination diagrams for the sample years 2002-2004 showing the temporal response of the beetle assemblages to the significant re-creation management treatments. Figure 'A' represents the companion beetle species scatter plot to figure 'B'. The change with time of the beetle assemblages in response to the management treatments in figure 'B' is emphasised by the connection of the centroids of year*treatment interaction with arrows, from 2002*treatment (start of first arrow) to 2004*treatment (end of final arrow). Only species with high axis 1 or 2 scores are shown (thin line vectors) are shown in figure 'A'. Note year has been abbreviated to the last two digits so that 2002 and 2004 are respectively 02 and 04. The label for the interaction between treatment and year for 2003 has not been shown on the biplot figure. Treatment abbreviations are: Cont. = Control, no re-creation management; Grass = simple grass-only seed mixture; Hay (H) = high rate of hay spreading; Brush (L) = Low rate of application for brush harvested seeds. Specific names are given in full, abbreviations for generic names are: Chrysomelidae: Lo. = Longitarsus; Ch. = Chaetocnema; Sp. = Sphaeroderma; Phy. = Phyllotreata. Curculionidae: Ph. = Phyllobius; Tr. = Trichosirocalus; Si. = Sitona. Apionidae: Ce. = Ceratapion; Pr. = Protapion.

When the individual factor levels of the GRASS-ONLY ×LOCAL SEED interaction were considered there were far fewer significant effects that had been seen for the plants. Significant effects were found only for the 'Control × year', 'High rate hay spreading × year', 'Grass-only × year' and 'Low rate brush harvesting with grass seed mix × year'. In contrast to the biplots seen for the plants, the phytophagous beetles showed much greater shifts in assemblage structure over time (Figure 6).

Success of Re-Creation

While the RDA biplots provide an indication of how close in structure the assemblages of the plant and beetles are to the donor site, the analyses of Jaccard's similarity to the same donor sites provides a more rigorous test of re-creation success. For the plants there was, however, a significant year × LOCAL SEEDS interaction, in addition to significant individual effects of year and the LOCAL SEEDS treatment (Table 3). This interaction was characterised by the control having the lowest similarity relative to the target community in 2002, and while this did increase, it remained lower than those plots that had received either hay spreading or brush harvested seeds by 2004. For plots where hay spreading or brush harvesting of seeds was applied similarity to the donor site did increase slowly over the three year period, this being greatest local provenance seeds had been applied by brush harvesting at a high rate (Figure 7 a). Plant similarity to the donor site was also affected by the GRASS-ONLY treatment, where similarity was greatest where the grass-only seed mix had not been sown.

For the phytophagous beetles similarity to the donor site was affected by the significant treatment interactions of year × GRASS-ONLY and year × LOCAL SEEDS, as well as the individual terms of year, GRASS-ONLY and LOCAL SEEDS (Table 3). Overall similarity for the beetles increased with time since re-creation was initiated. All factor levels of the interaction between year × LOCAL SEEDS increased in similarity to the donor site over the three year period, however, this rate of increase was greatest for those plots that had received hay spreading at a high rate of application (Fig 7b). For the year × GRASS-ONLY interaction those plots not sown with the grass-only seed mix increased in similarity to the donor site ever year from 2002 – 2004. In contrast, the similarity to the donor site peaked for plots that were sown with a grass-only seed mix in 2003, although this similarity had declined by the final year of the study in 2004.

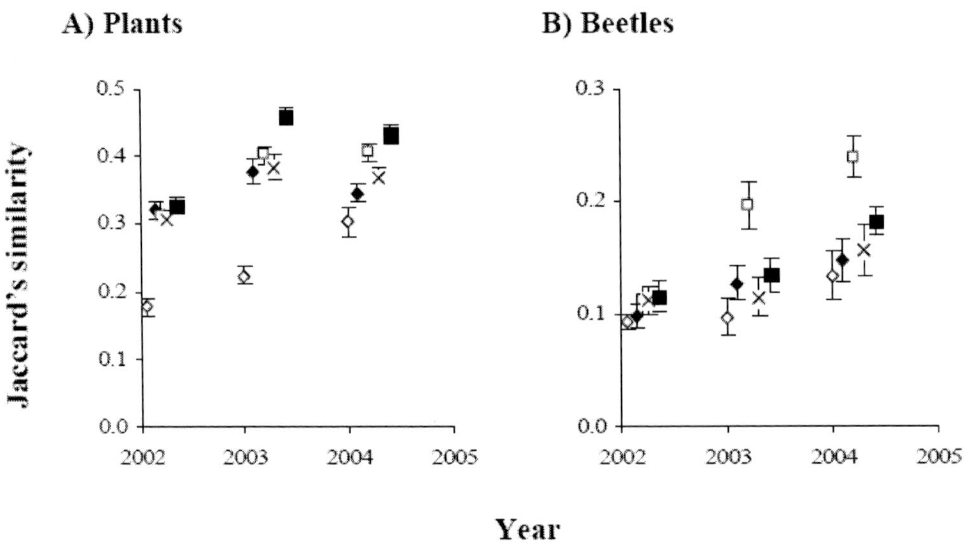

Figure 7. Temporal change in re-creation success in response to management for the plant (A) and beetle (B) assemblages. Re-creation success is measured by the Jaccard's similarity (± SE) between the experimental treatments for 2002, 2003 and 2004 and the target chalk grassland. Higher similarity to this target grassland community represents increased re-creation success. Experimental treatments were the four methods of local provenance seed addition and a control. Where the levels of the seed addition treatment were: ◊ = control; ♦ = low rate of hay spreading; □ = high rate of hay spreading; × = brush harvested seeds at a low rate; ■ = brush harvested seeds at a high rate. Note that for clarity an offset 0.05 years has been used for consecutive seed addition methods from 2002 –2004.

Discussion

As it is normal to characterise grassland habitats on the basis of their plants, it is perhaps unsurprising that goals for re-creation have focused almost exclusively on re-establishing the desired botanical composition (Bakker and Berendse 1999; Walker et al. 2004). The assumption has been that successful re-creation of the plant communities will be followed by other components of these habitats, such as the invertebrates (Woodcock et al. 2008; Woodcock and McDonald In press). In the case of the invertebrates, this assumption reflects their dependence on the plant communities for the provision of both host plants and as structural refuges (Mortimer et al. 1998; Morris 2000; Mortimer et al. 2006). While there is certainly a relationship between the establishment of plants and invertebrates during habitat re-creation, this relationship is complex (Woodcock *et al.* 2008). By understanding how more than one trophic level responds to re-creation management, the potential for successfully replicating target communities - in this case a species-rich chalk grasslands – should be increased (Forup and Memmott 2005; Woodcock et al. 2008).

Grass-Only Seed Mix

The practice of using seed mixes from commercially available sources has been shown to be effective in both promoting the rapid establishment of sward cover and reducing the occurrence of pernicious weeds (Stevenson et al. 1995). Potentially, such seed mixes may contain both grass and forb components (Stevenson et al. 1995; Pakeman et al. 2002; Walker et al. 2004). In this study we have considered a simple grass-only seed mix, as this is common practice in UK agri-environment schemes. This grass-only seed mix has proved popular to farmers as, in addition to being effective in controlling pernicious weeds, it is relatively inexpensive (DEFRA 2005). However, the impact of this seed mix was to reduce overall plant species richness, independent of whether this approach was used in combination with hay spreading or brush harvesting techniques used to introduce a broader spectrum of local provenance seeds. This suggests that while the grass-only seed mix may be effective in reducing undesirable weed establishment (Stevenson *et al.* 1995), it was also limiting the establishment of desirable chalk grassland species (Crofts and Jefferson 1999; Willems 2001). The impact of the grass-only seed mix was also seen for beetle species richness, which tended to be lower where the grass-only seed mix had been sown in combination with either of the methods used for introducing local provenance seeds. While the effect of the grass seed mix on the beetles was small, its impact reflects the strong dependence of the beetles on the establishment of key forb host plants, a factor reduced where the grass-only seed mix was sown (Woodcock *et al.* 2006; Woodcock *et al.* 2008).

The impact of the grass-only seed mix on the plants was limited not only to the species richness of the sward, but also affected its botanical composition. Unsurprisingly, this was characterized by strong associations with the sown grass species *F. ovina*, *C. cristatus*, *A. capilaris* and *Dactylis glomerata* L. where the grass-only seed mix had been used, either in combination with the local provenance seeds or on its own. However, where the grass-only seed mix was sown on its own, plant community structure was largely dominated by these grasses, and so distinct from when sown in combination with local provenance seeds.

However, the use of the grass-only seed mix acted as a competitive filter limiting forb establishment when sown in combination with either hay spreading or brush harvesting methods. For example, *Centaurea nigra* L. (Asteraceae) was successful in becoming established only when introduced in combination with the grass-only seed mix (Pakeman *et al.* 2002).

The response of beetle assemblage structure to the grass-only seed mix shows parallels to that of the plants, although in contrast there were far fewer significant responses to the treatment interactions with local provenance seed treatments. The parallel was most apparent in terms of a divergence in beetle assemblage structure where management had either included or not included the grass-only seed mix. Typically, the beetle assemblages associated with the grass-only seed mix contained species that were found ubiquitously across all treatments, with few beetles being strongly associated with this management practice. Given that the majority of UK phytophagous beetles found within grasslands are dependent on forbs as host plants, rather than grasses (Bullock 1992), the poor establishment of forbs where the grass-only seed mix was sown would explain this trend (Mortimer *et al.* 2006; Woodcock *et al.* 2008).

Local Provenance Seeds

The use of local provenance seeds during grassland re-creation had several advantages over the use of commercially available seed mixes, such as those described above (Stevenson *et al.* 1995; Pakeman *et al.* 2002; Walker *et al.* 2004). Firstly, seeds will be sourced from local biotypes of plants and so adapted to local environmental conditions. It has been suggested that this will improve survival of these plants once established as well as also contributing to the preservation of local genetic diversity (Jones et al. 1999; Gustafson et al. 2004; Walker et al. 2004). Additionally, methods used in the collection of local provenance seeds will harvest species present within the swards of the donor sites which are not commercially available (Morgan and Collicutt 1994; Jones et al. 1999).

Relative to the grass-only seed mix, the benefits of using hay spreading and brush harvesting techniques was in general seen in terms of the establishment of forb species harvested from the donor site chalk grassland. However, both the method of sourcing local provenance seeds and the application rates affected the success of re-creation. In particular the use of low rates of application for the local provenance seeds tended to result in benefits to the plant community that did not persist for the duration of the study. For example, plant species richness was always higher than the control treatment after the initial application of local provenance seeds, although after three years this initial difference only remained where high application rates has been used. High rates of application of local provenance seeds are therefore likely to be a limiting factor to enhancing plant species richness (Edwards *et al.* 2007).

Beyond simple enhancement of species richness, differences in community structure for the plants where local provenance seeds has been sown was determined primarily by whether or not they had been applied in combination with the grass-only seed mix. Such an overriding impact on the structure of the plant communities in response to the addition of the grass-only seed mix reflects the competitive dominance of some of these sown grass species (Willems 2001; Bakker *et al.* 2003). However, the relatively small differences in community structure between the hay spreading and brush harvested seed treatments, either with or without the grass-only seed mix, seem to be characterized by underlying differences in the composition of seeds collected by these two methods (Morgan and Collicutt 1994; Jones et al. 1999; Edwards et al. 2007). These relatively small differences should be weighed against the relative scarcity of the machinery required for brush harvesting (Morgan and Collicutt 1994). Whether the costs associated with such specialized machinery are warranted is arguable, although the overall benefits associated with this technique relative to the more conventional hay spreading methods were apparent, at least in terms of the enhancement of plant species richness.

The success of re-creation was measured relative to an existing example of a species-rich chalk grassland (Edwards *et al.* 2007; Woodcock *et al.* 2008). For both the plants and the beetles, the methods used for harvesting local provenance seeds played an important role in influencing the long term success of chalk grassland re-creation. For the plants, similarity to the target grassland increased for the control, hay spreading and brush harvesting methods over the three years period after re-creation management was initiated. The high rates of application of both hay spreading and brush harvesting resulted in the highest similarity to the target grassland, while the control had the lowest. This predictable pattern reflected expected benefits associated with high sowing rates (Stevenson *et al.* 1995; Jones *et al.* 1999; Edwards *et al.* 2007). Although high application rates of local provenance seeds were clearly superior for the plants, botanical similarity to the target grassland under the low application rates remained superior to the control treatment. This was in contrast to the beetles which were reliant on high application rates of hay spreading or brush harvested seeds if similarity to the target grassland was to remain higher than the control over the same three years. Even where high rates of application had been used for the local provenance seeds, the brush harvesting method was only slightly superior to the control in replicating the beetles of the target grassland. Why high rates of brush harvesting were less effective in establishing chalk grassland beetle assemblages than hay spreading is not immediately clear. The probable explanation is that brush harvesting may have failed to introduce key forb species that supported large number of beetles typical of chalk grasslands (Edwards *et al.* 2007).

Conclusion

For the attempt to re-create chalk grasslands this study has shown that there are parallels between the plants and the beetles in how they respond to management used to introduce seeds. In both cases the use of a grass-only seeds mix appears to have had an overall detrimental effect in terms of either enhancing species richness or the overall success of re-creation. Similarly, high rates of application of local provenance seeds have proved crucial to the success of re-creation for both the plants and beetles, although while the plants benefit from both methods, hay spreading proved to be the superior method for the beetles. Introducing the widest possible breath of plant species is likely to be crucial to promoting re-creation, and in this sense the use of multiple seed harvesting methodologies, potentially combined with multiple donor sites and several harvesting times throughout the year would potentially be of greatest benefit (Bakker and Berendse 1999; Willems 2001; Walker et al. 2004). However, such a combination of approaches may be unrealistic given financial restrictions common to most re-creation attempts (Walker *et al.* 2004). Independent of this restriction, consideration of the beetles in addition to the plants has served to highlight management practices that would be likely to promote re-creation of a functional system with more than one trophic level.

ACKNOWLEDGEMENTS

This project was funded by the UK Department for the Environment, Food and Rural Affairs (project BD1441). English Nature gave access to the sites at Aston Rowant. Victoria Chapman, Tracy Gray and Corin Wilkins assisted with botanical sampling. Thanks to Darren Mann and James Hogan at the Hope Entomological Collection, University Museum Oxford.

REFERENCES

Bakker JD, Wilson SD, Christian JM, Li XD, Ambrose LG and Waddington J (2003). Contingency of grassland restoration on year, site, and competition from introduced grasses. *Ecological Applications*, 13, pp. 137-153.

Bakker JP and Berendse F (1999). Constraints in the restoration of ecological diversity in grassland and heathland communities. *Trends in Ecology and Evolution*, 14, pp. 63-68.

Blackstock TH, Rimes CA, Stevens DP, Jefferson RG, Robertson HJ, Mackintosh J and Hopkins JJ (1999). The extent of semi-natural grassland communities in lowland England and Wales: a review of conservation surveys 1978-96. *Grass and Forage Science*, 54, pp. 1-18.

Bullock JA (1992). *Host Plants of British Beetles: A List of Recorded Associations.* Essex, UK: Cravitz Publishing Company Ltd.

Crofts A and Jefferson RG (eds) (1999) *The lowland grassland management handbook. 2nd Edition.* English Nature and Wildlife Trusts, Peterborough.

DEFRA (2005). *Environmental Stewardship: Higher Level Stewardship Handbook.* London, UK: Department for the Environment, Food and Rural Affairs.

Edwards A, Mortimer SR, Lawson CS, Westbury DB, Harris SJ, Woodcock BA and Brown VK (2007). Hay strewing, brush harvesting of seed and soil disturbance as tools for the enhancement of botanical diversity in grasslands. *Biological Conservation*, 134, pp. 372-382.

Forup ML and Memmott J (2005). The restoration of plant-pollinator interactions in hay meadows. *Restoration Ecology*, 13, pp. 265-274.

Gibson CWD and Brown VK (1991). The nature and rate of development of calcareous grassland in southern Britain. *Biological Conservation*, 58, pp. 297-316.

Gustafson DJ, Gibson DJ and Nickrent DL (2004). Conservation genetics of two co-dominant grass species in an endangered grassland ecosystem. *Journal of Applied Ecology*, 41, pp. 389-397.

Jones GH, Trueman IC and Millet P (1999). The use of hay strewing to create species-rich grasslands (i). General principles and hay strewing versus seed mixes. *Land Contamination and Reclamation*, 3, pp. 104-107.

Krebs CJ (1999). *Ecological Methodology*. California: Addison Wesley Longman.

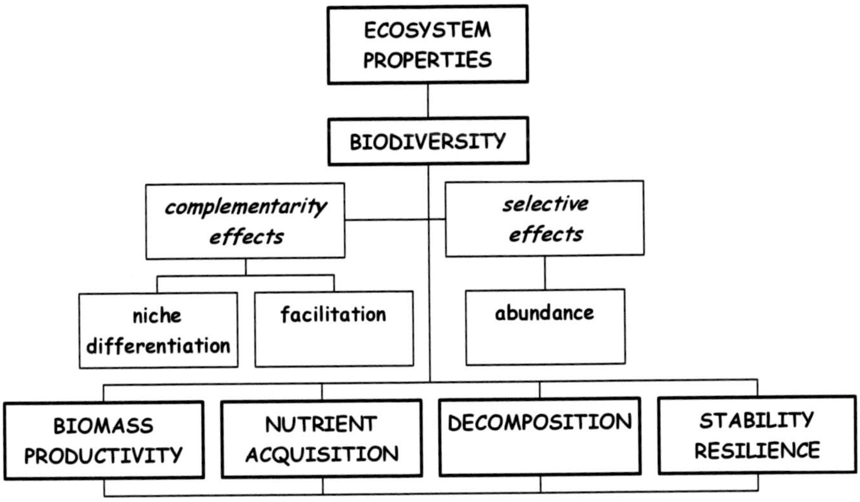

Figure 1. Most commonly listed ecosystem properties (bold squares), with biodiversity affecting other properties through the mechanisms underlying vegetation processes (light squares).

Theoretical models that predict responses of ecosystem processes to biodiversity loss can be generally grouped into two classes: first, there are the effects generated from niche differentiation and/or facilitation between species, which can increase the performance of communities above the expected performance of individual species, collectively named as complementarity effects; and second, there are the selective effects, such as interspecies competition, which result in a dominance (high relative abundance) of species with particular traits (Figure 1) (Loreau and Hector 2001). Compositional effects therefore trace back typically to a few functionally dominant species, whereas richness effects are independent of species composition and are believed to result from increased facilitation and niche complementarity at high species richness (Worm and Duffy 2003).

Both productivity and the use of resources (Figure 1) increase asymptotically according to the original biodiversity of a community. On average, increasing diversity incorporates a greater variation in species traits, and therefore affects interspecies competitive interactions. In the case of multispecies competition for resources at the ecosystem level, however, greater resource use is associated with higher plant diversity. This would reduce ecosystem nutrient losses, leading to long-term increases in ecosystem carbon and nutrient stores, which would also increase productivity. Rapid rates of species extinction and species loss from managed ecosystems may therefore have significant effects on their productivity (Tilman and al. 1997).

Much of the experimental, descriptive and theoretical work has shown that local richness often has a unimodal relationship to productivity, whereas regional richness is often a linear function of productivity. This apparent contradiction appears, in part, because studies have simulated how changing species pool sizes affect local productivity, with a positive richness–productivity relationship in many studies with typically mesophitic assemblages, and declining richness often seen in transitions to eutrophic systems. Cross-site comparisons, on the other hand, address how the productivity potential of the environment adjusts the local richness, given a particular species pool. These points highlight the causality between biodiversity and productivity; namely, increasing potential diversity (the species pool size) can enhance production, whereas increasing potential productivity (resource flux) adjusts local diversity (Worm and Duffy 2003).

A meta-analysis of more than 300 studies reporting on the nutritional needs of plants with an emphasis on nitrogen and phosphorus, for example, showed that phosphorus is rarely available as a great excess relative to nitrogen, and therefore the addition of one of these elements will rapidly provoke a limitation in the other. At the organism level, many plant species have adaptations that allow them to obtain a nutrient that would otherwise be difficult to acquire, in the form of mycorrhizae or symbiosis with nitrogen-fixing bacteria (Davidson and Howarth 2007). The diversity of arbuscular mycorrhizal fungi from the roots of a single plant species and from a single sampling location that showed only seven of 49 phylotypes to be closely similar to known sequences (>99% identity), however, came as a surprise to the research community. Functional diversity within species of arbuscular fungi also appears to be very high, although despite their importance, little is known about the diversity of arbuscular fungi in natural habitats (Vandenkoornhyse and al. 2002; Munkvold and al. 2004). Numerous studies have nevertheless indicated that below-ground mycorrhizal fungal diversity is a major factor that contributes to ecosystem functioning, presumably through its influence on plant fitness due to changes in hormonal balance, nutrient capture, and resistance to stress. The consequence of this is greater plant species composition and productivity, and lower variability (Gogala 1991; Francis and Read 1994; van der Heijden and al. 1998; Smith and Read, 1997).

Several studies that have tested the effects of species diversity on ecosystem properties have been mainly concerned with the effects of plant diversity on primary production and nutrient retention in temperate grasslands, both of which are under the direct control of the plant. They have, however, frequently failed to detect significant effects on below-ground decomposition processes (Figure 1), potentially because these processes are under microbial control. Current evidence suggests that in governing soil process rates, the properties of individual plant species are more important than plant diversity (Wardle and al. 1998; Loreau and al. 2001).

Effects of biodiversity on ecosystem stability (Figure 1) are difficult to demonstrate (Tilman and al. 2006). Field observations and laboratory studies that were designed to test the hypothesis of biodiversity effects on ecosystem stability have, in general, led to the conclusion that greater ecological diversity is associated with greater ecosystem stability despite lower species stability, as predicted by multispecies competition models (Loreau and al. 2001; Tilman and al. 2006). In a long-term experimental field test, for example, greater numbers of plant species led to a greater temporal stability of the annual above-ground plant production of the ecosystem. The ecosystem stability was also positively dependent on the root mass as a measure of the biomass of perennating species (Tilman and al. 2006). At first glance, it is less clear whether this dependence on diversity arises from the need for recruitment of a few key species from within the regional species pool, or whether is it due to the need for a rich assortment of complementary species within a particular ecosystem (Loreau and al. 2001). Species, that dominate under a given set of environmental conditions serve to maintain the ecosystem function under those conditions, while minor species will be functionally similar to the dominant species, but with different environmental requirements and tolerances. These species, therefore, provide resilience to perturbations (persistence) that will allow the maintenance of function under changing conditions. Therefore, the plant functional diversity is an important plant trait that supports the flow of ecosystem goods and services, whereas a functional similarity between the dominant and minor species, and among the minor species, may be equally important to ensure resilience to perturbations in the functions of an ecosystem under changing environmental conditions (Figure 2) (Walker and al. 1999).

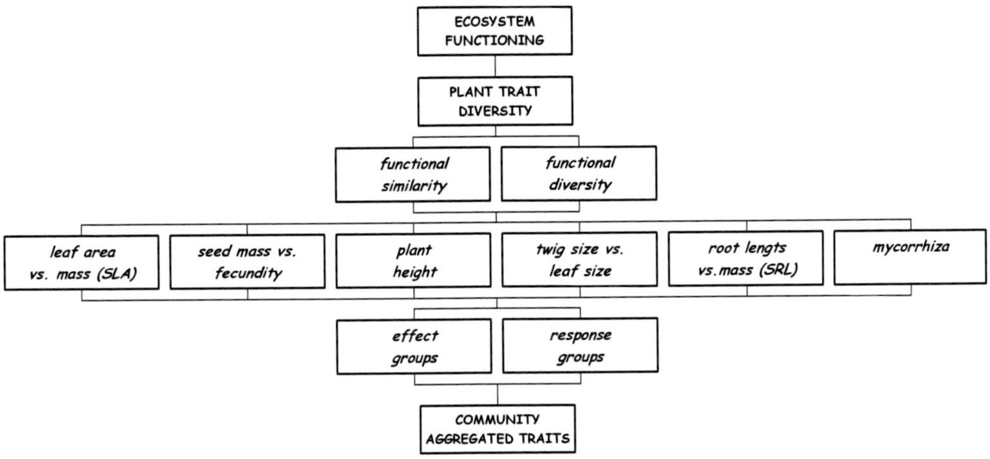

Figure 2. The most common plant functional traits in contemporary functional-ecology studies.

3. Plant Functional Diversity and Traits

Biodiversity has two main components at the species level, the richness and the composition. As such, it is generally agreed that human impacts on the environment from local to global scales cause not only a general decline in diversity, but also predictable functional shifts, as sets of species with particular traits are replaced by other sets with different traits (Loreau and al. 2001). Recent studies that were aimed at quantifying leaf economics across the plant species of the World lead to the formation of a global plant trait network.. They have revealed a universal spectrum of correlated leaf traits that affect global patterns of nutrient cycling and primary productivity and can thus be used to calibrate vegetation–climate models. The spectrum runs from a quick return to a slow return of nutrients and dry mass on leaves. The correlation patterns are seen in species from the arctic to the tropics, and they are largely independent of growth form or phylogeny. This generality suggests that unidentified fundamental constrains control the return of photosynthates on investments of nutrients and dry mass on leaves (Wright and al. 2004; Shipley and al. 2006a).

On the basis of their traits, plants can be grouped into functional types as non-phylogenetic groups of species that show close similarities in their responses to environmental and biotic constraints. They are considered as reflecting adaptations to variations in the physical environment, and trade-offs (ecophysiological and/or evolutionary) among different functions within a plant (Díaz and Cabido 1997; Duckworth and al. 2000; Lavorel and al. 2007). These can aid in our understanding of the ecological processes, such as the assembly and stability of communities, and the succession, and facilitate the detection and prediction of responses to environmental change that occur on a wide range of scales. The members of a plant functional type share similar morphological, physiological and/or life-history traits, while the differences between members within one functional type are smaller than those between functional types (Duckworth and al. 2000). Measuring functional diversity is therefore about measuring the diversity of functional traits, where these functional traits are the components of an organism phenotype that influence the landscape of an ecosystem (Figure 2) (Díaz and Cabido 1997; Díaz and al. 1998; Petchey and Gaston 2006). This approach of grouping species according to their function within an ecosystem has been used both to investigate fundamental ecological issues for the prediction of the functional consequences of biotic change caused by human or other types of disturbance (Walker and al. 1999, Díaz and Cabido 1997; Díaz and al. 1998; Regvar and al. 2006) and to investigate remediation (van der Putten and al. 2000).

Specific knowledge about how particular organisms interact with their environment, with each other, and how traits vary over environmental gradients, are essential to determine specific traits to use. This selection, however, also depends on the importance of a particular trait in the function of interest, thus giving it an informative value within the hypotheses that need to be tested. The critical points in the development of predictive measures of functional diversity are: (i) the choice of functional traits with which organisms are distinguished; (ii) how the diversity of that trait information is summarized to provide a measure of functional diversity; and (iii) the validation of these measures of functional diversity through quantitative analyses and experimental tests. One of the ways for the selection of traits is to select those that maximize the explanatory power of the functional diversity (Petchey and Gaston 2006).

In a synthesis of empirical and theoretical studies, it was therefore proposed that at least four axes of plant specialization should be considered in studies of functional trait diversity (Figure 2). These four axes are: (i) the specific leaf area (SLA = the ratio between leaf area and leaf biomass), as the leaf life-span trade-off is associated with the turnover time of plant parts; (ii) the trade-off between fecundity and seed mass, which addresses the opportunities for plant establishment and success in the face of hazards (seed mass and fecundity are negatively correlated, even after correcting for plant size); (iii) the potential plant height, that carries several trade-offs and adaptive elements, but that captures multiple constraints, such as the density and height of the shade-producing competitors, the water economy, and the response to disturbance; and (iv) the coupled variation between twig size and leaf size, which determines the texture of the canopy (Westoby 1998; Westoby and al. 2002; Lavorel and al. 2007). Recent studies that have focused on the functions of root traits have indicated how the trade-offs among them can be used as proxies for below-ground function, and how these trade-offs relate to above-ground traits. In general, low specific root length (SRL = the ratio between root length and root biomass) is associated with thick, dense roots with low nitrogen and high lignin concentrations (Lavorel and al. 2007; Biondini 2007) and should be therefore considered in functional classifications (Figure 2). In addition, the incorporation of mycorrhizal associations into current functional-type classifications (Figure 2) is a valuable tool for the assessment of plant-mediated control of carbon and nutrient cycling. Indeed, a study of the existence of possible links between plant carbon cycling traits among 83 British plant species of known mycorrhizal type revealed that plant species forming ericoid mycorrhiza have consistently low inherent relative growth rates, low foliar N and P, and poor litter decomposability. In contrast, plant species forming ectomycorrhiza had intermediate relative growth rates, higher foliar N and P, and intermediate to poor litter decomposability, while those forming arbuscular mycorrhiza have comparatively high relative growth rates, high foliar N and P, and rapid litter decomposition (Cornelissen and al. 2001).

4. Plant Traits at the Ecosystem Level: Community Aggregated Traits

Although different plant species often influence different functions, it seems unlikely that a single species or group of species can control the functioning of an ecosystem, thus leaving other species as redundant. Instead, ecosystem multifunctionality does depend on a greater number of species, and so studies that have focused on individual processes in isolation will have consequently underestimated the levels of biodiversity required to maintain multifunctional ecosystems (Hector and Bagchi 2007).

Several studies using randomly assembled communities have shown that the primary production has a positive relationship with plant species and functional-group diversity. Once plants are at a site, the biomass (abundance) of each will be proportional to the total amount of resource units that each species can capture at that point in time (Yachi and Loreau 1999; Loreau and al. 2001). Soil microbes are important regulators of plant productivity, especially in nutrient-poor ecosystems, where plant symbionts are essential for the acquisition of limiting nutrients. Mycorrhizal fungi and nitrogen-fixing bacteria provide an estimated 5% to 20% of the grassland and savannah nitrogen, up to 80% of the temperate and boreal forest nitrogen, and up to 75% of the phosphorus that is acquired by plants annually (van der Heijden and al. 2008). In estimating the relevance of those interactions for individuals and communities, and under fluctuating environmental conditions, it was seen that each mycorrhizal type is associated with an ecosystem and soil environment with distinctive characteristics, in which selection has favoured the development of particular attributes. It is well established that plant species with ericoid mycorrhizae predominate in the more humus soils of high latitudes and altitudes, those with ectomycorrhizal species predominate in forest ecosystems with surface litter accumulation at intermediate altitudes and latitudes, and those with arbuscular mycorrhizae dominate herbaceous and woody plant communities on mineral soils at lower latitudes (Read 1991). These differences arise largely due to the trade-off between costs and benefits of both of the partners, since plant species benefit most from their mycorrhizal associations when they are growing in habitats that lie outside of the tolerance limits of non-mycorrhizal plants (Fitter 1991; Smith and Read 1997; Keddy 2007). In addition, the recent functional metagenomic profiling of microbial processes of nine biomes has revealed that most of the functional microbial diversity is maintained in all of the communities, whereas the relative occurrence of the metabolisms varied, and the differences between metagenomes predicted the biogeochemical conditions of each environment (Dinsdale et al. 2008).

Comparative studies have begun to reveal the extent to which functional substitutions alter ecosystem properties, such as productivity, nutrient cycling, decomposition rates and resilience to perturbation (Loreau and al 2001). A multiple linear regression analysis was also applied in a comparative approach to predict plant competitive ability according to plant traits, in terms of the relative ability of each plant species to suppress the growth of a common indicator species (phytometer). Using data from 12 measured variables of 44 wetland species it was shown that plant biomass explained 63% of the variation in competitive ability, and plant height, canopy diameter, canopy area and leaf shape explained most of the residual variation (Gaudet and Keddy 1988). The plant dominance, which occurs through the ecological "selection" of species with particular traits, and the complementarity among species with different traits are two ways by which phenotypic diversity is mapped onto ecosystem processes. These two mechanisms, however, may also be viewed as two poles on a continuum, from pure dominance to pure complementarity. With our knowledge now, we cannot reject the hypothesis that a few dominant species suffice to provide the functional diversity that is necessary to explain the level of primary production seen in grassland ecosystems at the small spatial and temporal scales that have been considered in recent studies. However, species that are functionally redundant for an ecosystem process at a given time may no longer be redundant over a larger time scale. Therefore, in the long term, biodiversity provides an "insurance", or a buffer, against environmental fluctuations because different species respond differently to these fluctuations, thus leading to more predictable aggregate community or ecosystem properties (Yachi and Loreau 1999; Loreau and al. 2001).

Community stability can also be defined as the inverse of the temporal variability of an assemblage, such as that measured by the coefficient of variation in an aggregate biomass (Worm and Duffy 2003). A number of empirical and experimental studies have demonstrated decreased variability of ecosystem processes with increased diversity. As diversity increases, the variability in the populations of individual species can actually increase as a result of the destabilizing influence of strong species interactions within the ecosystem, while the variability of the aggregate ecosystem properties decreases because of the stabilizing influence of the responses of asynchronous species to intrinsic or extrinsic environmental fluctuations (Loreau and al. 2001). A mathematical model based on eight functional traits of 30 herbaceous species and the community-aggregated values of these traits in 12 sites through a 42-year chronosequence of secondary succession has been applied that predicted 94% of variance in the relative abundance. Thus it was shown that the relative abundance of an individual species within a species pool is a function of how closely the functional traits of that individual species agree with the community-aggregated traits (Figure 2) (Shipley and al 2006b).

In addition, a comparative approach involving 20 dicotyledonous herbaceous species was used to explore the possible relationships between several plant ecophysiological traits, plant litter decomposition and plant effects on soil properties. Decomposition rates of the plant stem and leaf litter were negatively related to plant mass, time to flowering, and vegetative growth rate, and they were positively related to the nitrogen content of the stem. Root decomposition was also related to several traits. Multiple regression analyses then showed that 74% and 84% of the variation across species for stem and litter decomposition, respectively, is predicted by the plant traits. The effects of a plant species on soil properties, including microbial biomass and activity, pH, nitrate concentration, and total nitrogen, were often closely related to the various plant traits. These results thus suggest that plant traits: (i) are powerful predictors of decomposition; (ii) have potential as alternative predictors of litter quality characteristics; and (iii) are important in determining the effects of plant species on soils (Wardle and al.1998).

5. Plant Traits and Disturbance

Disturbances can be defined as natural or land-use-related events that remove biomass or individuals. The environment can be considered to act as a filter in the prevention of the establishment of species lacking a specific combination of traits. The species that ultimately comprise the community are those that survive these filter effects (Keddy 1992; Duckworth and al. 2000; Lavorel and al. 2007). This arises simply as a result of the non-random demographic processes of dispersal, growth, survival, and reproduction of individuals within and between species, as determined by the functional traits they possess, where the interspecies variations of these functional traits are greater than their intraspecies variations. Species sorting by the environment is a process of selection over short (ecological) time scales, which results in changes in species composition over an environmental gradient (Keddy 1992; Díaz and al. 1998; Shipley and al 2006b; Lavorel and al. 2007).

The rules of community assembly provide a means to predict future states of communities. Functional classification often has two relatively distinct goals, one of which is to investigate the effects of species on ecosystem properties (functional effect groups), and another of which is to investigate the responses of species to changes in the environment, such as disturbance, resource availability and climate (functional response groups) (Figure 2). Most studies have focused on functional effect groups, rather than using groupings based on species responses. However, a merging of these two perspectives is needed to better understand the effects of biodiversity on the properties of an ecosystem (Keddy 1992; Hooper and al. 2002).

Early syntheses of changes in species traits along nutrient gradients recognized that species from nutrient-rich habitats tend to be inherently fast growing. Rapid resource capture and fast turn-over of organs leads to the poor internal conservation of resources, while the reverse is true for nutrient-poor habitats. Soil disturbance, therefore, favours plants with a suite of traits that goes beyond the ruderal syndrome, including a prostrate stature with either the stoloniferous architecture in perennial grasses or the flat rosettes in forbs, and a high fecundity and a small seed pool. Nutrient and/or water limitation tends to select for a conservative competitor strategy, with leaf traits promoting resource conservation, such as low specific leaf area, high tissue density, long life-span, and low nutritive value. As a consequence of these primary traits, the predominant plants in these environments will compete with their neighbours by sequestering nutrients, possibly with the help of mycorrhizal associations, and by accumulating poorly degradable litter, and sometimes through allelopathy or nutrient immobilization (Westoby 1998; Westoby and al. 2002; Lavorel and al. 2007).

Under intense anthropogenic pressure, patterns of biodiversity become increasingly variable. There are three main factors that limit species richness: harshness of the environment, competitive exclusion, and species-pool limitation. Species loss from ecosystems is usually caused by increasing effects of these three factors. The current experimental evidence indicates that the least productive species are those that have recently been excluded from temperate European grasslands. This also means a reduction in the available diaspore pool on a landscape scale, and could result in increased species-pool limitation in other communities (Lepš 2004). As productivity is often a non-linear, concave function of the number of species or functional groups, an increase in spatial variability of biodiversity can cause dramatic decreases in the mean productivity of the ecosystems. Thus the impact of the loss of biodiversity on productivity may be larger than current estimates indicate (Benedetti-Cecchi 2005). In addition, the strengths of interactions among plants are expected to co-vary with other key traits along environmental gradients (Lavorel and al. 2007), supporting an expected increase in the importance of mycorrhizal interactions as a result of the increased importance of the trade-offs between costs and benefits between both partners in extremely disturbed sites. Current results support the hypothesis of mycorrhizal associations as a plant strategy for exploitation of nutrient resources and resistance to stress, with its most intense development in periods of increased plant demands (Read 1991; Regvar and al. 2006; Vogel-Mikuš and Regvar 2006; Pongrac and al. 2007).

6. Pollution, Monitoring and Phytoremediation Techniques

Soil contamination occurs when either a solid or liquid substance mixes with the soil and becomes physically or chemically attached to soil particles or trapped in the spaces between them. This can result in an actual or calculated threat to human health or the environment. In sufficient quantities, chemicals become soil contaminants. Thus, they can harm humans, plants and animals, and they can leach contaminants into the ground water at unacceptable levels, and cause unacceptable degradation of the soil resources. In remediation practice, however, decision makers often face an intriguing question: "How clean is clean?". This frequently turns out to be difficult to answer (Belluck and al. 2006).

From a regulatory or clean-up perspective, a contamination is often defined as a concentration exceeding a particular value to which a level of risk has been assigned. Because risk varies with each metal and the associated exposure pathways, the definition of the degree of contamination is specific for each contaminant. Regulatory limits also differ depending on site-specific factors and specific land-use restrictions. In addition, the regulatory limits for metal concentrations in soil vary considerably according to State and even according to site (Raskin and Ensley 2000) and they frequently do not address categories that would be of a comparable level, although the limiting concentrations may be of the same order of magnitude. For example, there are criteria that have been defined for residential direct-contact soil clean-up according to the New Jersey Department of Environmental Protection and Risk. These are in the range of critical concentrations that are defined by the government of Slovenia as those at which the damaging effects on human health and the environment makes the polluted soil not suitable for the raising of plants intended for human or animal consumption (Table 1). These definitions have however largely differing consequences for phytoremediation management practices and local residents.

Once it has been determined that a soil is unacceptably contaminated, the necessary clean-up can be achieved through: (i) soil excavation, treatment and/or disposal; (ii) *in-situ* soil treatment; (iii) soil containment to prevent soil contaminant movement; or (iv) limiting exposure to the contaminated soil. In the past, soil remediation was primarily carried out by the physical removal of soils from contaminated sites, which went for landfilling, incineration or chemical stripping of the contaminants from soil that results in soils sufficiently clean to leave on site.

Table 1. Soil clean-up criteria established by the New Jersey Department of Environmental Protection, with risk concentrations for the metal limits for dangerous substances in the soil defined in R. Slovenia (NJDEP, 1996, from Raskin andEnsley 2000; OJ. RS, No. 68/96)

	NJDEP (1996) Clean-up criteria ($mg\ kg^{-1}$)		OJ RS No. 68/96 Risk concentrations ($mg\ kg^{-1}$)	
	Residential	Non-residential	[1]Warning	[2]Critical
Cd	1	100	2	12
Cu	600	600	100	300
Ni	250	2.400	70	210
Pb	400	600	100	530
Zn	1.500	1.500	300	720
Cr	n.d.	n.d	150	380
Hg	n.d.	n.d	2	10
Co	n.d	n.d	50	240
Mo	n.d	n.d	40	200
As	20	20	30	55

[1] The concentration that indicates the probability of damaging effects on human health or the environment for certain types of soil use.

[2] The concentration at which due to damaging effects on human health or the environment the polluted soil is not suitable for the raising of plants intended for human or animal consumption or for the retaining or filtering of water.

nd: no data.

However, due to significant damage having frequently been caused to a site, an evaluation of alternative methods that were less invasive but that provided similar clean-up results was required. With the development of phytoremediation technologies, site clean-up managers have new options that can allow for site clean-up without necessarily disrupting soil profiles and function (Belluck and al. 2006). Phytoremediation is defined as the use of vegetation for the *in-situ* treatment of contaminated soils, sediments and water. It is best applied to sites with a shallow contamination of metals, nutrients or organic pollutants that are amenable to one of its applications: rhizofiltration, bioremediation of the rhizosphere, phytotransformation, phytoextraction, and in extreme cases phytomining or phytostabilisation (Figure 3). In addition to providing very competitive remediation technologies when it comes to costs, these are aesthetically pleasing and have a high public acceptability (GWRTAC 1997; Salt al. and 1998; Raskin and Ensley 2000; Brooks 2000; Khan and al. 2004).

Rhizofiltration refers to the use of plant roots to absorb, concentrate and precipitate the metal contaminants in surface or groundwater (Table 2). The roots of these plants can absorb large quantities of lead and chromium from soil water or water that is passed through the root zone of the densely growing vegetation. It can also be applied to radionuclide contamination, nutrients and organic pollutants (GWRTAC 1997; Raskin and Ensley 2000; Brooks 2000).

Rhizosphere bioremediation increases the soil organic carbon, bacteria and mycorrhizal fungi, all of which are factors that encourage the degradation of organic chemicals in the soil (Table 2).

2. Phytoremedial technologies and their applications for remediation of different types of pollution (adopted from GWRTAC 1997; Salt al. and 1998; Raskin and Ensley 2000; Brooks 2000)

Phytotechnology	Rhizofiltration	Rhizosphere bioremediation	Phytotransformation	Phytoextraction	Phytomining	Phytostabilisation
Remediation of						
water	√	n.a.	√	n.a.		n.a.
soil	n.a.	√	√	√	√	√
sediment	√	√	√	√		√
Pollutants						
metals	√	√	√	√	√	√
organics	√	√	√			√
radionuclids	√			√		√
Plants						
herbs	√		√	√	√	√
grasses	√	√	√	√		√
woody plants	√	√	√			√
aquatic plants	√	√	√	n.a.		n.a.
hyperaccumulating	√		√	√	√	√

n.a.: not applicable.

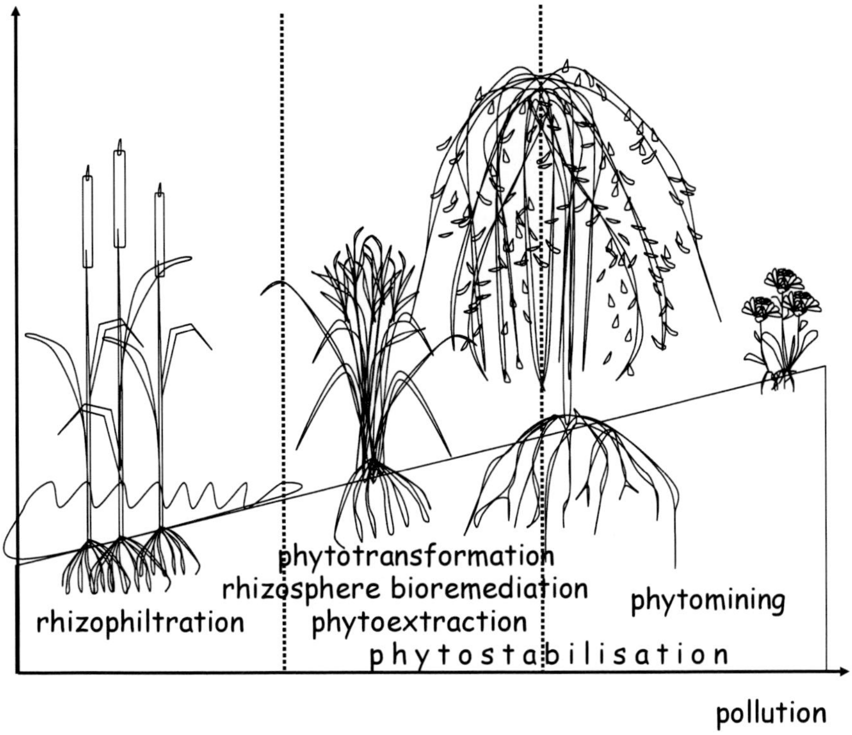

Figure 3. A schematic representation of phytoremedial technologies and their applications in relation to the levels of pollution.

The role of plants is either direct, with their exudates helping, stimulating or degrading the soil contaminants enzymatically, or indirectly through sustaining the growth of other soil microbes via the leaking of up to 20% of photosynthates and/or the aerating of the rhizosphere (GWRTAC 1997; Salt al. and 1998; Khan 2005).

Phytotransformation refers to the uptake of organic and nutrient contaminants from the soil and groundwater and their subsequent transformation by the plants (Table 2). For its environmental application, it is vitally important that the transformed metabolites that accumulate in the vegetation are non-toxic, or at least significantly less toxic than the parent compound. This also includes *phytovolatilization*, whereby volatile chemicals or their metabolic products are released into the atmosphere through plant transpiration (GWRTAC 1997; Salt al. and 1998; Brooks 2000).

Phytoextraction refers to the use of metal-accumulating plants that translocate and concentrate the metals from the soil into their roots and above-ground parts (Table 2). Chelate-assisted phytoextraction, in particular, is more developed than continuous phytoextraction, and it is being implemented commercially. This can be applied to metal and radionuclide contamination, including sites with mixed wastes. An important issue in phytoextraction is whether the metals can be economically recovered from the plant tissue, thus becoming the process of *phytomining* (Table 2), or whether disposal of the waste is required. As a general rule, the readily bioavailable metals for plant uptake include cadmium, nickel, zinc, arsenic, selenium and copper, with those moderately bioavailable being cobalt, manganese and iron; lead, chromium and uranium are not very bioavailable (GWRTAC 1997; Salt al. and 1998; Raskin and Ensley 2000; Brooks 2000; Ernst 2005).

Phytostabilization aims to immobilize the toxic contaminants in soils and sediments by vegetation, therefore preventing the spread of contamination by windblown dust, which represents an important pathway for human exposure to several hazardous contaminants (Table 2). This also contributes to the hydraulic control of water due to transpiration, thus preventing the leaching of contaminants into the groundwater. It can also be applied to low-level radionuclide contaminants, and especially where the half-lives are not too long. In the case of contamination with pollutants that are easily translocated to the shoots, like cadmium, care needs to be taken in the assessment of the risk to the food chain. Phytostabilization is especially applicable to sites where the contaminants are below the levels for regulatory action, and to severe metal contamination of sites, where the removal or treatment are not practical (GWRTAC 1997; Raskin and Ensley 2000; Brooks 2000) and where the potential of other phytoremediation technologies appears less promising.

The main constraints of phytoremediation include the difficulties involved with treating wastes greater than three metres in depth, the long periods of time required to remediate below action levels, the difficulties in establishing plant cover due to high toxicity at sites, and the possible migration of contaminants off-site due to their spread with the plant material or water flow. In addition, as in any other treatment systems, the maintenance including irrigation and agronomic inputs are required. The technology is currently being tested at several sites and it remains to be seen if it is effective at full scale and whether it will become a major new technology in the future (GWRTAC 1997; Raskin and Ensley 2000; Brooks 2000).

7. Plant Traits as Predictors of Performance in Phytoremediation

Stress is a major component of natural selection in soil ecosystems. Organisms can deal with different stress factors to different extents, which determines the limits of their ecological amplitudes (Roelofs et al. 2008). As the literature indicates, plants growing on metalliferous soils cannot prevent metal uptake, but only restrict it, and hence they accumulate metals in their tissues to varying degrees (Baker 1981). Functional genomic tools were used to reveal stress responses in different ecologically relevant soil organisms. Adaptation to stress factors appears to evolve through enhanced constitutive transcription of stress responsive genes in plants and animals (Roelofs et al. 2008). Three distinct categories of plant responses to metal(loid) levels have been defined: (i) metal hypotolerance, which is synonymous with hypersensitivity or sensitivity, and which describes mutants and transgenic plants that are more sensitive than their wild-type to one or several meta(loids) simultaneously; (ii) basal metal tolerance, which includes plant species or ecotypes that can regulate their distribution of metals at the cellular and whole-plant level in such a way as to survive and reproduce on non-metal-enriched soils or substrates; and (iii) metal hypertolerance, which is synonymous with metal resistant or metallicolous species or ecotypes, which can survive and reproduce on highly metal-enriched soils (Ernst and al. 2008). In reports of metal tolerance of plant species, populations and genotypes, the existence of such different degrees of metal tolerance is insufficiently recognized (Ernst 2006). Resistance to metals can be achieved by either of two strategies: avoidance, by which a plant is protected externally from the influences of the stress, and tolerance, by which a plant survives the effects of the internal stress. Hence, excluders, indicators and accumulators are frequently seen as plant survival strategies on metal-enriched soils (Baker 1987).

Tolerance is a phenomenon that involves the response of plants to external metals, and does not appear to be influenced by internal metals (Mcnair and al. 1999). In general, root growth responds more rapidly to metal exposure than shoot growth, and therefore the root tolerance index (TI = root growth in metal solution/ root growth in control solution) is frequently used as a measure of tolerance. However, short-term tests are insufficient for the evaluation of metal tolerances of a plant as a whole, because the survival of seedlings does not ensure the survival at later vegetative states and does not guarantee reproduction (Ernst and al. 1992; Ernst 2006). It has long been known that tolerance and hyperaccumulation are genetically independent characters (Macnair and al. 1999; Ernst 2006). Thus, metal tolerance as a plant trait should be considered before plant species can be used in any phytoremedial action.

We are frequently faced with multi-metal contaminations at specific sites, and therefore multi-metal tolerant plants are highly appreciated for phytoremedial efforts. Unfortunately, such plants are particularly rare. They are characterised by low growth rates and very restricted reproduction, as a consequence of the high metabolic costs of natural selection for the new beneficial alleles. They therefore produce less biomass compared to their non-metal-tolerant ancestors, which may also result from their adaptation to other environmental constraints, such as low nutrient and water supplies. An estimate of 20% integrated costs has been suggested relating to plant survival on metal-enriched soils that results in changes in the morphology and physiology among populations, independent of the genetics of metal tolerance (Brooks 2000; Raskin and Ensley 2000; Ernst 2006).

Metal accumulation of plants on metal-enriched soils combines two components: the intrinsic demand of the plant metabolism, and the impact of the external metal supply on metal uptake, with its consequences for storage in roots and for translocation from roots to shoots (Ernst 2006). Wide variations exist in the extent to which accumulated metals can be transported from the root system to the shoot. Therefore, transfer factors can be calculated (TF = shoot metal concentration/ root metal concentration), with a TF >1 being typical of an accumulator/ hyperaccumulator plant species; similarly, a TF <<1 is typical of an excluder species (Baker and al. 1994; Dahmani-Muller and al. 2000). Metal exclusion, however, cannot be regarded as an avoidance strategy at the whole plant level, since the uptake by the root has to meet the demand of a plant for its primary metabolic processes and its defence function (Ernst 2006). Efficient root-to-shoot translocation, on the other hand, is regarded as a key trait of hyperaccumulating plants as it is an important tolerance mechanism, which is potentially related to limited vacuolar sequestration of the roots (Xing and al. 2007).

Bioaccumulation factors (BAF = metal concentration in plant parts/ metal concentration in solution or soil) are used to allow for different metal concentrations in culture solutions/ soils (Baker and al. 1994). Some plants can accumulate exceptionally high concentrations of metals in their above-ground tissues. Thresholds for plant hyperaccumulation are set at 10,000 mg kg^{-1} for Zn and Mn, 1,000 mg kg^{-1} for Ni, Pb, Co, Cu and Se, and 100 mg kg^{-1} for Cd (Baker 1987; Brooks 2000; Raskin and Ensley 2000). Therefore, metal hyperaccumulating plants with particularly high BAFs are most appreciated for phytoextraction and phytomining (Brooks 2000; Raskin and Ensley 2000; McGrath and Zhao 2003; Audet and Charest 2007a). However, when testing the efficiency of phytoextraction, repeated studies are needed, along with a careful selection of sampling and metal analysis techniques, as the metals may be increasingly dissolved after their extraction due to equilibration processes (Keller and Hammer 2004). Some metal hyperaccumulating plant species can extract metals from the total metal pool, and therefore information on the bioavailable metal pools is not sufficient (Brooks 2000; Vogel-Mikuš and al. 2005). Root-associated bacteria and fungi, including the mycorrhizal fungi, have been shown to contribute to biological weathering, in addition to physical disintegration and chemical decomposition of rocks (Hofland and al. 2004; Calvaruso and al. 2006). According to the element defence hypothesis, high BAFs, and therefore high metal concentrations in plants, can protect plants from different herbivores, which can also be beneficial for remediation activities. The selection of such plants, however, should be carefully considered in the framework of the environmental land use and the possible consequences of the intake of metals into the food chain, since human intake of metals generally has largely detrimental effects (Poschenreider and al. 2006; Järup 2003).

As stress factors, metals cause a physiological strain on plants that is seen as reduced vigour, or in extreme cases, total inhibition of plant growth (Baker 1981). In an analysis designed to determine the relationships between Cd hyperaccumulation and plant vegetative and reproductive traits of natural *Thlaspi caerulescens* populations, it was shown that plants originating from populations with high Cd hyperaccumulation abilities had better growth, through developing more and bigger leaves, having taller stems, and producing more fruits and heavier seeds. The same study also demonstrated, that 75% of plant Cd concentrations can be accounted for by the concentrations in the soil, while all of the residual variance was covered by the concentrations of other metals in the plants. Hence, metal accumulation/ hyperaccumulation is also an important plant trait in metal-enriched soils that should be considered in phytoremedial actions (Basic and al. 2006).

8. The Value of Metallophytes in Phytoremediation: Lessons Learned from a Case Study

Successful ecosystem restoration requires a fundamental understanding of the ecological characteristics of the component species, together with the knowledge of how they assemble, interact and function as communities (Pywell and al. 2003). Succession is one of the most studied processes in ecology, and succession theory provides strong predictability. As such, it can be regarded as the sequential (or directional) change in species number and composition, and in the community structure, including the dominance relations and the community type on a site. Competition between individual plants can explain species replacement during succession, which involves shifts in the relative competitive abilities of different types of plants along the environmental gradient (van der Putten and al. 2000; Duckworth and al. 2000; Wiegleb and Felinks 2001).

The evolution of tolerance to one or more metals has enabled plant species to colonise natural metal-enriched soils. Anthropogenic metal emissions, on the other hand, have generally forced plants from non-metalliferous soils to evolve *de-novo* metal tolerances, which is predominantly realised in the grasses. As a result, the evolution of high metal tolerance is not homogenously distributed over taxonomic groups. Adaptation to high metal levels and hence the evolution of the necessary metabolism can be realised by enhancing the tolerance of the population to the higher metal levels, with a concomitant evolution of metal tolerance mechanisms at the cell, tissue, organ and whole plant levels; this aspect has been extensively reviewed elsewhere (Brooks 2000; Raskin and Ensley 2000; Ernst 2006; Vogel-Mikuš and Regvar 2006). As a consequence, in open sites of post-mining landscapes there are a great variety of vegetation types that have a highly individualistic natures and that are floristically complex and contain several special variants with a lot of exclusive species. They are therefore not easily compared with the vegetation types of undisturbed areas (Brooks 2000; Ernst 2006; Wiegleb and Felinks 2001). Community studies on contaminated soils have yielded compelling evidence of constant plant associations that develop on distinct types of metal-polluted soils; for example, on calamine soils that contain excess Zn, on serpentine soils that contain large amounts of Ni and Cr, and on the copper-belt soils of central Africa (Cu). These can thus also be used as indicators of an underlying mineralization base (Baker 1981, Baker 1987; Brooks 2000; Ernst 2006). They can provide a magnificent diversity of metallophyte floras, from which plant species suitable for remediation should be selected in relation to contemporary phytoremediation aims. It is within metallophyte floras where metal hyperaccumulating species can be found, and they are usually endemic to a given type of substrate (Brooks 2000). One of these community studies led to the discovery of the Zn/Cd hyperaccumulating *Thlaspi praecox* from a multimetal polluted site in Žerjav in northern Slovenia (southern Alps) that can accumulate up to 14,590 mg kg^{-1} Zn, 5,960 mg kg^{-1} Cd, and 3,500 mg kg^{-1} Pb (Vogel-Mikuš and al. 2005).

From an ecotechnological viewpoint, succession research can either be viewed as an accompanying element of re-vegetation studies for the re-establishment of productive ecosystems, or as an instrument to deliberately direct succession in a desired fashion. The extent to which succession can be altered has been questioned in several studies (Wilcox 1998; van der Putten and al. 2000; Wiegleb and Felinks 2001). Primary colonisation of large mining areas is strongly influenced by the vegetation of the surrounding neighbourhood. The main dispersal mechanism is wind blow, and the vitality and survival of the diasporas includes their germination rates and early mortality rates, which are of utmost importance (Wiegleb and Felinks 2001). Early colonising species are typically annual plants that are eventually replaced by later-colonizing perennial dominated communities. The superior long-term competitive ability of many perennial species arises from their well developed root systems and their ability to endure lower levels of resources than the annual species (Wilcox 1998). Sowing later-stage successional species can essentially change the initial stage of vegetation development on abandoned land. The diversity of plant species at a certain stage can affect the course of succession at least temporarily. However, local conditions and species identities affect the suppressive capacity of plant communities towards other species, which complicates the generalization of any relationship between the diversity of communities and their susceptibility to invasion (van der Putten and al. 2000).

In a multimetal polluted site in northern Slovenia, the vegetation and flora of an area affected by post-mining and smelting were screened to select the most suitable plant species for facilitated phytostabilization, as the other activities were not applicable due to the high pollution levels, the steep slopes, and the land use purposes in the vicinity of the non-residential industrial area. The consequences of the centuries-long activities resulted in the degradation of the local flora at the most affected sites, accompanied by severe erosion. The profound effects of the pollution can, however, be seen on a much larger scale, with serious consequences for local residents, and particularly for their children. Two predominant grass species (*Sesleria caerulea* and *Calamagrostis varia*) were selected for the first-stage phytostabilization of the eroded slopes, while willow species (*Salix caprea* and *Salix appendiculata*) were selected for the later-stage stabilisation activities, as the only woody species present at the most polluted and degraded locations. Unfortunately, the severely reduced germination capacity of the selected, apparently tolerant grass species prevented their use in practice, and therefore commercial grass mixtures were applied. This indicated the vegetative propagation of the selected grass species, rather than seed dispersal, as an important trait contributing to their dominance on this site. In contrast, significantly higher germination capacity of a co-dominant perennial hyperaccumulating *T. praecox* indicated metal tolerance mechanisms in this species that provided a decisive competitive advantage, favouring its generative reproduction. Altogether, these data supported the initial floristic composition model, in which the underlying dynamics of the succession are an expression of the life histories of the propagules within the soil (Regvar and al. 2006; Regvar and al. 2000; Vogel-Mikuš and al. 2005; Eržen and Janet 2005; Wilcox 1998).

Within the initially formed stands, biotic mechanisms, such as facilitation and inhibition, contribute in differential ways to the further vegetation development. The obviously very successful colonization of various grass species is enhanced by their property of forming facultative mycorrhizal associations (Wiegleb and Felinks 2001). In addition, metal hyperaccumulating plants belonging to the Brassicaceae family were shown to form arbuscular mycorrhiza, including *Biscutella laevigata* and the hyperaccumulating *T. praecox* (Orlowska and al. 2002; Regvar and al. 2003, Vogel-Mikuš and al. 2005; Regvar and al. 2006). A functional analysis of plant species of distinct mycorrhizal type was calculated according to the relative abundances of the plant species (Braun-Blanque) along the successional gradient of plots with differing metal concentrations in Žerjav and revealed the dominance of plant species with arbuscular mycorrhizae on all of the plots. This thus indicated the importance of the arbuscular mycorrhizae for the establishment of the early plant community structure in these metal-enriched soils. Non-mycorrhizal plant species were most numerous on plots where vegetation cover was not complete, whereas ectomycorrhizal plant species were most numerous in the less polluted locations, consistent with their shrubby and early-woodland successional stages (Regvar and al. 2006).

A meta-analytical literature survey of the possible roles of mycorrhizal colonisation of plants in metal-enriched soils has revealed an enhanced uptake of mycorrhizal plants compared to non-mycorrhizal plants at low metal concentrations, and a reduced metal bioavailability at high metal levels, presumably due to the adsorption of metals, and hence resulting in increased plant biomass and enhanced tolerance through metal stress avoidance (Audet and Charest 2007b). These findings are in line with the reduced metal uptake seen to be accompanied by an enhanced absorption of nutrients in mycorrhizal *T. praecox* plants under controlled conditions. On the basis of this and other studies, mycorrhizal colonisation was suggested as a survival strategy for plants in metal-enriched soils (Vogel-Mikuš and al. 2006; Vogel-Mikuš and Regvar 2006).

Some plant species are consistently more mycorrhizal than others. Despite frequently observed seasonal variations, a consistent hierarchy is maintained. Differences in mycorrhizal colonisation levels of *T. praecox* as greenhouse-grown and as field-collected from a metal polluted site were accompanied by significant changes in metal and nutrient compositions, thus satisfying the mineral needs of the plant. Changes in mycorrhizal colonisation seen along the metal pollution gradient, however, can as well be attributed to the differing plant developmental stages at the differing locations, rather than to metal pollution effects, since the myorrhizal frequencies seen in autumn samples were of comparable levels (Peat and Fitter 1993; Regvar and al. 2006; Vogel-Mikuš and al. 2006; Pongrac and al. 2007). Altogether, these results support the hypothesis that the development of mycorrhizal colonisation in plants on metal-enriched soils results from a trade-off between both partners, which is affected by both the nutrient demands of the plant and the soil pollution levels (Figure 4). Therefore, the remediation of the rhizosphere should accompany other phytoremediation technologies in the contemporary adaptive phytoremedial approaches used.

All of the above-mentioned data indicate profound consequences for phytoremedial activities. Metal-tolerant grass species with extensive root systems provide good soil cover and prevent wind and/or water erosion of the metals, and therefore serve phytostabilization strategies well. Immobilization of the metals within the rhizosphere is accomplished by the precipitation of the metals within the soil, their adsorption onto the root surface and/or their uptake and accumulation within the roots. Arbuscular mycorrhizal fungi also contribute to the immobilization of metals in the soil beyond the plant rhizosphere, and thereby improve phytostabilization (Göhre and Paszkowski 2006). Therefore, in phytostabilisation, the lower metal concentrations of the mycorrhizal plants can contribute to increased plant fitness and a decreased risk of metal transfer to the food chain.

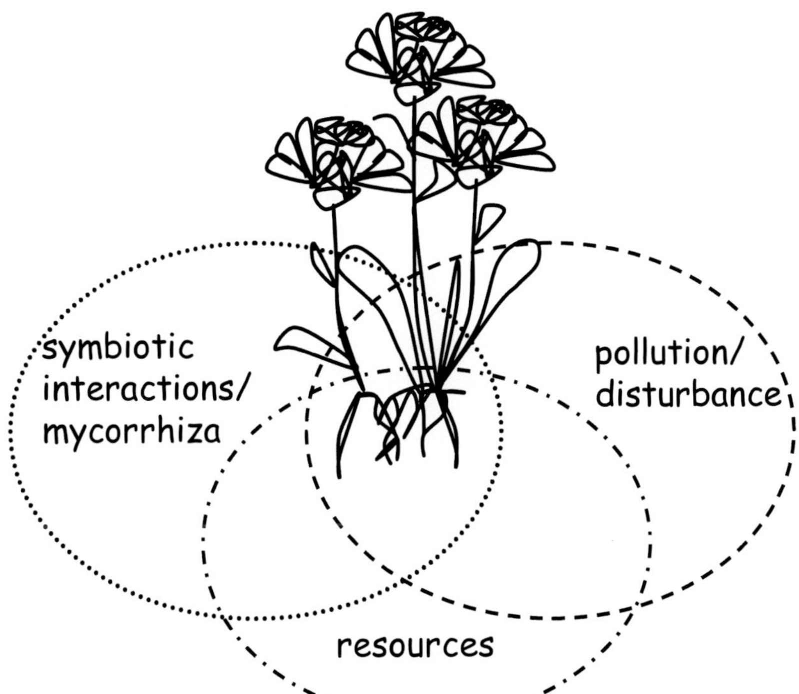

Figure 4. Schematic representation of the trade-offs between the costs and benefits of both partners in mycorrhizal symbiosis as affected by the acquisition of resources and stress tolerance under differing levels of pollution (disturbance), as determinants of the strength of interaction.

The consequences for phytoextraction, however, are less straightforward. Fungal isolates from metal-enriched soils usually have a higher metal tolerance than isolates from non-contaminated locations, whereas higher metal uptake is usually seen after inoculation with non-tolerant isolates, when compared to tolerant ones (Leyval and al, 1997; Gaur and Adholeya 2004). Thus, the prospect of having a combination of hyperaccumulating plant hosts with mycorrhizal isolates that are less efficient in restricting the metal uptake (Regvar and al. 2003; Göhre and Paszkowski 2006) is tempting. However, in the case of lower fungal tolerance levels, it appears predictable that a lot of testing would be needed before a successful plant–fungus combination for the range of metal levels at polluted sites can be found, if at all. After inoculation with the tolerant strains, on the other hand, lower metal accumulation can be expected, when compared to nontolerant ones, which does not appear very useful for phytoextraction at first glance, but provides a better chance of a successful combination, particularly for the more heavily polluted sites. In addition, at multimetal contaminated sites the plants frequently face a problem for their absorption of sufficient quantities of essential nutrients due to high concentrations of the "cheating" non-essential metals, as seen for *Thlaspi caerulescens* (Rosens and al. 2003) and *T. praecox*. This can be alleviated by mycorrhizal colonisation (Pongrac and al. 2007), which consequently results in greater plant fitness, which can frequently prove decisive for successful phytoremedial activities.

CONCLUSION

There is a compelling need to protect natural plant communities and to restore them in degraded areas. These activities must be guided by practical conservation tools, sound scientific principles, and clear priorities. With roughly one-third of the flora of the World facing extinction, scientists and conservation managers will need to work rapidly and collaboratively, and to recognize each other's strengths and limitations (Keddy 2005). To benefit from the synthesizing of experimental results across different continents and environments, the potential indicators of the ecology of different species need to be measured easily and consistently worldwide (Westoby and al. 2002). To have further progress in a predictive approach that will allow us to reveal the more general principles in studies of ecosystem functioning, four themes in functional trait research that need to be emphasized were suggested recently: (i) functional traits; (ii) environmental gradients; (iii) biotic interactions; and (iv) performance currencies (explaining how trait variations affect performance) (McGill and al 2006). It has become clear that no single functional classification is applicable to all situations, and that there is a multidimensional continuum of possible functional classifications, with the functional groups defined being dependent on the questions and scale addressed by any study (Pendry and al. 2007).

Thus, although a variety of remediation technologies are available for site remediation, no single technology is appropriate for all contaminant types and for the variety of site-specific conditions that exist at different contaminated sites. In addition, the contaminant source, the source control measures, and the potential impact of the potential remedial measures will determine the choice of remediation strategy and technology. Therefore, often more than one remediation technology is needed to effectively address most contaminated site problems (Khan and al. 2004). Also, as phytoremediation is too new a practice to be approved by regulatory agencies in *pro forma* reviews, design teams need to work with regulatory personnel from an early stage and with high frequency to obtain a satisfactory solution for all of the parties at the site. The main questions that need to be answered are: (i) Can phytoremediation remediate the site to a sufficient standard and reduce the risk to human health and the environment? (ii) What time-scale will this be achieved in? (iii) Will it create any toxic intermediates or products? (iv) Will it be as cost-effective as alternative methods? and (v) Will the public accept the technology? (GWRTAC 1997; Raskin and Ensley 2000).

Despite the abundance of theory and guidance, restoration goals are not always achieved, and the correct pathways towards targets are not highly predictable. This is understandable, since each restoration project has many constraints and unique challenges (Zedler 2005). To reach an informed remediation decision, risk managers must integrate both scientific and non-scientific findings, to generate final soil remediation values that credibly support both the numerical values of the soil remediation used for a site and the acceptability of residual chemicals and their risks that will remain with the contaminated soil site following its clean-up (Belluck and al. 2006). From the theoretical predictions and their correspondence with the experimental data, it can be concluded that in a future management context, the decision-making should be structured according to a solid analysis of the site (the possible developmental pathways) and in relation to the justified conservation aims (the desired developmental pathways). The natural dynamics relate to abiotic (soil dynamics, erosion, geomorphology), biotic (colonization, migration, succession) and combined (soil formation) processes (Wiegleb and Felinks 2001). To improve the restoration progress, the site restoration should be designed in distinct phases that will allow for learning and for adjustments to the decisions during the processes for the contemporary adaptive phytoremediation (Zedler 2005). As a concept, contemporary ecosystem management should be based on several principles, among which there are the scientific terms of ecosystem health, ecosystem integrity, biodiversity and sustainability that need to be precisely defined and clearly separated from other values and priorities (Lackey 1998). Following plant traits seem to be one of the options for a more standardized assessment of restoration success using species diversity components. Political and social factors, economics, an efficient and cost-effective clean-up, and stakeholder acceptability are among the many aspects that are also important in the decision making (Belluck and al. 2006). The good news is that the managers who are faced with protecting and/or restoring landscapes already have a wide variety of theoretical as well as practical applicative research studies in their hands, although unfortunately there appears to be no universal rules as to how the degraded landscapes can be restored. Therefore, an integrative approach of the parties with the accumulated knowledge from different fields and a variety of skills will be needed for contemporary adaptive phytoremediation activities.

REFERENCES

Audet, P. and Charest, C. (2007a). Heavy metal phytoremediation from a meta-analytical perspective. *Environ. Pollut.,. 147*, 231-237.

Audet, P. and Charest, C. (2007b). Dynamics of arbuscular mycorrhizal symbiosis in heavy metal phytoremediation: meta-analytical and conceptual perspectives. *Environ. Pollut., 147*, 609-614.

Baker, A.J.M. (1981). Acumulators and excluders, strategies in the response of plants to heavy metals. *J. Plant Nutr.*, *3*, 643-654.

Baker, A.J.M. (1987). Metal tolerance. *New Phytol.*, *106 (Suppl.)*, 93-111.

Baker, A.J.M., Reeves, R.D. and Hajar, A.S.M. (1994). Heavy metal accumulation and tolerance in British populations of the metallophyte *Thlaspi caerulescens* J. and C. Presl (Brassicaceae). *New Phytol,*. *127*, 61-68.

Basic, N., Keller, C., Fontanillas, P., Vittoz, P., Besnard, G. and Galland, N. (2006). Cadmium hyperaccumulation and reproductive traits in natural *Thlaspi caerulescens* populations. *Plant Biology*, *8*, 64-72.

Belluck, D.A., Benjamin, S.L. and David, S. (2006). Why remediate? In J-L. Morel, G. Echevarria, and N. Goncharova (Eds.), *Phytoremediation of Metal-Contaminated Soils: NATO Science Series IV: Earth and Environmental Sciences Vol. 68.* (pp.1-23).Dodrecht, NL: Springer.

Benedetti-Cecchi, L. (2005). Unanticipated impacts of spatial variance of biodiversity on plant productivity. *Ecology Letters*, *8*, 791-799.

Biondini, M. (2007). Plant diversity, production, stability, and susceptibility to invasion in restored northern tall grass priaries (United States). *Restoration Ecology*, *15*, 77-87.

Brooks, R.R. (2000). *Plants That Hyperaccumulate Heavy Metals.* Wallingford, UK: CAB International.

Calvaruso, C., Turpault, M.-P., and Frey-Klett, P. (2006). Root-associated bacteria contribute to mineral weathering and to mineral nutrition in trees: a budgeting analysis. *Appl. Environ. Microb.*, *72*, 1258-1266.

Cornelissen, J.H.C., Aerts, R., Cerabolini, B., Werger, M.J.A. and van der Heijden, M.G.A. (2001). Carbon cycling traits of plant species are linked with mycorrhizal strategy. *Oecologia*, *129*, 611-619.

Davidson, E.A. and Howarth, R.W. (2007). Nutritents in synergy. *Nature*, *449*, 1000-1001.

Dahmani-Muller, H., van Oort, F., Gélie, B. and Balabane, M. (2000). Strategies of heavy metal uptake by three plant species growing near a metal smelter. *Environ. Pollut.*, *109*, 231-238.

Díaz, S., Cabido, M. and Casanoves, F. (1998). Plant functional traits and environmental filters at a regional scale. *J. Veg. Sci.*, *9*, 113-122.

Díaz, S. and Cabido, M. (1997). Plant functional types and ecosystem function in relation to global change. *J. Veg. Sci.*, *8*, 463-474.

Dinsdale, E.A., Edwards, R.A., Hall, D., Angly, F., Breitbart, M., Brulc, J.M., Furlan, M., Desnues, C., Haynes, M., Li, L., McDaniel, L., Moran, M.A., Nelson, K.E., Nilsson, C., Olson, R., Paul, J., Brito, B.R., Ruan, Y., Swan, B.K., Stevens, R., Valentine, D.L., Thurber, R.V., Wegley, L., White, B.A. and Rohwer, F. (2008). Functional metagenomic profiling of nine biomes. *Nature*, *452*, 629-632.

Duckworth, J.C., Kent, M. and Ramsay, P.M. (2000). Plant functional types: an alterntative to taxonomic plant community description in biogeography? *Prog.Phys. Geog.*, *24*, 515-542.

Ernst, W.H.O. (2005). Phytoextraction of mine wastes – options and impossibilities. *Geochemistry*, *65*, 29-42.

Ernst, W.H.O. (2006) Evolution of metal tolerance in higher plants. *For. Snow Landsc. Res.* *80*, 251-274.

Ernst, W.H.O., Verkleij, J.A.C. and Schat, H. (1992). Metal tolerance in plants. *Acta Bot. Neerl.*, *41*, 229-248.

Ernst, W.H.O., Krauss, G.-J., Verkleij, J.A.C. and Wesenberg, D. (2008). Interaction of heavy metals with the sulphur metabolism in angiosperms from an ecological point of view. *Plant Cell Environ.*, *31*, 123-143.

Eržen, I. and Janet, E. (2005). Blood lead concentrations in three-year-old children living of the Upper and Lower Mežica valley. *Slovenian Journal of Public Health*, *44*, 18-25.

Fitter, A.H. (1991). Costs and benefits of mycorrhizas: implications for functioning under natural conditions. *Experientia*, *47*, 350-355.

Francis, R., and Read, D.J. (1994). The contributions of mycorrhizal fungi to the determination of plant community structure. *Plant Soil*, *159*, 11-25.

Gaudet, C.L. and Keddy, P.A. (1988). A comparative approach to predicting competitive ability from plant traits. *Nature 334*, 242-243.

Gaur, A. and Adholeya, A. (2004). Prospects of arbuscular mycorrhizal fungi in phytoremediation of heavy metal contaminated sites. *Current Science*, *86*, 528-534.

Gogala, N. (1991). Regulation of mycorrhizal infection by hormonal factors produced by hosts and fungi. *Experientia*, *47*, 331-340.

Göhre, V. and Paszkowski, U. (2006). Contribution of the arbuscular mycorrhizal symbiosis to heavy metal phytoremediation. *Planta*, *223*, 1115-1122.

GWRTAC (1997). *Phytoremediation* (Publication No. GWRTAC-TE-98-01). Pittsburgh, USA: Ground-Water Remediation Technologies Analysis Center, 425 Sixth Avenue, Regional Enterprise Tower.

Hector, A. and Bagchi, R. (2007). Biodiversity and ecosystem multifunctionality. *Nature*, *448*, 188-191.

van der Heijden, M.G.A., Klironomos, J.N., Ursic, M., Moutoglis, P., Steitwolf-Engel, R., Boller, T., Wiemken, A., and Sanders, I.R. (1998). Mycorrhizal fungal diversity determines plant biodiversity, ecosystem variability and productivity. *Nature*, *396*, 69-72.

van der Heijden, M.G.A., Bradgett, R.D., and van Strallen, N.M. (2008). The unseen majority: soil microbes as drivers of plant diversity and productivity in terrestrial ecosystems. *Ecology Letters*, *11*, 296-310.

Hofland, E., Kuyper, T.W., Wallander, H., Plassard, C., Gorbushina, A.A., Haselwandter, K., Holmström, S., Landeweert, R., Lundström, U.S., Rosling, A., Sen, R., Smits, M.M., van Hees, P.A.W., and van Breemen, N. (2004). The role of fungi in weathering. *Front. Ecol. Environ.*, *2*, 258-264.

Hooper, D.U., Solan, M., Symstad, A., Díaz, S., Gessner, M.O., Buchmann, N., Degrange, V., Grime, P., Hulot, F., Mermillod.Blondin, F., Roy, J., Spehn, E. and van Peer, L. (2002). Species diversity, functional diversity, and ecosystem functioning. In: M. Loreau, S. Naeem, P. Inchausti (Eds.) *Biodiversity and Ecosystem Functioning: Synthesis and Perspectives*. Oxford, UK: Oxford University Press.

Järup, L. (2003). Hazards of heavy metal contamination. *British Medical Bulletin*, *68*, 167-182.

Keddy, P.A. (1992). Assembly and response rules: two goals for predictive community ecology. *J. Veg. Sci.*, *3*, 157-164.

Keddy, P. (2005). Putting the plants back into plant ecology: six pragmatic models for understanding and conserving plant diversity. *Annals of Botany*, *96*, 177-189.

Keddy, P.A. (2007). *Plants and Vegetation*. Cambridge, UK: Cambridge University Press.

Keller, C. and Hammer, D. (2004). Metal availability and soil toxicity after repeated croppings of *Thlaspi caerulescens* in metal contaminated soils. *Environ. Pollut.*, *131*, 243-254.

Khan, A.G. (2005). Role of soil microbes in the rhizospheres of plants growing on trace metals contaminated soils in phytoremediation. *J. Trace Elem. Med. Bio.*, *18*, 355-364.

Khan, F.J., Husain, T. and Hejazi, R. (2004). An overview and analysis of site remediation technologies. *J. Environ. Manage*, *71*, 95-122.

Lavorel, S., Diaz, S., Cornelissen, J.H.C., Garnier, E., Harrison, S.P., McItyre, S., Pausas, J.G., Pérez-Harguindeguy, N., Roumet, C. and Urcelay, C. (2007). Plant functional types: are we getting any closer to the holy grail? In: J.G. Canadell, D. Pataki and L. Pitelka (Eds.) *Terrestrial Ecosystams in a Changing World* (The IGBP Series, pp149-164). Berlin, Germany: Springer-Verlag.

Lackey, R.S. (1998). Seven pillars of ecosystem management. *Landscape and Urban Planning,40(1-3)*, 21-30.

Lepš, J. (2004). What do biodiversity experiments tell us about consequences of plant species loss in the real world? *Basic and Applied Ecology*, *5*, 529-534.

Leyval, C., Turnau, K., and Haselvandter, K. (1997). Effect of heavy metal pollution on mycorrhizal colonization and function, physiological, ecological and applied aspects. *Mycorrhiza*, *7*, 139-153.

Loreau, M. and Hector, A. (2001). Partitioning selection and complementarity in biodiversity experiments. *Nature*, *412*, 72-76.

Loreau, M., Naeem, S., Inchausti, P., Bengtsson, J., Grime, J.P., Hector, A., Hooper, D.U., Huston, M.A., Raffaelli, D., Schmid, B., Tilman, D. and Wardle, D.A. (2001). Biodiversity and ecosystem functioning: current knowledge and future challenges. *Ecology*, *294*, 804-808.

van der Maarel, E. (2005). *Vegetation Ecology*. Oxford, UK: Blackwell Publishing.

Macnair, M.R., Bert, V., Huitson, S., Samitou-Laprade, P. and Petit, D. (1999). Zinc tolerance and hyperaccumulation are genetically independent characters. *Proc. R. Soc. Lond. B*, *266*, 2175-21¸79.

McGill, B.J., Enquist, B.J., Weiher, E. and Westoby, M. (2006). Rebuilding community ecology from functional traits. *Trends Ecol. Evol.*, *21*, 178-185.

McGrath, S.P. and Zhao, F.-J. (2003). Phytoextraction of metals and metalloids from contaminated soils. *Current Opinion in Biotechnology*, *14*, 277-282.

Mooers, A.Ø. (2007). The diversity of biodiversity. *Nature*, *445*, 717-718.

Munkvold, L., Kjøller, R., Vestberg, M., Rosendahl, S. and Jakobsen, I. (2004). High functional diversity within species of arbuscular mycorrhizal fungi. *New Phytol.*, *164*, 357-364.

OJ. RS, No. 68/96. Decree on the Limit, Warning and Critical Concentration Values of Dangerous Substances in Soil.

Orlowska, E., Zubek, Sz., Jurkiewitcz, A., Szarek-Lukaszeewska, G., Turnau, K. (2002). Influence of restoration on arbuscular mycorrhiza of *Biscutella laevigata* L. (Brassicaceae) and *Plantago lanceolata*, (Plantaginaceae) from calamine spoil mounds. *Mycorrhiza*, *12*, 1-17.

Peat, H.J. and Fitter, A.H. (1993). The distribution of arbuscular mycorrhiza in the British flora. *New Phytol. 125*, 845-854.

Pendry, C.A., Dick, J., Pullan, M.R., Knees, S.G., Miller, A.G., Neale, S. and Watson, M.F. (2007). In search of a functional flora-towards a greater integration of ecology and taxonomy. *Plant Ecol.*, *192*,161-167.

Petchey, O.L. and Gaston, K.J. (2006). Functional diversity: back to basics and looking forward. *Ecology Letters*, *9*, 741-758.

Pongrac, P., Vogel-Mikuš, K., Kump, P., Nečemer, M., Tolra, R., Poschenrieder, C., Barceló, J. and Regvar, M. (2007). Changes in elemental uptake and arbuscular mycorrhizal colonisation during life cycle of *Thlaspi praecox*. *Chemosphere*, *69*, 1602-1609.

Poschenreider, C., Tolrà, R. and Barcheó, J. (2006). Can metals defend plants against biotic stess?. *Trends in Plant Science*, *11*, 288-295.

Prach, K. and Pyšek, P. (2001). Using spontaneous sucession for restoration of human-disturbed habitats: Experience from Central Europe. *Ecol. Eng.*, *17*, 55-62.

Pywell, R.F., Bullock, J.M., Roy, D.B., Warman, L., Walker, K.J., and Rothery, P. (2003) Plant traits as predictors of performance in ecological restoration. *J. App. Ecol.*, *40*, 65-77.

van der Putten, W.H., Mortimer, S.R., Hedlund, K., van Dijk, C., Brown, V.K., Lepš, J., Rodrigues-Barrueco, C., Roy, J., Diaz Len, T.A., Gormsen, D., Korthals, G.W., Lavorel, S., Santa Regina, I. and Smilauer, P. (2000). Plant species diversity as a driver of early succession in abandoned fields: a multi-site approach. *Oecologia*, *124*, 91-99.

Raskin, I. and Ensley, B.D. (2000). *Phytoremediation of toxic metals*. NY, USA. John Whiley and Sons, Inc..

Read, D.J. (1991). Mycorrhizas in ecosystems. *Experientia*, *47*, 376-391.

Regvar, M. and Vogel-Mikuš, K. (2008). Recent advances in understanding of plant responses to excess metals: exposure, accumulation and tolerance. In N.A. Khan, S. Singh. and S. Umar (Eds.) *Sulfur Assimilation and Abiotic Stress in Plants* (pp. 227-251). Berlin, D: Springer-Verlag.

Regvar, M., Vogel, K., Kugonič, N., Vodnik, D., Jug, M. & Kret, M. (2000). Toward a comprehensive model for revegetation of the heavy metal polluted site in Žerjav. In: L. Santaloce and A. Massacci (Eds.) *Intercost workshop on bioremediation: [Action 831 - Action 837 - Action 838]*. Roma, IT: Istituto di Biochimica ed Ecofisiologia Vegetali, 140-142.

Regvar, M., Vogel, K., Irgel, N., Wraber, T., Hildebrandt, U., Wilde, P. and Bothe, H. (2003). Colonization of pennycresses (*Thlaspi* spp.) of the Brassicaceae by arbuscular mycorrhizal fungi. *Journal of Plant Physiology*, *160*, 615-626.

Regvar, M., Vogel-Mikuš, K., Kugonič, N., Turk, B. and Batič, F. (2006). Vegetational and mycorrhizal successions at a metal polluted site: Indications for the direction of phytostabilisation? *Environ. Pollut.*, *144*, 976-984.

Roelofs, D., Aarts M.G.M., Schat H. and van Straalen, N.M. (2008). Functional ecological genomics to demonstrate general and specific responses to abiotic stress. *Funct. Ecol.*, *22*, 8-18.

Sala, O.E. (2001). Price put on biodiversity. *Nature*, *412*, 34-36.

Rosens, N., Verbruggen, N., Meerts, P., Ximénez-Embún and Smith, J.A.C. (2003). Natural variation in cadmium tolerance and its relationship to metal hyperaccumulation for seven populations of *Thlaspi caerulescens* from Western Europe. *Plant Cell Environ.*, *26*, 1657-1672.

Salt, D.E., Smith, R.D. and Raskin, I. (1998). Phytoremediation. *Ann. Rew. Plant Physiol. Plant Mol. Biol.,49*, 643-668.

Shipley, B., Lechowicz, M.J., Wright, I., Reich, P.B. (2006a). Fundamental trade-offs generating the worldwide leaf economic spectrum. *Ecology*, *87*, 535-541.

Shipley, B., Vile, D. and Garnier, E. (2006b). From plant traits to plant communities: a statistical mechanicistic approach to biodiversity. *Science*, *314*, 812-814.

Smith, S.E. and Read, D.J. (1997). *Mycorrhizal Symbiosis*. London, UK: Academic Press.

Tilman, D., Lehman, C., Thomson, K.T. (1997). Plant diversity and ecosystem productivity: Theoretical considerations. *Proc. Natl. Acad. Sci. USA*, *94*, 1587-1861.

Tilman, D., Reich, P.B. and Knops, J.M.H. (2006). Biodiversity and ecosystem stability in a decade long grassland experiment. *Nature*, 441, 629-632.

Vogel-Mikuš, K. and Regvar, M. (2006). Arbuscular mycorrhiza as a tolerance strategy in metal contaminated soils: prospects in phytoremediation. In D. Rhodes (Ed.), *NewTtopics in Environmental Research* (pp. 37-56). NY, USA: Nova Science Publishers, Inc..

Vogel-Mikuš, K., Drobne, D. and Regvar, M. (2005). Zn, Cd and Pb accumulation and arbuscular mycorrhizal colonization of pennycress *Thlaspi praecox* Wulf. (Brassicaceae) from the vicinity of a lead mine and smelter in Slovenia. *Environ. Pollut.*, *133*, 233-242.

Vogel-Mikuš, K., Pongrac, P., Kump, P., Nečemer, M. and Regvar, M. (2006). Colonisation of a Zn, Cd and Pb hyperaccumulator *Thlaspi praecox* Wulfen with indigenous arbuscular mycorrhizal fangal mixture induces changes in heavy metal and nutrient uptake. *Environment. Pollut.,139*, 362-371.

Vandenkoornhuyse, P., Baldauf, S.L., Leyval, C., Straczek, J., Young, J.P.W. (2002). Extensive fungal diversity in plant roots. *Science*, *295*, 2051.

Walker, B., Kinzig, A. and Langridge, J. (1999). Plant attribute diversity, resilience, and ecosystem function: the nature and significance of dominant and minor species. *Ecosystems*, *2*, 95-113.

Walker, B.H. and Landgridge, J.L. (2002). Measuring functional diversity in plant communities with mixed life forms: a problem of hard and soft attributes. *Ecosystems*, *5*, 529-538.

Wardle, D.A., Barker, G.M., Bonner, K.I., and Nicholson, K.S. (1998). Can comparative approaches based on plant ecophysiological traits predict the nature of biotic interactions and individual plant species effects in ecosystems? *J. Ecol.*, *86*, 405-420.

Westoby, M. (1998). A leaf-hight-seed (LHS) plant ecology strategy scheme. *Plant Soil, 199*, 213-227.

Westoby, M., Falster, D.S., Moles, A.T., Vesk, P.A. and Wright, I.J. (2002). Plant ecological strategies: some leading dimensions of variation between species. *Annu. Rev. Ecol. Syst. 33*, 125-159.

Wiegleb, G. and Felinks, B. (2001). Primary succession in post-mining landscapes of Lower Lusatia – chance or necessity. *Ecol. Eng.*, *17*, 199-217.

Wilcox, A. (1998). Early plant succession on former arable land. *Agriculture, Ecosystems and Environment*, *69*, 143-157.

Worm, B. and Duffy J.E. (2003). Biodiversity, productivity and stability in real food webs. *Trends Ecol. Evol.*, *18*, 628-632.

Wright, I.J., Reich, P.B., Westoby, M., Ackerly, D.D., Baruch, Z., Bongers, F., Cavender-Bares, J., Chapin, T., Cornelissen, J.H.C., Diemer, M., Flexas, J., Garnier, E., Groom, P.K., Gulias, J., Hikosaka, K., Lamont, B.B., Lee, T., Lee, W., Lusk, C., Midgley, J.J.,

Navas, M.-L., Niinemets, Ü., Oleksyn, J., Osada, N., Poorter, H., Poot, P., Prior, L., Pyankov, V.I., Roumet, C., Thomas, S.C., Tjolker, M.G., Veneklaas, E.J., and Villar, R. (2004). The worldwide leaf economics spectrum. *Nature*, *428*, 821-827.

Xing, J.P., Jiang, R.F., Ueno, D., Ma, J.G., Schat, H., McGrath, S.P. and Zhao, F.J. (2007). Variation in root-to-shoot translocation of cadmium and zinc among different accessions of the hyperacumulators *Thlaspi caerulescens* and *Thlaspi praecox*. *New Phytol.*, *178*, 315-325.

Yachi, S. and Loreau, M. (1999). Biodiversity and ecosystem productivity in a fluctuating environment: the insurance hypothesis. *Proc. Natl. Acad. Sci. USA*, *96*, 1463-1468.

Zedler, J.B. (2005). Ecological restoration: guidance from theory. *San Francisco Estuary and Watershed Sciece, 3, (Iss.2, Art. 4)*, 1-31.

In: Grasslands: Ecology, Management and Restoration
Editor: Hans G. Schröder
ISBN 978-1-60692-023-7

Chapter 7

VEGETATION AND FIRE DYNAMICS IN SOUTHERN BRAZIL DURING THE LATE QUATERNARY AND THEIR IMPLICATION FOR CONSERVATION AND MANAGEMENT OF MODERN GRASSLAND ECOSYSTEMS

Hermann Behling[*1] ***and Valério D. Pillar***[2]

[1]Albrecht-von-Haller-Institut für Pflanzenwissenschaften, Georg-August-Universität Göttingen, Abteilung für Palynologie und Klimadynamik, Untere Karspüle 2, 37073 Göttingen, Germany

[2]Departamento de Ecologia, Universidade Federal do Rio Grande do Sul, Porto Alegre, Brazil

ABSTRACT

In different regions of southern Brazil, often a mosaic of grassland and forest is found, but little is known how these vegetation pattern has been formed. Palaeoecological background information is needed for management and conservation of the highly diverse mosaic of forest and grassland in southern Brazil. Questions on the origin of grasslands, its development, dynamic and stability, its response to environmental change such as climate, and the role of human impact, are essential. Further questions on its natural stage of vegetation or its alteration by pre- and post-Columbian anthropogenic activity are also important. To answer these questions, palaeoecological and palaeoenvironmental data based on pollen and charcoal analysis of radiocarbon dated sedimentary archives from southern Brazil are used to provide an insight into past vegetation changes that allow us to improve our understanding of the modern vegetation and to develop conservation and management strategies for the strongly affected grassland ecosystems in southern Brazil.

[*] E-mail: Hermann.Behling@bio.uni-goettingen.de

INTRODUCTION

Subtropical grassland ecosystems in southern Brazil present very high levels of biodiversity and have been the predominant land cover type in parts of the region. Despite the strong alteration of the landscapes by deforestation, agriculture and silviculture, in some less disturbed regions a mosaic of Campos (grassland) and forests is found, which have still a somewhat natural aspect. The existence of these mosaics in southern Brazil has puzzled naturalists and ecologists for a long time, because the modern wet climatic conditions would implicate a landscape covered by forest. Early researchers such as Lindman, traveling across the region in late 19th century, observed that forests should be able to expand over these grasslands, and attributed their presence to the transitional situation between tropical forests to the north and temperate grasslands to the south (Lindman 1906). Based primarily on plant geographical evidence, Rambo (1956a, 1956b) and Klein (1975) put forward that the grasslands were the older vegetation type with forest expansion, being a more recent process after changes to more humid climate conditions. Also Hueck (1966) questioned how the southern Brazilian grassland could exist under present humid forest climatic conditions.

Understanding the origin of the grasslands is of prime importance to their conservation and management. In case the mosaics of grasslands and forest are a consequence of human induced deforestation, vegetation management would likely be focused on reestablishment of complete forest vegetation. If grasslands are original and have prevailed in the past due to different climatic regimes, it could be reasonable to conserve these 'natural' relicts and their high biodiversity.

Sustainable management and conservation of modern mosaics of Campos and forest could be assisted by the knowledge of pre-historical and historical reference conditions for past human impact, fire magnitude and frequency of burning, and the applied use of this information in developing long-term management and monitoring plans.

Results from several palaeoecological investigations based on fossil pollen and microfossil charcoal of sediment cores from peat bogs provide important background information on past vegetation and fire dynamics (Figure 1). Most of the sites were sampled on the southern Brazilian highlands were a mosaic of grassland and *Araucaria* forest is found (e.g., Behling et al. 2004, Behling and Pillar 2007) and a few on the lowland campos region (Behling et al. 2005, 2007). Key questions to be addressed include: How was the formation of the diverse mosaic of forest and grassland vegetation in southern Brazil? Was its development caused by pre-Columbian slash and burn activity and deforestation of the post-Columbian settlers, or rather a natural process of climate driven forest expansion constrained by grassland fires? Are fires natural or of anthropogenic origin? What is the effect of climate change, fire and human impact on the overall biodiversity of the region through the time? How should we management and conserve the modern species rich forest and grassland ecosystems?

With examples from the southern Brazil, we want to demonstrate how essential it is to have knowledge on past vegetation and environment dynamics to understand the modern grassland ecosystem in a holistic way and that this background information has to be considered in developing conservation, restoration and management strategies for grassland ecosystems.

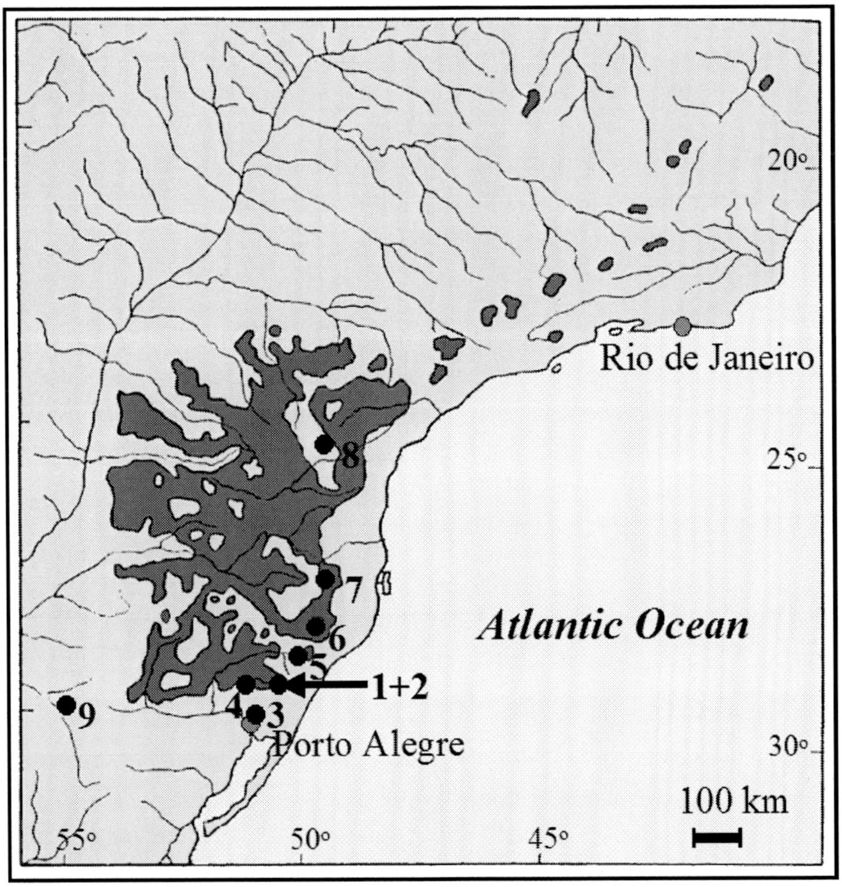

Figure 1. Map showing the distribution of Araucaria forest and Campos in southern Brazil (adapted after Hueck 1953, 1966) and the location of the sites mentioned in the text: (1) Cambará do Sul, (2) Aparados da Serra, (3) Morro Santana, (4) São Francisco de Paula (5) Serra do Rio Rastro, (6) Morro da Igreja, (7) Serra da Boa Vista, (8) Serra Campos Gerais and (9) São Francisco de Assis.

Modern Vegetation and Climate in Southern Brazil

Southern Brazil is a transitional zone between tropical and temperate climates and the natural vegetation is a mosaic of grassland, shrubland and different forest types (Hueck 1953, 1966, Klein 1978, 1979, 1984, Rambo 1953, 1956b, Por 1992, Boldrini 1997, Pillar and Quadros 1997, Overbeck at al. 2007). As regards grassland vegetation, on the highland plateau in Paraná, Santa Catarina and Rio Grande do Sul states, grasslands form mosaics with *Araucaria* forest (Figs. 1, 2). The grasslands of the southern and western parts of Rio Grande do Sul State are often included in the same type of grasslands that extend into Argentina and Uruguay (Burkart, 1975; Soriano et al., 1992).

Boldrini (1997) estimated a total number of 3,000 grassland plant species in Rio Grande do Sul State alone. The grassland vegetation is often regionally differentiated into "campo limpo" ("clean grassland"), which contains a predominance of herbs of the Poaceae, Cyperaceae and Asteraceae family, and "campo sujo" ("dirty grassland"), which in addition to grasses and sedges contains shrubs, primary of the families Asteraceae (*Baccharis* spp.) and

Apiaceae (e.g. *Eryngium* spp.) (Klein, 1978). Boldrini (1997) describes physiognomic regions of *Campos* vegetation in Rio Grande do Sul State, considering local floristic variations associated with climate, topographic variation and soil heterogeneity. However, a good portion of the variation in grassland physiognomy and in the composition of the dominant species seems to be determined by grazing and fire regimes (Pillar and Quadros, 1997). Under low grazing intensity tall grasses dominate (*Andropogon* spp., *Aristida* spp., *Schizachyrium* spp.) mixed with shrubs such as *Baccharis* spp., *Vernonia* spp. (Asteraceae) and *Eryngium horridum* (Apiaceae).

Campos de altitude (high elevation grasslands) occur on mountains, at elevations above 1600 m in southern Brazil and above 1800 m in southeastern Brazil. This grassland contains species of the Poaceae and Cyperaceae families, but also includes small shrubs of Melastomataceae, Ericaceae, Eriocaulaceae, Asteraceae and Verbenaceae (Safford 1999a, 1999b). Campos de altitude is rich in endemic species and occurs on several mountain peaks in S and SE Brazil (Ferrão and Soares 1989).

The present-day grassland vegetation is highly affected by different agricultural land-use practices and more recently by large-scale afforestation by exotic species, (mainly *Pinus* and *Eucalyptus*). Until 1996, an estimated decrease of 25% in the total area of natural grasslands has occurred due to a strong expansion of agricultural activities (Nabinger et al. 2000). More recent mapping of natural vegetation remnants in the south and west of Rio Grande do Sul State alone (MMA 2007) indicates ca. 50% loss of natural grassland habitats by the year 2002.

The climate of the southern highland region is defined as subtropical warm temperate and humid without marked dry periods (Nimer 1989). The southern highlands form an orographic barrier for southeasterly to northeasterly winds. As a consequence, the rainfall is high with an average annual rainfall of over 2000 mm. The mean annual temperature is 14.5 °C. The lowest recorded temperature for the eastern highland is -6.5 °C (Nimer 1989). In general, the climate of southern Brazil is influenced by the South Atlantic anticyclone, a semi-permanent high-pressure system which transports moist tropical air masses over the continent from easterly and northeasterly directions during the whole year. Disturbances are related to polar cold fronts, which when meeting the tropical air masses, produce strong rainfall in southern Brazil (Nimer 1989, Hastenrath 1991). Higher rainfall in southern Brazil is also related to El Niño events (Martin et al. 1993, Ratisbona 1976).

Late Quaternary Vegetation, Climate and Fire Dynamics in Southern Brazil

Several palaeoecological studies from the *Araucaria* forest and Campos regions of the southern Brazilian highlands have been carried out in the last decades (Behling 2002) (Figure 2). Data from the states of Paraná (Serra Campos Gerais: Behling 1997), Santa Catarina (Serra do Rio Rastro, Morro da Igreja, Serra da Boa Vista: Behling 1995) and Rio Grande do Sul (Aparados da Serra: Roth and Lorscheitter 1993; São Francisco de Paula: Behling et al. 2001; Cambará do Sul: Behling et al. 2004) have proved that extensive areas of Campos vegetation existed on the highlands through glacial, early and mid Holocene times. The dominance of Campos vegetation was attributed to cold and dry glacial, and warm and dry early Holocene climates.

Figure 2. Photo showing a mosaic of *Araucaria* forest and grassland on the southern Brazilian highland (Photo Behling).

A dry season lasting probably about three months per year was characteristic for the early and mid Holocene period (Behling 1997, 2002). Initial expansion of *Araucaria* forests started by migration from the gallery forests along the rivers about 3210 calibrated radiocarbon years before present (cal yr BP), which indicates a turn to somewhat wetter climates. A marked expansion of *Araucaria* forests started on the highlands, replacing Campos vegetation in Santa Catarina State about 930 cal yr BP ago, and in Paraná State (Serra Campos Gerais) about 1400 cal yr BP ago, reflecting a very humid climate without a marked seasonal dry period.

Pollen data are also available from the lowland Campos (grassland) region near the town of São Francisco de Assis in the western Rio Grande do Sul State (Behling et al. 2005). The region was naturally covered by Campos throughout the recorded glacial and Holocene period under cold and relatively dry and warm and dry condition, respectively. Initial expansion of gallery forest after 5170 cal yr BP indicates a change to wetter climatic conditions. Maximum extent of gallery forest after 1550 cal yr BP reflects the wettest recorded period.

Detailed high resolution data are available from a peat core from a shallow basin about 7 km distance to the village Cambará do Sul (Figure 3) spans the last 42,840 uncalibrated radiocarbon years before present (uncal yr BP). A more complete palaeoenvironmental data set and reconstruction, including records of selected single pollen and spore taxa have been published in Behling et al. (2004). The pollen record from Cambará do Sul also documents the dominance of a diverse Campos vegetation between 42,840 uncal and 11,500 cal yr BP. The highland region of northwestern Rio Grande de Sul State was probably almost treeless.

The evidence of a few single pollen grains representing *Araucaria* forest and Atlantic rain forest vegetation in the late Pleistocene sediments were most likely wind transported over some distance from possible forest refugia in deep and protected valleys in the highlands or from the 6-7 km distant slopes of the Serra Geral mountains or from the lowland vegetation of the coastal region, respectively.

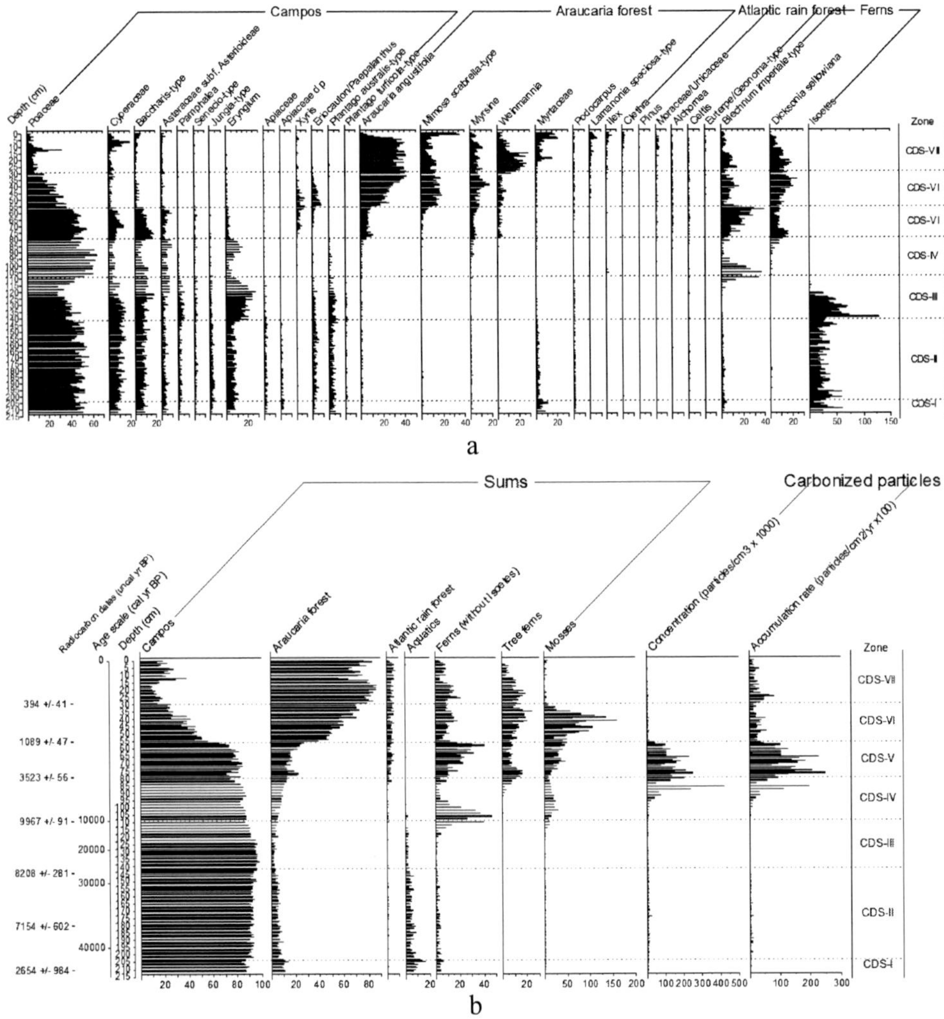

Figure 3. Percentage (a) and summary (b) pollen percentage diagram for Cambará do Sul, including the radiocarbon dates, age scale (cal yr BP), records of different taxa and ecological groups (Campos, *Araucaria* forest, Atlantic rainforest, Ferns), concentration and influx of carbonised particles, and pollen zones. For more details see Behling et al. 2004).

The widespread treeless Campos vegetation, indicate cold and dry climate conditions. Repeated frosts events and minimum austral winter temperatures below -10 °C, which do not permit the growth of *Araucaria* on the highlands is suggested. Mean annual temperature depression was probably similar to the southeastern Brazilian highland which was in the range of 5-7 °C between about 26,000 and 17,000 uncal years ago during the Last Glacial Maximum (LGM) period (Behling and Lichte 1997). During the LGM, the driest period recorded, pollen of *Eryngium* was abundant indicating drier conditions, and the shallow lake at that time was not permanent, indicating a seasonally dry climate. Based on the Cambará do Sul record, it is suggested that seasonal climatic conditions developed after 26,900 uncal years ago. Seasonal climate with a long annual dry period since the LGM period prevailed until the late Holocene.

Campos vegetation still dominated the landscape during early and mid Holocene (11,500 to 4320 cal yr BP). Species of the *Araucaria* forest increased slightly but were still rare, indicating that populations migrated into the study region, probably along small streams. Atlantic rain forest taxa became more common, suggesting an expansion on the Serra Geral coastal slopes closer to the study site. Extensive grasslands coupled with the rare occurrence of *Araucaria* forest taxa, suggest a dry climate. Changes in the composition of the Campos vegetation, reflect a change to a warm and dry climate. The annual precipitation must have been lower than 1400 mm and the climate was seasonal, with a dry season lasting about 3 months. Apparently the climatic conditions did not favor expansion of *Araucaria* forest in the study area during the early and mid Holocene.

Only during the first part late Holocene period (4320 to 1100 cal yr BP), *Araucaria* forests expanded in the region around Cambará do Sul, forming a net of gallery forests along streams, while regionally grassland vegetation dominated. *Araucaria* forests included populations of *Myrsine,* and less frequently individuals of arboreal such as *Mimosa scabrella*, Myrtaceae, *Podocarpus* and *Ilex. Dicksonia sellowiana,* a tree fern was already common in the gallery forests. Since the late Holocene the Atlantic rain forest and probably also cloud forest species were well established on the upper coastal slopes, located about 6 – 10 km distance to the study site. During the second period of the late Holocene (1100 – 430 cal yr BP), a remarkable strong expansion of *Araucaria* forest taxa took place, primarily *Araucaria angustifolia* and *Mimosa scabrella,* within 100 years replacing Campos vegetation. During the uppermost late Holocene period (430 cal yr BP - modern) still *Araucaria* forests kept expanding, reducing the Campos area in the study region. The expansion of *Araucaria* forests, including tree ferns, since 4320 cal yr BP and specially after 1100 cal yr BP is apparently related to a change to a wetter climate, with higher rainfall rates and a shorter annual dry season, or no marked dry season.

The results from Cambará do Sul indicate a complete replacement of the original grassland by forest in the surrounding of the site. Other above mentioned records such as Serra Campos Gerais or Serra da Boa Vista document the formation of a mosaic of forest and grassland during the late Holocene.

The results of pollen and charcoal analysis at lowland site at Morro Santana (Behling et al. 2007) indicate the occurrence of a mosaic of taxonomically highly diverse forest and Campos vegetation that were under the influence of fire during the late Holocene period. Between 1230 and 580 cal yr BP, Campos taxa were well represented, while forest taxa were relatively less abundant. The Campos vegetation was primarily composed of species of the Poaceae, Cyperaeceae and Asteraceae families, with the genera *Eryngium* and *Eriocaulon* also being important taxa. Some Cyperaceae may also originate from the studied swamp. Forest taxa, such as Moraceae/Urticaceae, Myrtaceae and *Dodonaea*, were rare. At about 580 cal yr BP, there is a change in the composition of the Campos vegetation by the increase of *Baccharis* and the decrease of *Eryngium,* among others. This trend is indicated by phase transitions in the vegetation composition around the same period and as well between 380 and 300 cal yr BP. This can be interpreted as an indicator of a decrease in fire frequency.

Baccharis, which was negatively correlated to charcoal concentration, is the most important genus of grassland shrubs in the present-day vegetation, leading to development of shrubland in the absence of fire over longer periods (Müller et al. 2007). Species of the genus *Eryngium*, positively correlated to charcoal concentration, have been characterized as disturbance specialists profiting from frequent fires in abundance and reproductive output (Fidelis et al. 2008), and thus should diminish in importance with greater fire intervals or – as grassland species – with forest encroachment. During the following period until the present, forests expanded continuously, especially by the increase of the pioneer *Myrsine.* The marked higher representation of *Myrsine*, which was negatively correlated to charcoal concentration, may indicate a change in the disturbance regime.

Pre- and Post-Columbian Fire Dynamics and Human Impact

The role of fire including its origin, if natural by lightning or by man as a tool for hunting, and its effect on the vegetation needs to be understand. A few charcoal records are available from southern Brazil. The records Serra Campos Gerais in Paraná State and the São Francisco de Assis in Rio Grande do Sul State (Figure 1) demonstrate that fires were quite rare during glacial periods and had little effect on the glacial grasslands (Behling 1997, Behling et al. 2004). At the beginning of the Holocene a marked increase in fire frequency is found in both records.

In the detailed Cambará do Sul record (Figure 3), the relatively low amount of carbonised particles (Figure 2, data, calculated as accumulation rate (particles/cm^2/yr), documents also that natural grassland fires were rare during the glacial periods (between 42,840 uncal to 11,500 cal yr BP). Fire became frequent for the first time at about 7400 cal yr BP, but not with the beginning of the Holocene as in the records Serra Campos Gerais and São Francisco de Assis. This is probably in response to the onset of occupation by indigenous peoples (Dillehay et al. 1992), who may have used fire for hunting, coupled with seasonal climatic conditions leading to accumulation of flammable biomass. The different timing of the marked increase in fire frequency, which suggests an earlier occupation than on the southernmost highland region, is a good indication that frequent fires during the Holocene period are not natural, but of anthropogenic origin. Fire was facilitated by the presence of Poaceae species; it is likely that the presence of tall grasses during this period led to accumulation of large amounts of highly flammable biomass in the growing season (Pillar and Quadros 1997). The great quantity of Poaceae and the decrease of some other Campos taxa suggest that frequent fires may also have been a driving factor changing the floristic composition of the Campos (Bond and Wilgen 1996).

Carbonised particles became less frequent in the sediments of Cambará do Sul at about 1100 cal yr BP. The expansion of *Araucaria* forests reduced the area of Campos near the peat bog and fires became rare near the site during the entire late Holocene period since 1100 years ago. The somewhat higher accumulation rates of carbonised particles showing that fire were still frequent in the wider region. Frequent fires at the end of the Holocene are also documented from other records where present day vegetation is still a mosaic of grassland and forest (Behling 1997, Behling et al. 2005, Behling et al. 2007).

In addition to the impact of early anthropogenic fires, the Cambará do Sul record show an increase of Poaceae pollen (13.5 cm core depth) and then Cyperaceae pollen (9.5 cm) at about 170 cal yr BP, (AnnoDomini (AD) 1780) and about 100 cal yr BP (AD 1850), respectively. This indicates a post-Columbian disturbance of the *Araucaria* forest, probably by cattle within the forest. Cattle were introduced in the highland Campos in the first decade of the 18th Century by the Jesuits from the Missions of the east side of the Uruguay river (Porto 1954). The village of Cambará do Sul was founded in 1864. The first farmers had free-range cattle likely entering into the forest somewhat earlier. First pine pollen grains (11 cm core depth) have been found at about 130 cal yr BP, AD 1820. This coincides with the first German settlements in the lower slopes of the Serra Geral in Rio Grande do Sul, around AD 1825, which introduced the exotic taxa. A decrease in *Araucaria* was detected between 30 and 15 cal yr BP (AD 1920 and AD 1935), signaling the start of intense selective logging in the region. An enormous reduction of *Araucaria angustifolia* (decrease of pollen from 41 to 2%, starting at 3.5 cm core depth) begins at about 10 cal yr BP, AD 1940, with intensified selective logging of *Araucaria* trees during the last 50-60 years, but still not in the area surrounding the peat bog. Due to the cattle in the forest and the logging of the *Araucaria* trees, other tree species, especially *Mimosa scabrella*, and Myrtaceae, *Lamanonia speciosa* and *Ilex* became more frequent in forming a secondary vegetation. Also the tree fern *Dicksonia sellowiana*, trunks are used for orchid and other ornamental plant cultivation, were removed from the *Araucaria* forest during the last about 50-60 years.

It is therefore reasonable to conclude that the increase in charcoal amounts during the Holocene more likely indicates anthropogenic fire, caused first by Amerindians and later by European settlers. Furthermore, the significant correlation between charcoal concentration and pollen taxa found for the Morro Santana record (Behling et al. 2007), which were negative for taxa characterizing shrub and forest encroachment and positive for taxa characterizing the opposite, supports the conclusion that the charcoal originated mainly from grassland fires and not from fires after cutting and subsequently burning forest areas for agriculture. If the latter were the case we would expect no correlation between charcoal and pollen taxa, since the areas burn for cultivation of forest soils were likely small.

Our results indicate a general trend towards increase of woody vegetation cover in Morro Santana starting sometime between about 620 and 540 cal yr BP (AD 1330 and 1410), with faster changes, like phase transitions between 380 and 300 cal yr BP (AD 1570 to 1650) and between about 60 and 20 cal yr BP (AD 1890 to 1930). What factors might have caused these changes? In the case of Morro Santana (Behling et al., 2007) and the period described by the sediment core, climate and human induced disturbances cannot be disentangled. This climate driven process of woody vegetation expansion, was likely also affected by human induced changes in disturbance regimes, which may be linked to early Amerindian population displacement caused by the first Portuguese incursions in the territory and the introduction of cattle from the west during the 17th Century. Impact of the Guarani on the grasslands is unclear, but should not have been high, as they used forest areas, not grassland, for agriculture. They may have caused occasional fires. Cattle grazing, principal activity of European settlers in the region, affects vegetation directly by biomass removal and trampling, controlling plants unable to resprout such as seedlings of forest woody species, and indirectly affecting fire intensity and frequency by changing flammable biomass accumulation (Pillar and Quadros 1997). The lack of a clear signal in the pollen and charcoal record for the first officially established cattle farming settlements around AD 1740 indicates that this had minimal effect on vegetation and fire frequency, because feral cattle may have been affecting vegetation since much earlier, being the main cause of the phase transition in pollen composition that ended in 300 cal yr BP (AD 1650). The sharp transition in pollen composition after about AD 1960 may indicate a change in land use coinciding with the ending of cattle grazing in the area due to the proximity to new urban settlements.

The record from Morro Santana in the Porto Alegre area in southern Brazil provides insight on vegetation, climate, fire dynamics as well as human activities during the last 1230 cal yr BP (Behling et al. 2007). The formation of a shallow swamp and the accumulation of sediments are related to a change to wetter conditions, as it has been documented from the southern Brazilian highland at the same time. The palynological results document the existence of grassland vegetation at the study area since the late Holocene, suggesting that present-day patches of Campos surrounded by forest can be seen as natural vegetation, and did not develop as a consequence of deforestation and cattle introduction. Under the wet late Holocene climatic conditions, forests expanded continuously since 580 cal yr BP.

History of Vegetation, Fire, Climate, Human Impact and Its Implication for Conservation

Here we demonstrate that palaeoecological and palaeoenvironmental background information from the southern Brazilian highland provide important implications for conservation and management regarding dynamics of vegetation, fire, and human impact including land use. Several pollen records document that the highly diverse patches of grassland are natural, remnants of a large glacial and early-mid Holocene area, and not former forest areas. This suggests that Campos should be protected and not be subject of reforestation as it is at the moment the case, where huge areas of original grassland are being replaced by exotic pine forests.

Pollen data show that Campos area became markedly reduced by the expansion of *Araucaria* forest, especially during the last about 1100 years caused by a change to wetter conditions. The natural expansion of *Araucaria* forest, which is nowadays in general not possible due to human activities, would contract the area of Campos. Recent studies show that by excluding grazing and fire present-day *Araucaria* forest tends to expand over grassland (Oliveira and Pillar 2004, Duarte and Pillar 2006).

The charcoal records indicate that natural fires ignited by lightning were rare in the grasslands of southern Brazil. The increase of fire frequency at the beginning of the Holocene is likely due to the occupation of the southern highland by first Amerindians or later after 7400 cal yr BP as in the Cambará do Sul area. For grassland management and conservation it has to be considered that anthropogenic fires played an important role during the Holocene and has certainly changed the floristic composition. Some plants and present-day plant communities may have adapted to fire (Overbeck et al. 2005). Further, the increase in charcoal coincided with a reduction in diversity in the Campos (Behling and Pillar 2007), but whether this could be seen as a causal relationship is not clear; climate change and mammal extinctions are important factors that may have affected diversity. The record shows that the later expansion of *Araucaria* forest was possible (or may have be facilitated) despite the high fire frequency and that fire in the surrounding of the study site Cambará do Sul became absent due to the expansion of the forest. Records from other sites, where a mosaic of grassland and forest is found, show still the occurrence of frequent fires.

Consequences of fire and grazing suppression at present in conservation areas with mosaics of forest and grassland in the south Brazilian highland should be carefully considered. The results show that with suppression of fire and grazing an active process of forest expansion, which had been halted by human disturbances, is re-established. If this management is maintained, in a few decades grassland in conservation areas will shrink and eventually disappear by forest expansion. We think that grasslands deserve conservation, and it is not because these ecosystems do not belong to present-day climate, and depend on human disturbances, that they should be condemned to extinction. Furthermore, fire and grazing suppression produce large accumulation of flammable biomass, increasing the risk of catastrophic, uncontrolled fires with unpredictable consequences on biodiversity. Grassland fire, by legislation, is prohibited in farms for management purposes; but in this case grazing by cattle maintains the grassland. The use of a limited number cattle for Campos management might be a good alternative rather than the use of frequent fire with its negative effects such as soil degradation, air pollution and the impact of uncontrolled fires. Large mammals which have lived in the grasslands of southern Brazil until the beginning of the Holocene might be an important factor to maintain the high diversity of the Campos similar to present-day cattle.

This example form southern Brazil shows that management for conservation of the mosaics of grassland and forest should take into account the history of vegetation and fire dynamics in southern Brazil during the late Quaternary. If management for conservation continues to include fire suppression, the present grassland patches will tend to disappear, as forest expansion will not be impeded, with losses to plant and habitat diversity. Grasslands have high biodiversity but are currently subject to severe land-use changes and thus reductions in area due to expansion of agricultural production and silvicultural plantations by exotic forest species, especially *Pinus* and *Eucalyptus*. In the few areas that are being protected (less than 0.5% of the biome; MMA 2000), management currently aims at excluding all human influence, including grazing and fire. This means that grassland areas within conservation units will disappear as a consequence of shrub encroachment and forest expansion. In these areas, fire could be used as a management tool to preserve grasslands or forest-grassland mosaics. A decision not to suppress fire, however, would require changes in conservation policies and legislation, since the state legislation prohibits the use of fire as management tool. Maintaining or reintroducing cattle grazing in conservation areas could be an alternative to fire, likewise impeding forest expansion over grasslands.

REFERENCES

Behling, H. (1995). Investigations into the Late Pleistocene and Holocene history of vegetation and climate in Santa Catarina (S Brazil). *Vegetation History and Archaeobotany* 4, 127-152.

Behling, H. (1997). Late Quaternary vegetation, climate and fire history in the *Araucaria* forest and campos region from Serra Campos Gerais (Paraná), S Brazil. *Review of Palaeobotany and Palynology* 97, 109-121.

Behling, H. (2002). South and Southeast Brazilian grasslands during Late Quaternary times: a synthesis. *Palaeogeography, Palaeoclimatology, Palaeoecology* 177, 19-27.

Behling, H., Bauermann, S.G. and Neves, P.C. (2001). Holocene environmental changes from the São Francisco de Paula region, southern Brazil. *Journal of South American Earth Sciences* 14, 631-639.

Behling, H. and Lichte, M. (1997). Evidence of dry and cold climatic conditions at glacial times in tropical Southeastern Brazil. *Quaternary Research* 48, 348-358.

Behling, H. and Pillar, V.D. (2007). Late Quaternary vegetation, biodiversity and fire dynamics on the southern Brazilian highland and their implication for conservation and management of modern *Araucaria* forest and grassland ecosystems. Philosophical *Transactions of the Royal Society of London Series B Biological Sciences* 362, 243-251.

Behling, H., Pillar, V.D. and Bauermann, S.G. (2005). Late Quaternary grassland (Campos), gallery forest, fire and climate dynamics, studied by pollen, charcoal and multivariate analysis of the São Francisco de Assis core in western Rio Grande do Sul (southern Brazil). *Review of Palaeobotany and Palynology* 133, 235-248.

Behling, H., Pillar, V.D., Müller, S.C. and Overbeck, G.E. (2007). Late-Holocene fire history in a forest-grassland mosaic in southern Brasil: Implications for conservation. *Applied Vegetation Science* 10, 81-90.

Behling, H., Pillar, V.D., Orlóci, L. and Bauermann, S.G. (2004). Late Quaternary *Araucaria* forest, grassland (Campos), fire and climate dynamics, studied by high resolution pollen, charcoal and multivariate analysis of the Cambará do Sul core in southern Brazil. *Palaeogeography, Palaeoclimatology, Palaeoecology* 203, 277-297.

Boldrini, I.I. (1997). Campos do Rio Grande do Sul: Caracterização Fisionômica e Problemática Ocupacional. *Boletin Instituto Biociencias/UFRGS* 56, 1-39.

Boldrini, I.B. (2002). Campos sulinos: caracterização e biodiversidade. In: Araújo, E.d.L., Noura, A.d.N., Sampaio, E.V.d.S.B., Gestinari, L.M.d.S., Carneiro, J.d.M.T. (eds.) *Biodiversidade, conservação e uso sustentável da flora do Brasil*, pp. 95-97. Universidade Federal Rural de Pernambuco.

Bond, W. J. and van Wilgen, B.W. (1996). *Fire and Plants*. London, Chapman and Hall.

Burkart, A. (1975. Evolution of grasses and grasslands in South America. *Taxon* 24, 53-66.

Dillehay, D.T., Calderón, G.A., Politis, G. and Beltrão, M.C. de (1992). Earliest hunters and gatherers of South America. *Journal of World Prehistory* 6, 145 – 204.

Duarte, L. da S., Dos Santos, M.M.G. Hartz, S.M. and Pillar, V.D. (2006). The role of nurse plants on Araucaria forest expansion over grassland in South Brazil. *Austral Ecology* 31, 520-528.

Ferrão, C. and Soares, J.P.M. (1989). Campos de altidude. Editora Index. Rio de Janeiro, 159pp.

Fidelis, A.T., Overbeck, G.E. and Pillar, V.D., Pfadenhauer, J. (2008). Effects of disturbance on population biology of the rosette species Eryngium horridum Malme in grasslands in southern Brazil. *Plant Ecology* 195, 55-67.

Hastenrath, S. (1991). *Climate dynamics of the tropics*. Kluwer Academic Publishers, Dordrecht Boston London, 488pp.

Hueck, K. (1953). Distribuição e habitat natural do Pinheiro do Paraná (*Araucaria angustifolia*). *Bolet. Fac. Filos. Ciênc. Univ. São Paulo, Bot.* 10,1-24.

Hueck, K. (1966). *Die Wälder Südamerikas*. Fischer, Stuttgart, 422 pp.

Klein, R.M. (1975). Southern Brazilian phytogeographic features and the probable influence of upper quaternary climatic changes in the floristic distribution. *Boletim Paranaense de Geociências* 33, 67-88.

Klein, R.M. (1978). Mapa fitogeográfico do estado de Santa Catarina. *Flora Ilustrada Catarinense*.

Klein, R.M. (1979). Ecologia da flora e vegetação do Vale Itajai. *Sellowia* 31, 1-164.

Klein, R.M. (1984). Aspectos dinâmicos da vegetação do sul do Brasil. *Sellowia* 36, 5-54.

Lindman, C.A.M. (1906). *A Vegetação no Rio Grande do Sul*. EDUSP/Itatiaia, São Paulo/Belo Horizonte.

Martin, L., Fournier, M., Mourguiart, P., Siefeddine, A. and Turcq, B. (1993). Southern Oscillation signal in South American palaeoclimatic data of the last 7000 years. *Quaternary Research* 39, 338-346.

MMA (Ministério do Meio Ambiente), (2000). Avaliação e ações prioritárias para conservação da biodiversidade da Mata Atlântica e Campos Sulinos. Relatório técnico. Brasília, DF.

MMA (Ministério do Meio Ambiente), (2007). Remanescentes de vegetação dos Campos Sulinos (do Pampa). Relatório técnico. Brasília, DF. URL:http://mapas.mma.gov.br/geodados/brasil/vegetacao/vegetacao2002/pampa/documentos/relatorio_bioma_Pampa.pdf accessed May 2008.

Müller, S., Overbeck, G.E., Pfadenhauer, J., Pillar, V.D. 2007. Plant functional types of woody species related to fire disturbance in forest-grassland ecotones. *Plant Ecology* 189, 1-14.

Nabinger, C., Moraes, A. and Maraschin, G.E. (2000). Campos in Southern Brazil. In: Lemaire, G., Hodgson, J., Moraes, A., Nabinger, C., Carvalho, P.C.F. (Eds.), *Grassland Ecophysiology and Grazing Ecology*. CAB International, pp. 355-376.

Nimer, E. (1989). *Climatologia do Brasil*. Fundação Instituto Brasileiro de Geografia e Estatística, Rio de Janeiro.

Oliveira, J.M. and Pillar, V.D. (2004). Vegetation dynamics on mosaics of Campos and Araucaria forest between 1974 and 1999 in Southern Brazil. *Community Ecology* 5, 197-202.

Overbeck, G., Müller S.C., Pillar V.D. and Pfadenhauer, J. (2005). Fine-scale post-fire dynamics in southern Brazilian subtropical grassland. *Journal of Vegetation Science* 16, 655-664.

Overbeck, G.E., Müller, S.C., Fidelis, A., Pfadenhauer, J., Pillar, V.D., Blanco, C.C., Boldrini, I.I., Both, R. and Forneck, E.D. (2007). Brazil's neglected biome: The South Brazilian Campos. *Perspectives in Plant Ecology, Evolution and Systematics* 9, 101-116.

Pillar, V.D. and Quadros. F.L.F. de (1997). Grassland-forest boundaries in southern Brazil. *Coenoses* 12, 119-126.

Por, F.D. (1992). *Sooretama the Atlantic rain forest of Brazil*. SPB Academic Publishing bv, The Hague, The Netherlands, 130pp.

Porto, A. (1954). *História das Missões Orientais do Uruguai*. 2nd ed. Livraria Selbach, Porto Alegre.

Rambo, B. (1956a). *A fisionomia do Rio Grande do Sul*. Livraria Selbach, Porto Alegre.

Rambo, B. (1956b). A flora fanerogâmica dos aparados riograndenses. *Sellowia* 7: 235-298.

Rambo, B. (1953). História da flora do planalto riograndense. *Anais Botânicos do Herbário "Barbosa Rodriges"* 5, 185-232.

Ramos-Neto, M.B. and Pivello, V.R. (2000). Lightning fires in a Brazilian Savanna National Park: rethinking management strategies. *Environmental Management* 26, 675-684.

Ratisbona, L. R. (1976). *The Climate of Brazil.* In: Schwerdtfeger, W. (ed.), *Climates of Central and South America*, Vol. 12, Elsevier, Amsterdam, 532pp.

Roth, L. and Lorscheitter, M.L. (1993). Palynology of a bog in Parque Nacional de Aparados da Serra, East Plateau of Rio Grande do Sul, Brazil. *Quaternary of South America and Antarctic Peninsula* 8, 39-69.

Safford, H.D. (1999a). Brazilian páramos I. An introduction to the physical environment and vegetation of the campos de altitude. *Journal of Biogeography* 26, 693-712.

Safford, H.D. (1999b). Brazilian páramos II. Macro- and mesoclimate of the campos de altitude and affinities with high mountain climates of the tropical Andes and Costa Rica. *Journal of Biogeography* 26, 713-737.

Soriano, A., León, R.J.C., Sala, O.E., Lavado, R.S., Deregibus, V.A., Cahuepé, O, Scaglia, A, Velazquez, C.A. and Lemcoff, J.H. (1992). Río de la Plata grasslands. In: Coupland, R.T. (Ed.), *Ecosystems of the world. Natural grasslands. Introduction and western hemisphere.* Elsevier, Amsterdam, pp. 367-407.

In: Grasslands: Ecology, Management and Restoration
Editor: Hans G. Schröder
ISBN 978-1-60692-023-7

Chapter 8

GRASSLAND CONTRIBUTION TO CLIMATE CHANGE IN THE BASQUE COUNTRY: OPPORTUNITIES TO MITIGATION

***P. Merino*[1]*, C. González-Murua*[2]*, A. Del Prado*[3]*, S. Menéndez*[2]*, M. Pinto*[1] *and J. M. Estavillo*[2]**

[1]Institute of Agricultural Research and Development
NEIKER-Tecnalia. Spain
[2]Department of Plant Biology and Ecology
University of the Basque Country, Spain
[3]Institute of Grassland and Environmental Research
IGER-North Wyke, UK

ABSTRACT

Agricultural activities are significant producers of nitrous oxide (N_2O) emission to the atmosphere (Johnson et al., 2007; Ugalde et al., 2007). That is, about 58% of total anthropogenic N_2O emissions are caused by agriculture (IPCC, 2007). The main cause of agricultural increases in N_2O is the application of N fertilizers and animal manures. Nitrous oxide is a long-lived greenhouse gas in the atmosphere with 296 times the global warming potential of CO_2. Denitrification is credited as the primary producer of N_2O (Johnson et al., 2005), which is generally favoured in poorly aerated soils with high NO_3^- concentrations (Van Groeningen et al., 2005). Currently, about 32% of the agricultural land in the EU is used for grassland production (Eurostat, 2005). The response of grassland to fertilizer N is greater than that of most other crops due both to its long growing season and to being harvested as vegetative growth (Whitehead, 1995). In the Basque Country, around 36% of agricultural land is occupied by grassland, and there is a high risk of N_2O losses by denitrification due to high rainfall, typically in excess of 1000 $mmyr^{-1}$, and warm temperatures, especially in spring and autumn when fertilizer is applied (Estavillo et al., 1994). Besides, the current dairying model is still developing towards an intensified dairy production system, resulting in increases in generation of animal manure, which may be also a significant source of harmful nutrient emissions into the environment if handled improperly. Also, the mountainous orography characteristic of the area together with the historically parcelled rural soil distribution and the current high industrial pressure has dramatically increased rural soil prices, which does not contribute to enhance farmland availability. Efficient use of nutrients is one of the major keys of sustainable agricultural production systems because inefficient nutrient use not only results in excessive and potentially harmful losses to the environment, it also negatively affects economic performance of production systems (Oenema and Pietrzak 2002).

This review focuses on N_2O emission especially in the Basque Country as related to grassland soil, considering the effect of management and mitigation options.

Keywords: denitrification, grassland, mitigation, nitrous oxide

INTRODUCTION

Nitrogen applied in fertilizers and manures is not always used efficiently by crops. To this respect, soil N availability can have a major impact on N_2O fluxes, thus, N fertilization increases N_2O fluxes relative to unfertilized controls (Smith et al., 1998).

There is still a big uncertainty in estimated nitrogen balances at different scales. The gap between input and output of N can be up to 40%. Denitrification is suspected to be the most important process that can explain this gap (Erisman and Sutton, 2007). N_2O is generated by the microbial transformation of nitrogen during nitrification and denitrification processes in soils and manures, and is often enhanced where available nitrogen exceeds plant requirements. Regarding environmental aspects, N_2O is involved in global warming and contributes to the destruction of parts of the ozone layer. It has a mean atmospheric residence time of more than 100 years (Prather et al., 2001). N_2O is mainly produced by denitrification, especially under wet conditions (Oenema et al., 2005). The climatic characteristics of the Basque Country provide the favourable conditions for the growth of grasslands and the development of livestock enterprises. The climate is Atlantic, with an average temperature of 14°C and an annual precipitation ranging between 1200 and more than 2000 mm (http://www.euskalmet.euskadi.net). The soils studied were typical of the region, and classified as dystric Gleysol for the top 10 cm (2.2% fine sand, 41.6% coarse sand, 30.4% loam, 25.7% clay for the top 10 cm). They are poorly drained soils, which affects soil water content and gas diffusivity in soil, promoting anaerobic conditions for denitrification. In fact, we found high rates of emissions on clay soils, suggesting that N_2O emission was regulated by soil aeration (Gregorich et al., 2005).

Initial studies in the Basque Country of Spain (Estavillo et al., 1994 and 1996) carried out using soil incubations determined that denitrification was the main way of N losses in grassland, being the greatest losses in spring and autumn following fertilization events, with denitrification rates up to 2 kg N ha^{-1} day^{-1}. These are seasons in which heavy rainfalls occur coinciding with fertilizer applications. Annual losses were about 10-15% of N applied, reaching in occasions 20-37% of the N applied (Estavillo et al., 1994).Values similar to this rate have also been observed in other studies carried out later in the area, with 1.56 kg N ha^{-1} day^{-1} (Estavillo et al., 2002). In fact, denitrification potential measurements carried out in different soil types (grassland, forest and crop) showed that grassland soils presented the highest denitrification potentials due to their high organic matter content (Virgel et al., 1995).

There was a linear relationship between potential denitrification in grassland and soil microbial carbon. To this respect, the importance of microbial N pool in grassland was evident: distribution of N showed that microbial biomass N content was 10 to 30 times larger than the mineral N and 1 to 5 times higher than that extracted by herbage, with values of 120, 6 and 28 kg N ha^{-1} respectively for the fifth cut carried out in autumn (Estavillo et al., 1997). From 1997 we have been monitoring N_2O fluxes in the field after fertilization events considering different treatments: fertilisation type, use of nitrification inhibitors and land use change. Field measurements took place the day the fertilizer was applied, the day after the fertilizer was applied daily for two weeks and then approximately weekly until fluxes diminished. A common methodology based on manually operated static chambers was used for N_2O flux measurements (Merino et al., 2001). To minimize the effects of diurnal variation, sampling was always carried out in the morning. Nitrous oxide emissions were measured using a closed air circulation technique in conjunction with a photoacoustic infrared gas analyzer (Brüel and Kjaer 1302 Multi-Gas Monitor). Fluxes were calculated from the linear concentration increase in the chamber headspace with time. Alternatively, gas measurements were carried out by gas chromatography equipped with an electron capture ^{63}Ni detector. An intercomparison of N_2O analyses was conducted for equipments calibration.

Cumulative N_2O emissions during the sampling period were estimated by averaging the rate of loss between two successive determinations, multiplying that average rate by the length of the period between the measurements, and adding that amount to the previous cumulative total. Table 1 shows cumulative N_2O values for the studies considered referred to the period of study and its conversion to CO_2 equivalents.

Usually higher percentage of N losses from the N applied were found from mineral fertilisation, although this trend was not consistent in all the studies. As mentioned above, the highest N_2O losses from mineral or organic fertilisation observed in our studies took place in spring and autumn. These losses coincided with high soil water contents; in fact, denitrification rates greater than 0.2 kg N $ha^{-1}day^{-1}$ occurred mainly after fertilization events if the soil was wet (soil water content > 35%) and soil nitrate was high (>7mg N kg^{-1} dry soil) (Estavillo et al., 1994). The difference between the different fertilizers would be that the increase in nitrate content in soil after mineral fertilization produced immediate losses of N by denitrification, whereas the nitrate content in soil after slurry application was lower. Thus, in the case of slurry fertilisation, losses occurred during longer periods of time, due to the mineralization of N between fertilization events. We have found that although nitrate availability determines the potential denitrification, the soil must be also we for denitrification to occur (Del Prado et al., 2006; Estavillo et al., 1994). Merino et al (2001) also observed that the interaction between nitrate and WFPS on N_2O emissions seems to indicate that the nitrate effect varies according to different ranges in WFPS. When these variables were considered in a purpose oriented experiment at laboratory scale, (Del Prado et al., 2006), we observed that under wet conditions (WFPS>60%) the fertilised soils gave cumulative N_2O emissions 3.5 times greater than the unfertilised ones; on drier conditions (WFPS<60%), this difference was a factor of about 1.5. Fertilised and unfertilised soils under wet conditions resulted in about 10 times greater N_2O emissions than soils with the same treatments but under dry conditions.

Grassland soils in particular have a high potential for mineralization and subsequent nitrification and denitrification (Merino et al., 2001). The microbial processes nitrification/denitrification involved in the production of N_2O have been studied by means of soil incubations with acetylene (Merino et al., 2001). In this sense, N_2O derived from cattle slurry applied to a grassland soil came in a greater proportion from nitrification rather than from denitrification, while for the mineral fertilisation most N_2O came from denitrification.

Table 1. Grassland derived N_2O emission measured at field expressed as cumulative emissions and CO_2 equivalents (CO_2e) for N_2O emissions following the recommendations of the Intergovernmental Panel on Climate Change (1996)

Treatment	Fraction* %	Season	Days	CO_2e Mg CO_2eha^{-1}	
M	5.6	Year	365	3.85	Merino et al., 2001
S	1.4			2.87	
M	1.55			0.60	
M+DCD	0.89			0.35	Merino et al., 2001
S	1.44	Autumn	20	0.59	
S+DCD	0.57			0.24	
M	6.0			3.66	
M+DMPP	2.5			1.54	
S	2.3	Autumn	60	1.43	Macadam et al., 2003
S+DMPP	0.91			0.55	
M	0.187			0.137	
Tillage+M(1)	0.195	Spring	5	0.143	Pinto et al., 2004
Tillage+M(2)	0.37			0.27	
S	8.5	Autumn	22	5.6	Merino et al., 2005
S	9.4	Spring	14	4.4	
M	4.58	Spring	59	2.16	Menéndez et al., 2006
S	15.97			7.54	
S	3.6			1.22	
S (solids)	4.1	Spring	15	1.41	Menéndez et al., 2008
S (liquid)	3.4			1.15	

*Values are expressed as percentage of the N applied emitted, which in the case of slurry is calculated with respect to the ammonium fraction applied.
(1)Tillage took place immediately before fertilisation.
(2)Tillage took place three days before fertilization.

High nitrification activity has been observed in the soil under study from both kinds of fertilisers (Merino et el., 2001; Merino et al., 2002; Pinto et al., 2004; Merino et al., 2005; Menéndez et al., 2006). Typically, a period of about 5 days following slurry and mineral fertilization was necessary to decrease soil ammonium content from fertilised treatments to the amount found in unfertilized treatments. This decrease might have been due to immobilization, nitrification or plan uptake. Probably all the processes were occurring simultaneously, but in most cases, nitrification was the main process involved as no favourable conditions for volatilization occurred (Merino et al., 2002) or a proportional increase in soil nitrate was accounted. Also, N_2O fluxes from mineral or slurry fertilisation have occurred mainly in the first week after application. Similarly, Comfort et al (1990) found that most of the N_2O emission occurred within the first 5 days following injection of manure into the soil when CO_2 evolution was greatest. On a global scale, CO_2 is the most important greenhouse gas contributing to global warming, although most of the contribution from agriculture to potential greenhouse warming does not originate from CO_2 but from N_2O and CH_4 (Izaurralde et al., 1997). This was confirmed after the measurements carried out in our experiments, where CO_2 emissions were evaluated in grassland soils. We observed enhanced CO_2 emissions during the first 4 days after the slurry application, although cumulative emissions after 59 days were not significantly different from the control (Menéndez et al., 2006).

In a study carried out in a grassland soil managed by a commercial farm, (Merino et al., 2001), N_2O emissions were greater if mineral fertilizer was used instead of slurry fertilizer, even if a higher amount of N was applied as slurry than as mineral fertiliser, with cumulative losses over a year of 7.9 and 5.9 kg Nha^{-1} respectively. Emissions were minimum two months alter fertilisation, with background levels for slurry and mineral fertiliser of 4 and 2 g N_2O-N $ha^{-1}d^{-1}$ respectively. Also in this case the seasons with the highest emission of N from the system were spring and autumn, with 4.3 for mineral and 2.0 kg N ha^{-1} for slurry respectively. The highest losses of N_2O by denitrification were found after the application of mineral fertilization, while slurry acted as a slow release fertilizer which supplies nitrate continuously at a small rate. This was the main difference observed between the slurry and the mineral fertilization with respect to N_2O emission. The determination of the kind of fertilization that produced greater N_2O losses resulted from the interaction of different factors, such as management (date of application, distribution of fertilisers, heterogenecity of cattle slurry), edaphoclimatic conditions. Flechard et al (2007) also observed from a study carried out most European climatological zones that N_2O measurements were extremely variable in time and in space at each site, depending on weather and management practices. In summary, those factors influencing soil microbial activity in soil, which in the case of slurry fertilisation is important due to the pulses of emissions occurring following mineralization of soil organic N. This slow release fertiliser effect observed in slurry fertiliser was simulated when nitrification inhibitors were applied together with mineral fertiliser or with cattle slurry. Adapting to social and policy demands of environmentally sustainable dairy production, some projects have been already developed in the territory regarding to study how different fertilizers (cattle slurry or inorganic fertilizer) amended with commercial nitrification inhibitors or slurry electroflotation techniques affect on nitrous oxide and CO_2 emissions on a temperate grassland. Nitrification inhibitors used in the field have been proposed as management alternatives to reduce both nitrate leaching and denitrification, providing greater N availability to the sward. It is likely that the use of nitrification inhibitors (NIs) will have increased potential in the long-term, although it currently shows limited potential (Smith et al., 2007). DCD (diciandyamide) and DMPP (3,4 dimethylpyrazol phosphate) are the nitrification inhibitors that have been evaluated in our conditions. They act by delaying the bacterial oxidation of ammonia to nitrite in the soil by depressing the activity of *Nitrosomonas* bacteria in the soil. These NIs have been known to successfully reduce N_2O emissions from mineral fertilisers (MacTaggart et al., 1997; Linzmeier et al., 2001). The mitigation potential of these inhibitors applied with slurry has seldom been studied, having been evaluated by our group even in terms of phytotoxic effects in clover (Macadam et al., 2003). Nitrification inhibitors applied in grassland soils efficiently delayed nitrification rates, keeping soil mineral N content as ammonium for a longer time (Merino et al., 2001; Merino et al., 2002, Macadam et al., 2003, Merino et al., 2005). The maintenance of soil mineral N in the ammonium form after NIs application led to a reduction in cumulative N_2O emissions, showing a percentage of reduction of about 40-60% with both inhibitors. Nevertheless, phytotoxic effects on white clover were observed with the application of DCD, which did not appear with the application of DMPP. When white clover was grown with DCD in the growth chamber, it showed the same visual symptoms of phytotoxicity as those observed previously in the field. They consisted mainly of chlorosis and further necrosis at the border of the leaves. DCD application caused a reduction in clover yield and a nutrient imbalance in leaves. These results led us to recommend the use of DMPP instead of DCD in grasslands as this inhibitor prevented N_2O emissions in identical amount, but without causing damage to clover. As mentioned above, DMPP has been widely studied if applied with mineral fertilisation, but few studies can be found considering its effect with slurry fertiliser. In this sense, and due to the decreasing mineral fertiliser amounts used by farmers in grasslands parallel to the increasing amounts of slurry applications in our region, application of DMPP with slurry was evaluated at different times of the year: spring and autumn. As previously mentioned, at this time of the year, periods of rainfall frequently coincide with warm temperatures, which may affect DMPP performance as a nitrification inhibitor. We proved that DMPP effect is lower during spring owing to the slightly higher mean soil temperature in that season, leading to a possible faster degradation of DMPP. In our field study in spring, DMPP had no more effect after day 14, in comparison to autumn, when DMPP delayed ammonium nitrification during 58 days (Merino et al., 2005). With respect to N_2O losses, 69% of N_2O emissions were decreased 22 days after slurry application in autumn. With respect to to CO_2, we found lower emissions from mineral fertilization than from control treatment and slurry fertilization, with 11.8, 15.9 and 15.3 Mg CO_2 ha^{-1} respectively. No effect of the nitrification inhibitor DMPP applied with mineral or with slurry fertiliser was found. Yield and botanical composition of the grass were not affected by the use of DMPP.

Adapting industrial processes of manure treatment is another action that has been evaluated in the Basque Country as a feasible way to decrease gaseous N emissions (Menéndez et al., 2008). To this respect, the electroflotation process, aimed to decrease the volume of slurries from intensive livestock farms was considered. The industrial process consists basically of an electrolysis of the slurry catalyzed by iron which leads to the flocculation of the solid particles, giving as a final result a solid and a liquid fraction. Thus, the solid and liquid fractions derived from the process were applied on a grassland soil to study the influence of N_2O and CO_2 emissions. As a result, the solid and liquid fractions of electroflotation can be considered useful products as fertilizers. Both products caused an increase in grassland yield with respect to the original untreated slurry, with grassland N extraction being even higher after the solid fraction application. Regarding environmental concerns, if applied under temperate conditions, they do not modify the risk of global warming, with N_2O and CO_2 emissions caused by their application being of the same magnitude as those caused by the application of the original untreated slurry.

Tillage is one of the management variables that may enhance or retard emissions of greenhouse gases from agriculture (Ugalde et al., 2007). Tillage influences interactions between soil structure and biota, which in turn influences the stability of nitrogen within the soil matrix. Tillage of grassland by farmers to grow maize, *Zea mays* (L), as a summer crop is a common practice in our region. Maize is sown immediately after tillage, and N fertiliser is applied at the same time to boost maize growth. As spring has been determined as a season when great losses of N_2O may take place, we conducted a field experiment in which the immediate effects on N_2O emission following tillage and mineral fertiliser application were studied (Pinto et al., 2004) at this time of the year. Also, production of N_2O was studied at different soil depths. In both the ploughed and unploughed treatments, the 0-10 cm layer was the major contributing layer to gaseous N production (Estavillo et al., 2002). A period of 3 days was necessary for the ploughed soil to reach the concentrations of the available mineral N originated from organic matter mineralization. This may have occurred as a consequence of mineralization following the incorporation of organic matter into the soil, which increases soil nitrogen availability and provokes important differences in the production and emission of N_2O (Table 1). As result, a recommended practice to reduce N loss from land use changes involving tillage would be to avoid immediate fertiliser addition as there is an extra N supply from mineralization of organic matter at this time.

Variability should have been reduced in this compilation by use of a common methodology, and by the fact that emissions are considered at the level of grassland sited in the same edaphoclimatic area, ensuring an improved estimation of N_2O emissions in the situations under study. Nevertheless, the highest variation was found for slurry treatment, the treatment that has received higher attention in our studies due to the environmental and economic advantages derived from its use at farm level. Besides, the uncertainty in the estimates coming from its use as fertiliser have required a deeper consideration by scientists. A wide variation was observed (Table 1), with ranges of 1.4 to 8.5 % of applied N for slurry fertilisation for 20 days in autumn.

The interaction of soil, climate and management systems needs further evaluation. Policies that support better agricultural land and fertiliser management practices are needed. Prudent management of N inputs, considering the type of fertiliser, time of application, use of nitrification inhibitors and land use and technology for slurries minimization can be an effective strategy to minimize N_2O emitted from grassland.

REFERENCES

Del Prado A, Merino P, Estavillo JM, Pinto M, González-Murua C (2006) N_2O and NO emissions from different N sources and under a range of soil water contents. *Nutrient Cycling in Agroecosystems* 74: 229-243.

Erisman JW and Sutton M. 2007. Reduced nitrogen in ecology and the environment. In: *Ammonia emissions in agriculture*. Edited by GJ Monteny and E Hartung. Wageningen Academic Publishers, The Netherlands 2007. pp 403.

Estavillo JM, Rodríguez M, Domingo M, Muñoz-RUeda A and González-Murua C (1994) Denitrification losses from a natural grassland in the Basque Country under organic and inorganic fertilization. *Plant and Soil* 162: 19-29.

Estavillo JM, Rodríguez M and González-Murua C (1996). Nitrogen losses by denitrification and leaching in grassland. *Fertilizer Research* 43: 197-201.

Estavillo JM, Rodríguez M, Lacuesta M and González-Murua (1997) Effects of cattle slurry and mineral N fertilizar applications on various components of the nitrogen balance of mown grassland. *Plant and Soil* 188: 49-58.

Estavillo JM, P Merino, M Pinto, S Yamulki, G Gebauer, A Sapek and W Corre (2002) Short term effect of ploughing a permanent pasture on N2O production from nitrification and denitrification. *Plant and Soil* 239: 253-265.

Flechard CR, Ambus P, Skiba U, Rees RM, Hensen A, van Amstel A, van den Pol-van Dasselaar A, Soussana JF, Jones M, Clifton Brown J, Raschi A, Horvath L, Neftel A, Jocher M, Ammann C, Leifeld J, Fuhrer J, Calanca P, Thalman E, Pilegaard K, Di Marco C, Campbell C, Nemitz E, Hargreaves KJ, Levy PE, Ball BC, Jones SK, van de Bula WCM, Groot T, Blom M, Domínguez R, Kasper G, Allard V, Ceschia E, Cellier P, Lavilla P, Renault C, Bizouard F, Abadía M, Williams M, Baronti S, Berretti F, Grosz B (2007) Effects of climate and management intensity on nitrous oxide emissions in grassland systems across Europe. *Agriculture, Ecosystems and Environment* 121: 135-152.

Gregorich EG, Rochette P, VandenBygaart AJ, Angers DA (2005) Greenhouse gas contribution and losses: erosion and deposition effects. Soil Tillage Research 47: 291-302.

Izaurralde RC, McGill WB, Bryden A, Graham S, Ward M and Dickey P (1997). Scientific challenges in developing a plan to predict and verify carbon storage in canadian prairie soils. In: *Management of carbon sequestration in soil.* Edited by R Lal, JM Kimble, RF Follet, BA Stewart. 1998. USA.

Johnson JMF, Reicosky DC, Allmaras RR, Sauer TJ, Venterea RT, Dell CJ (2005) Greenhouse gas contributions and mitigation potential of agriculture in the central USA. *Soil and Tillage Research* 83: 73-94.

Johnson JMF, Franzluebbers AJ, Weyers SL and Reicosky DC (2007). *Environmental Pollution* 150: 107-124.

Linzmeier W, Gutser R and Schmidhalter U (2001) Nitrous oxide emission from soil and from a nitrogen-15-labelled fertiliser with the new nitrification inhibitor 3,4-dimethylpyrazole phosphate (DMPP). *Biology and Fertility of Soils* 34: 103-108.

McTaggart I, Clayton H, Parker J, Swan L, Smith KA (1997) Nitrous oxide emissions from grassland and spring barley, following N fertiliser application with and without nitrification inhibitors. *Biol Fertil Soils* 25: 261-268.

IPCC Fourth Assessment Report: Climate Change 2007. http://www.ipcc.ch/ipccreports/ar4-wg3.htm http://epp.eurostat.ec.europa.eu

Macadam XMB, Del Prado A, Merino P, Estavillo JM, Pinto M and González-Murua C (2003) Dicyandiamide and 3,4-dimethylpyrazole phosphate decrease N_2O emissions from grassland but dicyandiamide produces deleterious effects in clover. *J. Plan Physiol.* 160: 1517-1523.

Menéndez S, Merino P, Pinto M, González-Murua C and Estavillo JM (2006) 3,4-Dimethylpyrazol phosphate effect on nitrous oxide, nitric oxide, ammonia and carbon dioxide emissions from grasslands. *Journal of Environmental Quality* 35: 973-981.

Menéndez S, Merino P, Lekuona A, Pinto M, González-Murua C and Estavillo JM (2008). The effect of cattle slurry electroflotation products as fertilizers on gaseous emissions and grassland yield. *Journal of Environmental Quality*. In press.

Merino P, Estavillo JM, Pinto M, Rodríguez M, Duñabeitia MK and González-Murua C (2001) Nitrous oxide emissions from grassland in an intensive dairy faro in the Basque Country of Spain. *Soil Use and Management* 17: 121-127.

Merino P, Estavillo JM, Graciolli LA, Pinto M, Lacuesta M, Muñoz-Rueda A and González-Murua C (2002). *Soil Use and Management* 18: 135-141.

Merino P, Menéndez S, Pinto M, González-Murua C and Estavillo JM (2005) *Soil Use and Management* 21: 53-57.

http://www.ingurumena.ejgv.euskadi.net/r4911293/es/contenidos/plan_programa_proyecto/plan_cambio_climatico/es_cc/adjuntos/borrador.pdf

Oenema, O. and Pietrzak, S. 2002. Nutrient management in food production: achieving agronomic and environmental targets. *Ambio* 31:159-168.

Oenema O, Wrage N, Velthof GL, Van Groeningen JW, Dolfing J and Kuikman PJ (2005) Trends in global nitrous oxide emissions from animal production systems. *Nutrient Cycling in Agroecosystems* 72: 51-65.

Pinto M, P Merino, A del Prado, JM Estavillo, S Yamulki, G Gebauer, S Piertzak, J Lauf and O Oenema (2004) Increased emissions of nitric oxide and nitrous oxide following tillage of a perennial pasture. *Nutrient Cycling in Agroecosystems* 70: 13-22.

Prather, M., Ehhalt, D., Dentener, F., Derwent, R., Dlugokencky, D., Holland, E., Isaksen, I., Katima, J., Kirchhoff, P., Matson, P., Midgley, P. and Wang, M. 2001. Atmospheric Chemistry and Green House Gases. p. 240-287. *In* J.T. Houghton et al. (ed.). *Climate Change 2001: The Scientific Basis.* Contribution of Working Group I to the Third Assessment Report of the Intergovernmental Panel on Climate Change Cambridge University Press, New York, E.E.U.U.)

Smith KA, Mctaggart IP, Dobbie KE and Conen F. 1998. Emissions of N2O from Scottish agricultural soils, as a function of fertilizer N. *Nutrient Cycling in Agroecosystems*, 52, 123-130.

Smith P, Martino D, Cai Z, Gwary D, Janzen H, Kumar P, McCarl B, Ogle S, O'Mara F, Rice C, Scholes B, Sirotenko O (2007) Agriculture. In Climate Change 2007: Mitigation. Contribution of Working Group III to the Fourth Assessment Report of the Intergovernmental Panel on Climate Change. B Metz, OR Davidson, PR Bosch, R Dave, LA Meyer (eds) Cambridge University Press, Cambridge, United Kingdom and New York, NY, USA.

Ugalde D, Brungs A, Kaebernick M, McGregor A, Slattery B (2007) Implications of climate change for tillage practice in Australia. *Soil Tillage and Research* (available on line).

Van Groeningen JW, Zwart KB, Harris D, Van Kessel C (2005) Vertical gradients of $\leq^{15}$Nand $\leq^{18}$O in soil atmospheric N2O-temporal dynamics in a sandy soil. *Rapid Communications in Mass Spectrometry* 19, 1289-1295.

Virgel S, Estavillo JM, Rodríguez M. 1995. Estudio de la desnitrificación potencial en diferentes suelos del País Vasco. Actas de la XXXV Reunión Científica de la Sociedad Española para el Estudio de los Pastos. Pp 259-262.

Whitehead DC (1995) Inputs of N to agricultural grassland. In *Grassland nitrogen.* Ed CAB International. pp 397.

In: Grasslands: Ecology, Management and Restoration
Editor: Hans G. Schröder
ISBN 978-1-60692-023-7

Chapter 9

THE ORIGIN AND AGE OF GRASSLANDS IN THE SOUTHERN KURIL ISLANDS

***N. G. Razzhigaeva*[1]*, L. A. Ganzey*[1]*, L. M. Mokhova*[1]*, N. F. Pshenichnikova*[1] *and N. A. Eremenko*[2]**
[1]Pacific Institute of Geography, Far East Branch
Russian Academy of Science, Vladivostok, Russia
[2]Reserve "Kurilskiy", Yuzhno-Kurilsk, Russia

ABSTRACT

The stages of grassland development on Kunashir and Iturup Islands (Southern Kuril Islands) and Lesser Kuril Ridge are reconstructed based on pollen analysis and radiocarbon dating of soil profiles. There are anthropogenic grasslands and meadows in wind shadows formed during cooling events in Late Holocene; these formations have been relatively stable during last 1500-2000 years. Grasslands on small islands in the south of the Lesser Kuril Ridge have existed last 4000-6000. Floristic components began to change in the Late Pleistocene. Grassland soils have multiple profiles. Grasslands have various floristic compositions. Herbaceously rich meadows are located on terraces and divides, with *Sasa* on some terrace surfaces and slopes. Shrub-grass associations developed on supersaturated substratum; herb-graminoid – on well drained areas. Grass associations typical for swamp accumulative lowlands are developed on low coastal plains.

The lithogenic component of landscapes is most dynamic on the islands and it quickly responds to climatic changes and changes in sea level. For example, substratum renewal has largely been a response to aeolian processes during small amplitude regressions in the Middle and Late Holocene. Sand buildup and the addition of volcanic ash with varied chemical compositions have changed water and mechanical soil characteristics that greatly influenced grass associations. Supersaturated sedge and sedge-herb meadows even on low isthmuses shifted to herb-graminoid and the role of xerophilous species increased.

Introduction

Grasslands on ocean islands are characteristically azonal landscape elements. Grassland communities are broadly distributed on islands located both in the boreal zone [6, 50] and on tropical islands in the Pacific Ocean [28]. Grasslands in a forest zone, at least dry land grasslands, are typically a result of anthropogenic activity and their origins are associated with the degradation and destruction of forest vegetation during human settlement and local economic development. The interest in the origin of grasslands communities on islands is in may ways connected with the stability of island landscapes. An additional question is what density of human population is necessary for island grassland communities to shift irreversibly and to set off catastrophic changes in vegetation cover. Paleogeographic study of Holocene deposits on Polynesian and Micronesian islands, where grasslands have developed on windy, dry coastlines show that grassland communities are natural, climate influenced formations that have existed for several thousand years and that these grasslands arose long before human settlement of the islands [27, 28].

Without addressing the difficult issue of the structural and functional organization of grassland ecosystems or their floristic composition, questions which have already been studied in detail [6, 7 and others], we want to discuss in this article the question of the age of grassland communities and to analyze the paleogeographic conditions during which they developed on temperate zone ocean islands. The Southern Kuril Islands (Kunashir, Iturup, Shikotan Islands, other islands of the Lesser Kuril Ridge) were selected as the research region. The availability of a large volume of paleogeographic information provides an opportunity to reliably reconstruct paleogeographic conditions in the late Pleistocene and Holocene [19].

Soil profiles (Figure 1) were selected as the key feature for a reconstruction of grassland communities because data on pollen spectra in soils are the most instructive for reconstructing local vegetation groups and they reflect the habitat conditions for individual species [9]. Although grass pollen does not move great distances, aerial pollen transfer is very important in the formation of the pollen spectra. L. M. Mokhova confirmed this when studying the structure of pollen rain and subfossil spectra for different layers of sediment on Kunashir Island [23]. A soil's pollen spectra should demonstrate the individual stages in the development of grassland communities. At the same time, the interpretation of the pollen spectra in soils present certain challenges since, as a rule, changes in fossil soils in the initial spectrum occur in response to aeration and moisture regime, to specificity of physical and chemical of soil processes and micro biotic factors [9]. Significantly less pollen is preserved in soils than is preserved in marshlands, alluvial and lake facie [47]. Interpretation of radiocarbon data obtained for humic acid soils reveals no single pattern. The renewal of humus occurs in paleosols close to the surface. Thus, the most reliable dating is obtained from buried soils that are shielded from the carbonic exchange zone [1].

Regional Setting

Iturup, Kunashir and Lesser Kuril Ridge are located on south Kuril Arc, stretching from South Kamchatka to Hokkaido Island. Iturup Island (44.45°– 45.55°N, 146.80°– 148.88°E) the largest island of Kuril Island Arc, is about 200 km long and from 5.5 to 46 km wide. Ekaterina Strait (6.3 km wide, up to 520 m depths) divides Iturup from Kunashir, Friz Strait (17.82 km wide, up to 890 m depth) – from Urup (Figure 1).

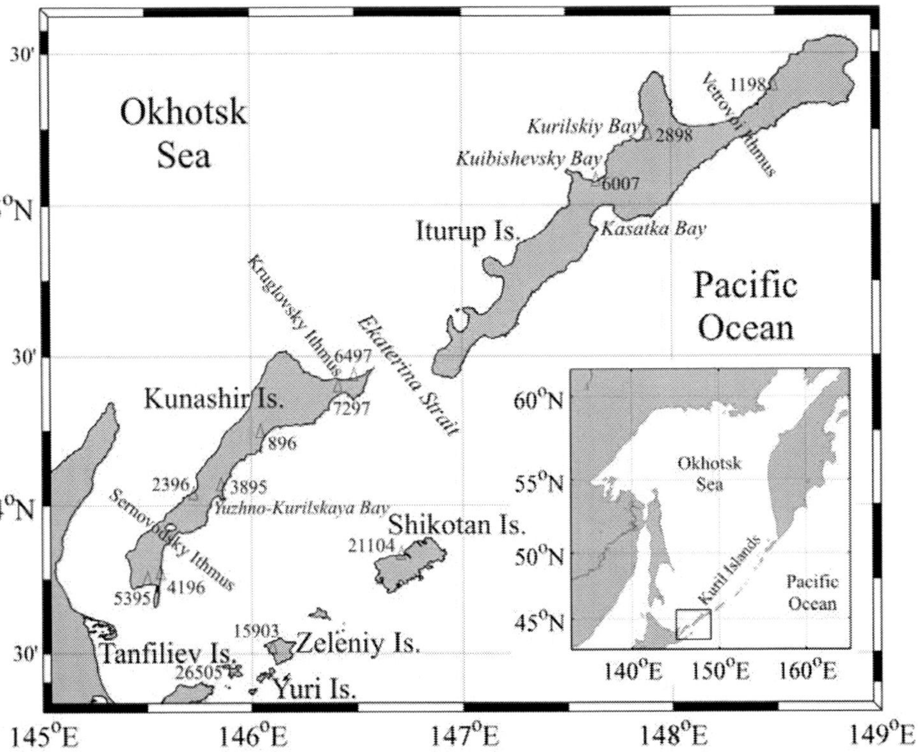

Figure 1. Location of the study area.

Kunashir Island (43°40'– 44° 30' N, 145°23'– 146°30' E) is divided by Izmena Strait (15.5 km wide) from northeastern Hokkaido, the island is about 123 km long, 7 to 35 km wide, and 1490 km^2 in area. The Pacific Ocean borders the islands on the east and the Sea of Okhotsk on the west. Lesser Kurils are located parallel Great Kuril Arc and divided from Kunashir Island by South Kuril Strait (48 km wide, depth up to 200 m). The islands are separated one by another by narrow shallow straits (1.9-5.6 km wide, up to 54 m depth), from Nemuro Peninsula - by strait with 0.7 km wide, depth up to 89 m. Shikotan island is largest and has 182 km^2 in area , 29 km long and 9 km wide. The islands located on south of the arc are small, less than 60km^2 in area, maximal length is 11.4 km, width – 9.2 km (Zeleniy Island).

Iturup and Kunashir Islands exhibit mountain relief (up to 500-1800 m) resulting from some volcanic groups, separated by low isthmuses (from 3-4 m to 60 m high). Shikotan has dissected low mountain relief (maximal elevation 412 m). Other islands located to the south from Shikotan are small flat island with elevation up to 40 m.

The islands have oceanic climate with small annual temperature amplitude, warm winter and cool summer. During the winter the north-west air masses from Asia produce severe cold and snow. During the summer moist cool Pacific air masses move south or east across the Kuril Islands toward the Asian low-pressure area, bringing with them extensive rainfall, fog and typhoons in August-September. Annual mean temperature is about 4.3°C on Iturup Island, 4.7°C on Kunashir, and 5.0-5.2°C on Lesser Kurils, mean monthly temperature differs from -7.2°C in February on Iturup Island, -6.7°C on Kunashir Island, -5.3…-5.9°C on Lesser Kurils and about 16°C in August on Itirup Island, 16.6°C on Kunashir, and 16.1-16.3°C on Lesser Kurils. Annual sum of mean daily temperatures (above 10°C) is 1451°C on Iturup, 1593°C on Kunashir, and 1562°C on Lesser Kurils. Annual rainfall is about 1040 mm on Iturup, 1255 mm on Kunashir, 1020-1240 mm on Lesser Kurils. The islands have large amount of days with strong winds (101-106 days with wind > 15 m/c) and lower snow cover (35 cm on Shikotan up to 68 cm on Iturup). Snow cover on small flat islands is unstable due to strong winds. N and NW winds prevail at October-March shifting to S at April-September. Amount of fog days are about 83, cloudy days – from 211 on Iturup up to 162 on Zeleniy Island [38, 39].

The Pacific Ocean with the Oyashio cold current borders the island on the east and the Sea of Okhotsk on the west. The warm Soya current penetrates up to Kurilsky Bay on Okhotsk side of the Iturup Island. Ocean currents are particularly important in influencing the regional climate of Southern Kuril Islands. The warm Soya current is most important, as it has a warming effect on the southern part of the Sea of Okhotsk. The Oyashio cold current brings cold water from north to south in the Kuril region. The warm Kuroshio Current produces synoptic eddies [18], which influence the southern part of the island.

Marine currents, mountain relief, and hot springs define a wide range of microclimatic conditions and a high diversity of ecotypes [2]. Vegetation period within the Sea of Okhotsk side is warmer than Pacific coast. Phenological data show that plants in Okhotsk side develop early than in Pacific side, the same phases of plant development begin on 8-10 days earlier [12]. This is explained by two causes: influence of warm Soya current and barrier role of mountain relief, protected from fog and cold winters forming by cold water of the Oyashio current.

Vegetation of north part of Iturup Island (northern from Vetrovoi Isthmus) belongs to Urup floristic district, central and southern part of Iturup Island and Kunashir are divided to South Kuril floristic district and Lesser Kurils are separated in Lesser Kuril floristic district [5]. South Kuril floristic district has the most diversity flora with high contents of heat-loving plants. Lesser Kuril floristic district differs by absence of some species, typical for Great Kuril Arc, such as *Pinus pumila, Quercus* and other trees. Coniferous and birch are dominate, larch forest is met on SE [48]. Grows of juniper are widespread on mountain slopes. Tree vegetation is absence on small islands on south Lesser Kuril Arc.

The moist climate, wide latitudinal range, and mountain relief combine to produce some botanical formations on the islands [2, 50]. Boreal coniferous forests of the Kunashiri-Iturup Formation with the dominant *Abies sachalienensis* are extensively distributed in northern Kunashir and south Iturup. Cool-temperate broadleaf and mixed coniferous-broadleaf forests of the Nemuro-Kunashiri Formation occupy the southern part of the Kunashir Island. Broad-leaved taxa include *Quercus crispula, Q. dentata, Acer pictum* and *A. ukurunduense* in association with *Kalopanax septemlobum, Phellodendron sachaliense, Cerasus sachalinensis, C. maximoviczii, Ulmus laciniata* and *U. propinqua.* Conifers are represented by *Abies sachalinensis, Picea microsperma, P. glehnii,* and *Taxus cuspidata.* Such thermophylous species as *Magnolia obovata, Betula maximowicziana, Alnus japonica, Fraxinus manshurica, Syringa amurensis, Actinidia arguta, Botrocarium controversum,* and some others are found only in this southern part of the Kuril Islands. The boundary between these two vegetation formations is located in the central part of the island.

MATERIALS AND METHODS

Grassland development reconstructions were based on the study of some soil profiles, located in different part of the islands from different floristic district (Figure 1). The study of the sections include pollen analysis, radiocarbon dating and tephrostratigraphy. The samples for the pollen analysis was treated with the standard KOH and acetolysis method, pollen grains were concentrated by means of the heavy-liquid flotation method [29]. Three pollen sums were calculated: total arboreal pollen, total nonarboreal pollen, and total spores. The percent of each taxon is calculated for these groups. ^{14}C dates were produced by liquid scintillation counting in the Geological Institute, RAS, Moscow (Table). The samples were treated with standard acid and alkali solutions. The calibrations of ^{14}C dates were made using the calibration program OxCal 3.9 [8, 45].

In paleoreconstructions we used non calibrated ^{14}C yr for more a correct correlation with the available paleogeographical data collected in surrounding regions. The correlation of ash layers is based on ^{14}C dates from under- and overlying deposits, on refractive indices and morphology of volcanic glass shards, and on mineral composition and chemical composition of volcanic glass (V.G. Khlopin's Radium Institute, St. Petersburg).

MODERN GRASSLANDS: DISTRIBUTION AND FLORISTIC FEATURES

Grassland landscapes are widely distributed in the Southern Kuril Islands. The most broadly dispersed grassland communities on the large islands are developed in areas with strong winds, in lowland isthmuses and at island extremes. Grasslands cover large areas in the north of Iturup Island, in the regions of Vetrovoi Isthmus and Medvezhii Peninsula where woody vegetation is stunted and is represented mostly as broken, narrow-leafed forests and larch forests, and where *Pinus pumila* thickets are found at higher elevations. Individual trees and small groves of *Quercus crispula* are encountered in certain areas [3]. Grasslands in the central portion of the island are distributed on Kuibyshevskii, Osenii, Dozornyi and Roka Isthmuses, on coastal lowlands at the head of bays, and they stretch like a narrow strip along the seashore on marine terraces and dunes. The northern (Lovtsov Peninsula and Kruglovsky Isthmus) and southern (area around the village of Golovnino) portions of Kunashir Island are almost devoid of woody vegetation and grassland communities develop on coastal lowlands and in dune fields. Thin forests on the Lesser Kuril Ridge cover only 23% of Shikotan Island and grasslands with shrub thickets are the primary landscape element (Photo 1). Alder, willow and marsh communities cover river valleys and inlet heads. Most of the small, flat islands have peatlands that cover the central portions of the islands, the river valleys and the lake shores. Woody vegetation is largely absent [10], grasslands are wide spread (Photo 2).

Grassland communities in the Southern Kuril Islands have a diverse floristic composition [2, 5, 14, 42, 43, 48, 50].

Table 1. C-ages of the soil profiles, Southern Kuril Islands.

Sample no.	Depth (m)	^{14}C-age (yr BP)	Calibrated age (2 σ)	GIN-no.
Kunashir Island				
1/5395	0.31-0.36	630±60	1280AD(95.4%)1420AD	8955
2/5395	0.53-0.58	3220±100	1750BC(95.4%)1200BC	8956
1/2396	0.45-0.50	930±100	890AD(2.0%)930AD 950AD(93.4%)1290AD	9623
2/2396	0.85-0.90	4090±70	2880BC(95.4%)2470BC	9624
1/896	1.50-1.60	2650±150	1250BC(95.4%)350BC	9638
2/896	3.0-3.10	3300±200	2200BC(95.4%)1000BC	9639
2/3895	0.40-0.43	1090±40	880AD(95.4%)1030AD	8952
4/3895	0.99-1.04	6590±50	5630BC(95.4%)5470BC	7883
1/4196	0.35-0.40	1010±70	680AD(95.4%)890AD	9616
2/4295	0.59-0.64	2350±40	850BC(95.4%)200BC	9617
3/4196	0.65-0.70	3020±50	1390BC(95.4%)930BC	9618
4/4196	0.77-0.82	3280±70	1740BC(2.5%)1710BC 1700BC(92.9%)1419BC	9618
5/4196	0.82-0.87	5630±50	4560BC(95.4%)4340BC	9620
1/6497	6.00-6.05	270±80	1400AD(95.4%)1800AD	9642
2/6497	6.90-6.95	630±140	1000AD(95.4%)1650AD	9643
1/7297	1.00-1.07	1780±40	130AD(93.6%)350AD 360AD(1.8%)390AD	9645
2/7297	1.30-1.35	3570±90	2200BC(95.4%)1650BC	9646
Zeleniy Island				
2/15903	0.38-0.48	1640±60	250AD(95.4%)560AD.	12557
3/15903	0.86-0.96	6130±130	5400BC(95.4%)4700BC	12558
Tanfiliev Island				
2/25605	0.52-0.57	4240±120	3350BC(95.4%)3450BC	13461
Iturup Island				
1/6007	0.02-0.07	650±40	1280AD(95.4%)1400AD	8951

Photo 1. Grasslands of Shikotan Island.

Photo 2. Grasslands of Zeleniy Island.

Grassland meadows, as a rule, are found in well drained areas on terrace like surfaces and on divides [5], and are often encountered in areas where snow blows off the land in winter [50]. A high level of species composition distinguishes these grassland meadows (Photo 3 and 4). Species characteristic of these areas include *Artemisia montana, Iris setosa, Poa macrocalyx, Festuca rubra, Saussurea rieder, Ptarmica macrocephala, Geranium erianhum, Dactylorhiza aristata, Fritillaria camschatcensis, Hemerocallis esculenta, Ligularia hodgsonii, Lilium pensylvanicum, Hypericum erectum, Carex scabrinervia* [5]. The species composition and the ratio of specific grassland vegetation species shift significantly from south to north [50]. The most diverse mountain grassland meadows on Shikotan Island (Photo 5) are forbic associations where *Juniperus sargentii* thickets are often encountered.

Photo 3. Grasslands of Yuri Island.

Photo 4. Grassland meadow of Shikotan Island near Krai Sveta Cape.

Grasslands with *Calamagrostis* communities including *Anemonastrum vilosissimum, Geranium erianhum, Sanguisorba tenuifolia, Adenophora tryphylla, Trientalis europea* are wide spread too.

On marine terraces in the central portion of Itutup Island grassland meadows contain bushes. Rare examples of *Larix kurilensis, Cerasus nipponika, Sorbus sambucifolia* are encountered. Grasslands on the northern portion of Iturup Island (Medvezhii Peninsula) resemble grassland associations on Urup Island. Coastal grasslands are distributed within terrace surfaces. *Ptarmica macrocephala, Anaphalis margaritacea, Bistorta viviparum, Vicia unijuga, Galium verum, Geranium erianthum, G. yezoense, Thalictrum minus, Anemonastrum vilosissimum, Acetosa lapponica, Thermopsis lupinoides, Cirsium kamtshaticum* are present on the island.

The grasslands with *Calamagrostis* assemblage are encountered where drainage on marine terraces and open cliffs is poor [5, 50]. As a rule, herbage is not thick and turf is weak. *Calamagrostis langsdorffii, Cirsium kamtschaticum, Chamaenerion angustifolium, Aruncus dioicus, Anaphalis margaritacea, Ligularia hodgsonii, Pedicularis resupinata, Thermopsis lupinoides, Trisetum sibiricum* are characteristic of these areas.

Photo 5. Grassland meadow of Western Shikotan Island.

Photo 6. Forb-sedge grassland on poor drainage surface, Zeleniy Island.

Along river valleys and lake shores and also near the outlets of hot springs there are reed grass and grassland meadows that in addition to *Calamagrostis langsdorffii,* one also encounters *Glyceria lithuanica, Carex cryptocarpa, Lathyrus pilosus. Hosta rectifolia, Stachys aspera, Onoclea sensibilis, Parathelypteris nipponica, Gentiana axillariflora, Carex maximowiczii* are typical in these communities [5]. Pure grasslands of *Calamagrostis langsdorffii* cover very limited areas along lakes and along river valleys, and often the area is marshy and gradual transitions between grasslands and herbaceous marshes are a common feature [50]. Forb-sedge grasslands are met within wet places on terraces (Photo 6).

A unique, broadly distributed landscape feature on the Southern Kuril Islands are communities consisting of various species of the genus *Sasa* that in certain locations are represented as pure thickets of *Sasa* (Photo 7). Development of these associations occurs in the absence of forest vegetation and the density and height of thickets depend on soil fertility, sunlight, and snow depth. These grassland types cover large portions of Kunashir Island, and present on Shikotan Island (Photo 8). Forb grasslands with *Sasa* (height less than 40 cm) are typical for Shikotan and Lesser Kurils (Photo 9).

Photo 7. Grasslands with *Sasa* communities of Kunashir Island.

Photo 8. Grasslands with *Sasa* communities of Shikotan Island.

Photo 9. Forb grasslands with *Sasa* of Tanfiliev Island.

Photo 10. Tall grass community with *Cacalia robusta,* and *Filipendula camtschatica,* Shikotan Island.

Unique grassland communities consisting of tall grass composed of thick vegetation that reaches 2-3 meters in height are characteristic of the Southern Kuril Islands [5, 48, 50]. These are encountered in the lower portion of ancient coastal accumulations and in creek valleys (Photo 10). Tall grass meadows on small islands in the south of the Lesser Kuril Range are found in river valleys closed off from the sea (Photo 11). *Petasites amplus, Reynoutria sachalinensis, Aconogonon weirichii, Cacalia robusta, Heracleum lanatum, Senecio cannabifolius, Filipendula camtschatica* are typical species in tall grass communities. *Urtica platyphylla, Veratrum grandiflorum, Saussurea fauriei, Cardiocrinum glehnii, Angelica ursine* are often found in these communities.

Photo 11. Tall grass community of Tanfiliev Island.

South Temperate flora species are often encountered in the understory: *Poa radula, Carex dissitiflora, Chrysosplenium grayanum, Laportea bulbifera, Erythronium japonicum, Trillium apetalum, Disporum sessile, Cremastra variabilis, Oreorchis patens, Polystichum tripteron, Trautvetteria japonica, Sanicula chinensis, Anthenoron filiforme* [5].

Marine grasslands develop on sandy stony beaches, coastal banks and dunes where halophytic and halophilous species are characteristic (Photo 12, 13). One encounters *Leymus mollis, Arctopoa eminens*, as well as species common to mixed grassland meadows: *Poa tatewakiana, P. maczocalyx, P. angustifolia, Festuca rubra, Hierochloe sachalinensis, Carex gmelinii, Angelica gmelinii, Ligusticum scoticum, Thermopsis lupinoides, Maianthemum dilatatum, Picris japonica, Arabis stelleri* [5]. Mixed grassland marine meadows with *Rosa rugosa* on Kunashir and Iturup Islands are characteristic for the Sea of Okhotsk coast [50].

Meadow soddy soils are distributed on the terrace like surfaces of Kunashir Island that are covered in grasslands [21, 44]. Several researchers separate out as grassland vegetation black, muck-humic soils that contrast to black alloophanous soils on Japanese Islands that have high fulvates in humus composition [17]. Soil profiles often include fossil soil horizons that correspond to different stages in the development of vegetation. As a rule, the lower portion of the soil profile is characterized by a thick, fossil humic horizon whose formation on Kunashir, Iturup and Zeleniy Islands dates to 4000-6000 ^{14}C yr BP. Only individual pollen grains and spores are found here, and in some instances they are entirely absent. Polygenetic soil profiles often include several layers of volcanic ash with various chemical and grain size compositions [32]. Volcanic ash from Ta-a (1739 AD), Ko-c2 (1694 AD) of Komagatake Volcano, Ma-b (about 1 ka) and Ma-f (about 6.5 ka) of Mashu Volcano, Ta-c (2.3-2.5 ka) of Tarumai Volcano is well expressed in the south of Kunashir Island (Photo 14) and on the islands in the Lesser Kuril Ridge. Marker ash layer B-Tm (969AD) of Baitoushan-Tomakomai Volcano from Northern Korea/China was found in some sections. These ash layers are wide spread within Eastern Hokkaido [13, 26]. In the north of Kunashir Island (at the extremes of Tyatya Volcano) and on Irurup Island there are layers of tephra from local volcanic sources. Thin, sod soils with grassland vegetation have developed in the upper portions of slopes and on divides. Poor developed soilts with well pronounced accumulative-humus horizon are formed under tall grass communities.

Photo 12. Grass community with *Senecio pseudoarnica*, and *Leymus mollis* on storm ridge of Dimitrov Bay, Shikotan Island.

Photo 13. Grasslands within dune field of Pervukhin Bay, Kunashir Island.

Photo 14. Soil profile with ash layers, South-Eastern Kunashir Island.

THE DEVELOPMENT OF GRASSLAND COMMUNITIES

The area around the village of Golovnino at the southern end of Kunashir Island is a region where grassland communities have developed. Herb-*Sasa* (west of the village of Golovnino), *Artemisia-Anaphalis*-herb-*Sasa* (Paltusov Cape) and *Sasa-Sieversia-Achillea-Carex*-herb assemblages (Ivanovskii Cape) are identified on terrace like surfaces [14]. The area has forest vegetation that dates to 2000 ^{14}C yr BP. Today the area is covered in grasslands [19]. South Temperate broad-leaved forests arose at the beginning of the Late Holocene (3790±70 BP, GIN-8443) in whose composition, given the dominance of *Quercus*, one would have found such thermophylic species as *Juglans, Ulmus, Corylus*, as well as *Carpinus* and *Fagus* that today are restricted to Hokkaido Island. V. M. Urusov holds that the three latter species have been absent from the Kuril Islands since the second half of the Pleistocene [48]. Climatic conditions were warmer than today and the average annual temperature was 2°C degrees higher [19]. With the cooling in the first half of the Late Holocene (2730±60 BP, GIN-8442), birch groves quickly colonized the area and many thermophylic species disappeared from the composition of oak forests.

West of the outlet of the Golovnina River, on a 12-15 meter high terrace like surface (section 5395), a profile of meadow-sod soil was studied at a polygenic soil profile that exhibits weak signs of podzolization. The top of the profile is represented by meadow soddy soil. Two buried soils appear in the lower portion of the profile. Contemporary vegetation is a herbaceous-mixed grass meadow with *Sasa*. Cover is 100%; height up to 50 cm. The pollen spectra from the soil cover demonstrate that bush-herbaceous groups covered this area in the second half of the Late Holocene. Pollen spectra from the buried soils fix a sharp drop in temperature: pollen composition is dominated by pollen similar in morphology to the *Betula* sect. *Nanae* pollen (Figure 2).

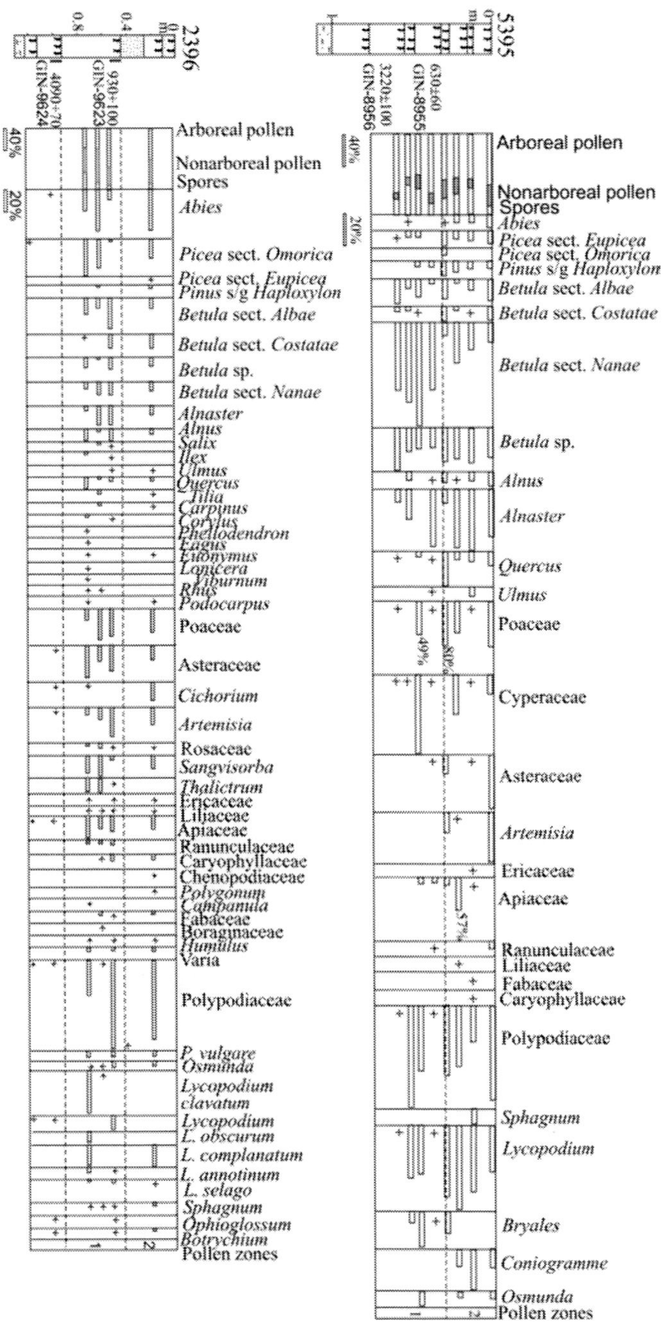

Figure 2. Percentage pollen diagram for soil profiles near Golovnino Village (section 5395) and near Lagunnoe Lake (section 2396), Kunashir Island.

Birch bushes are not observed in the south of the islands of the Kuril chain [4, 49]. It could be that the pollen found belongs to *Betula tatevakiana,* an endemic encountered on Hokkaido Island in marshes in the area of Nemuro Peninsula. The pollen might be a product of *B. fruticosa*, a species broadly distributed on the Japanese islands.

Bush-herbaceous groups developed on loamy substrate in supersaturated conditions, which is demonstrated by the high content of *Cyperaceae* pollen (up to 50%) and by the presence of *Bryales* that is gradual disappearing from the spectra. Sedge and sedge-grass associations are the most widely developed. *Alnaster*, which is broadly distributed in the contemporary vegetation groups on the coastline of southern Kunashir Island, begins to dominate in the upper portion of the profile. The large volume of *Quercus* pollen demonstrates that, apparently, this portion of the profile was formed in the lesser Holocene optimum, toward the end of the final two millenniums (^{14}C-data 630±60 BP, GIN-8955, probably renewed).

The pollen spectra obtained from the surface horizon of soil reflect a grassland community with a somewhat different composition. These are mixed herbaceous meadows that have a significant portion of *Artemisia*, *Cichoriaceae* pollen and that have developed in good drainage conditions. The horizon is formed of well sorted humus sand whose accumulation is associated with the onset of aeolian processes in the Little Ice Age. Aeolian cover from this period is especially well distributed on the south of the Sea of Okhotsk coastline of Kunashir Island [20, 36]. The fern *Coniogramme*, which is usually found in the cover of valley forests, is among the spores that are encountered.

Although grassland communities in this portion of the island have existed for a relatively long time (at least since 2500 ^{14}C yr BP), the rise of contemporary grasslands in the renewed, highly productive substratum occurred around 300 ^{14}C yr BP. Species remaining from Japanese gardens also make up part of the composition. This portion of the island is located in a zone influenced by volcanic ash from Hokkaido Island; this ash has enriched the soil with nutrients.

Grasslands on the Sea of Okhotsk side of the Yuzhno-Kurilskii Isthmus have existed since at least the lesser Holocene optimum (Figure 2). A grassland soil profile (section 2396) was studied. Three buried soils were uncovered in the lower portion of the profile. Mixed grass meadows are now present. Fossil soil appears at the base of the profile; it is dark loam that formed in warm climatic conditions between the end of the Middle and the beginning of the Late Holocene (^{14}C-data 4090±70 BP, GIN-9624). Rare coniferous pollen, grasses and spores are encountered in the fossil soil. Mixed grass meadows were distributed at this site in the second half of the Late Holocene; these were characterized by tall, diverse grass taxa (Figure 2). The slopes of Fregat and Otdelnaya Mountains were covered with dark coniferous, spruce-fir forests with pockets of birch. Remainders of South Temperate forests on the shore of Lagunnoe Lake date from the Holocene optimum. A thick soil profile formed in the Late Holocene with the addition of volcanic ash. The phase of active volcanic ash depositions on Kunashir Island is dated to 1700-1300 ^{14}C yr BP [19, 20]. Sand buildup improved drainage of the substratum and this enabled such species as Poaceae, Asteraceae, *Artemisia, Cichorium,* Caryophyllaceae to develop in dry habitats. Pollen from these species increased from 43% to 75%. The pollen spectra from a thin surface soil that is covered by a layer of finely sorted volcanic ash demonstrate that following the Little Ice Age the floristic composition of grassland communities did not undergo significant modification.

Grassland communities in the north of Kunashir Island, the Kruglovskii Isthmus and Lovtsov Peninsula are located in a zone affected by volcanic ash from Tyatya Volcano. These volcanic ash depositions exerted an especially strong influence on grassland communities on that part of the isthmus where volcanic ash fall was heavy. Volcanic ash has at time completely covered the soil. A shallow strait that was blocked from the Pacific Ocean by a barrier wall probably existed in the Middle Holocene, at the site where the isthmus was located in the end of Middle Holocene – beginning Late Holocene. The low isthmus located at this site was repeatedly closed by thick layers of tephra, the result of which is that the tephra grew to 20 meters in height. The radiocarbon age of fossil soils exposed in the creek valley that runs into Dlinnoe Lake demonstrates that the last, rather heavy eruptions of Tyatya Volcano occurred around 3500 and 2000 ^{14}C yr. BP (^{14}C-data 3570±90 BP, GIN-9646, 1780±40 BP, GIN -9645). Pollen spectra obtained from fossil soils and that are mixed with continental depositions reflect the broad development of mixed grass meadows on the isthmus (content of grass pollen is 28-29%), this, at least in the beginning of the Holocene. The vegetation of dry (*Artemisia,* Asteraceae, Caryophyllaceae) and wet (*Thalictrum,* Apiaceae, *Sanguisorba,* Ranunculaceae, *Polygonum, Rumex,* Primulaceae) habitats contain grass pollen. The percentage of pollen associations for dry meadows found in the surface soil atop tephra remains below 13%. The richness of the species composition of the grassland on the isthmus might be explained by the repeated renewal of the soil profile and by the large volume of nutrients entering with the tephra.

An analysis of fossil soils in dunes on the Sea of Okhotsk side of the Lovtsov Peninsula dates meadow vegetation to the Little Ice Age portion of the Holocene optimum (^{14}C-data from fossil soils in dunes 630±140 BP, GIN-9643, 270±80 BP, GIN-9642). In the pollen spectra, along with grass associations found in well drained areas (Asteraceae – up to 38%, *Artemisia* – up to 30%, Poaceae – up to 13%, *Cichorium, Sanguisorba*), the pollen spectra show that hydrophilous Apiaceae, Cyperaceae, *Thalictrum* and by other representatives of the family Ranunculaceae, *Polygonum,* Liliaceae, Iridaceae, Primulaceae, Polemoniacea played a large roll (up to 45%).

Development of stone birch forests with *Pinus pumila* and grassland communities on Iturup Island, features characteristic of areas north of Vetrovoi Isthmus occurred over an extended period of time. An important floristic boundary cuts across this region [3, 42, 48]. Studies of fossil soils in volcanic ash cover that are located in Parusnaya Bay at an elevation of 40 meters show that grassy communities were a sufficiently stable component of the landscape in the second half of the Late Holocene. The age of aeolian cover is estimated to correspond to the second or third generation volcanic aeolian sands for the Southern Kuils, whose formation took place during a cooling period between 1700-1300 BP and in the Little Ice Age [19, 36]. Pollen spectra from fossil soils at the base of the profile (Figure 3) demonstrate the broad development of alder forests on drained territory of Vetrovoi Isthmus in the middle of the Late Holocene. Arboreal pollen from Poaceae, and Poaceae-*Artemisia* assemblages in dry habitats that are mixed with forb-mix-grass associations dominate the upper horizons of the fossil soils. The pollen spectra from a contemporary soil profile demonstrate the distribution of *Pinus pumila,* whose grass pollen composition responds to mix grass meadow communities. Representatives of the genus *Botryhium* are noted among the ferns.

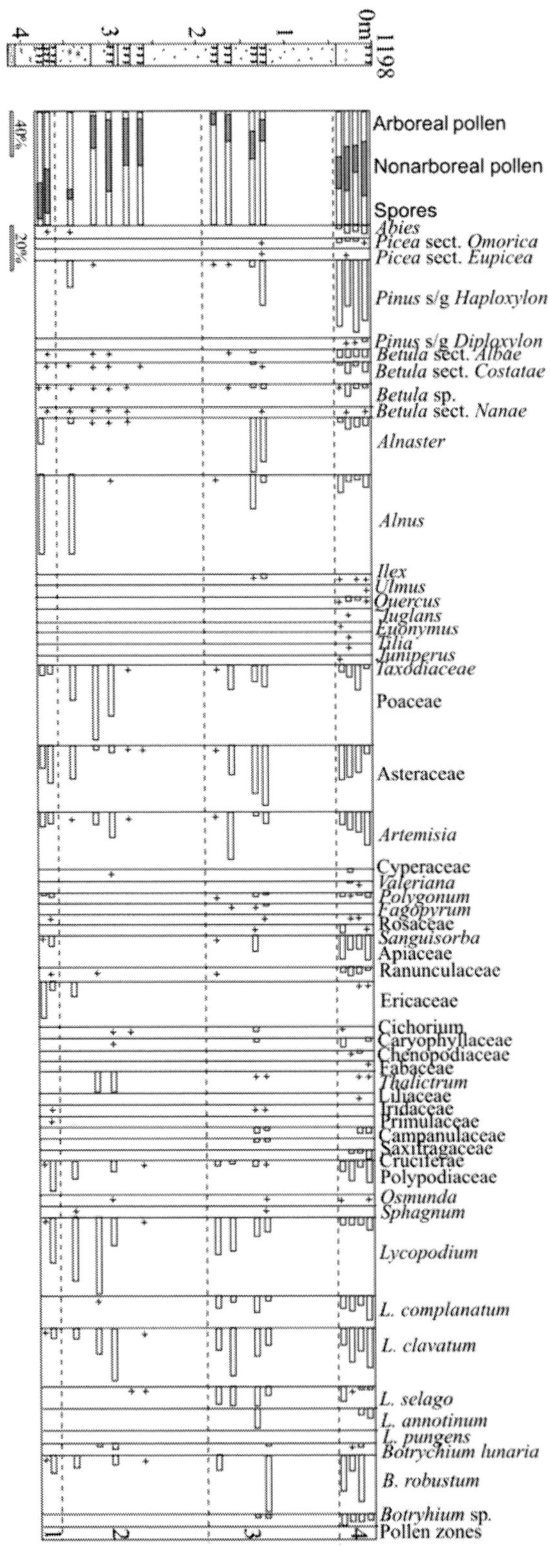

Figure 3. Percentage pollen diagram of eolian-paleosol cover on the coast of Parusnaya Inlet, Iturup Island (section 1198).

The abundance and large species diversity of mosses on Itirup Island contrasts to the grassland communities in the south of Kunashir Island.

The herbaceous communities on marshy, accumulative plains found along the coastlines also vary in age. The accumulative forms at the head of Yuzhno-Kurilskii Bay and in the lower drainage of the Rogachevka and Saratovka Rivers are the oldest examples on Kunashir Island. Their formation began in the second half of the Middle Holocene, which is demonstrated by the presence of dune ridges that were formed during the regression during the cooling period that occurred around 4500-4700 years ago and that now block low, marine terraces [20]. Herbaceous communities on the marshy plain in the Serebryanka River valley have a unique floristic composition and are distinguished by a high variety of species. The rich vegetation is possibly a result of habitat variety in areas with different saturation regimes, given the underlying substratum, and is related to the relatively long development period for vegetation cover in the alternating climatic conditions characteristic of the Middle and Late Holocene. Storm depositions in the inner portion of the isthmus are covered in dark coniferous forest that displaced South Temperate forests during the Holocene optimum. It is possible that some species in modern herbaceous communities are inherited from the Holocene optimum.

A low, accumulative valley in the lower drainage of the Saratovka and Rogachevka Rivers was formed at the end of the Middle Holocene. A ridge of dunes that stretches along the shore and that separates a low, marine terrace (section 896) began to form during the cooling at the cusp of the Middle and Late Holocene, this demonstrated by radiocarbon dating of the fossil and illuvial-humus soils (3300±200 BP, GIN-9639, 2650±150 BP, GIN-9638). Herbaceous communities at this site have existed since at least the beginning of the Late Holocene and have been in combination with dark coniferous forests that have birch and broad-leaved species covering surrounding slopes (*Ulmus, Quercus, Juglans, Carpinus, Tilia*). The pollen composition of herbaceous taxa reflects the development of mixed grass associations (Figure 4). A significant variety of mosses is found, which contrasts to grassland communities at the southern end of the island. The presence of *Potamogeton* pollen shows that lakes in the deflation basin have existed for at least 3000 years. The formation of soil with a loamy composition and that contains volcanic ash from the Tyatya Volcano, and that serves as a water-resistant horizon, is probably the reason for their formation amidst the dunes. The spread of grass pollen and spores across the profile shows that on the whole the structure of herbaceous communities in the Late Holocene did not change significantly.

The development of marshy-herbaceous communities occurred in younger, low marine terraces in the southeast of Kunashir Island. Accumulative deposits began to form around 3000 years ago, and at Veselovskii Spit - around 2000 years ago [19, 16]. Soil cover was studied on a terrace two meters above sea level (section 4196), one kilometer to the north of the outlet of Belozerskaya River. The profile revealed meadow sod and the buried, polygentic profile. The soil profile includes three buried soil horizons. Contemporary vegetation is a forbic mix-grass meadow. Middle Holocene age fossil soil appears at the base of the profile (^{14}C-data 5630±50 BP, GIN-9620). Pollen spectra from this soil horizon are primarily composed of grass pollen that belong to grass communities (Figure 5). The question of the preservation of woody pollen is, however, not entirely clear. Spores and pollen are not found at the top of this horizon. Fossil soil formed at the beginning of the Late Holocene is layered higher.

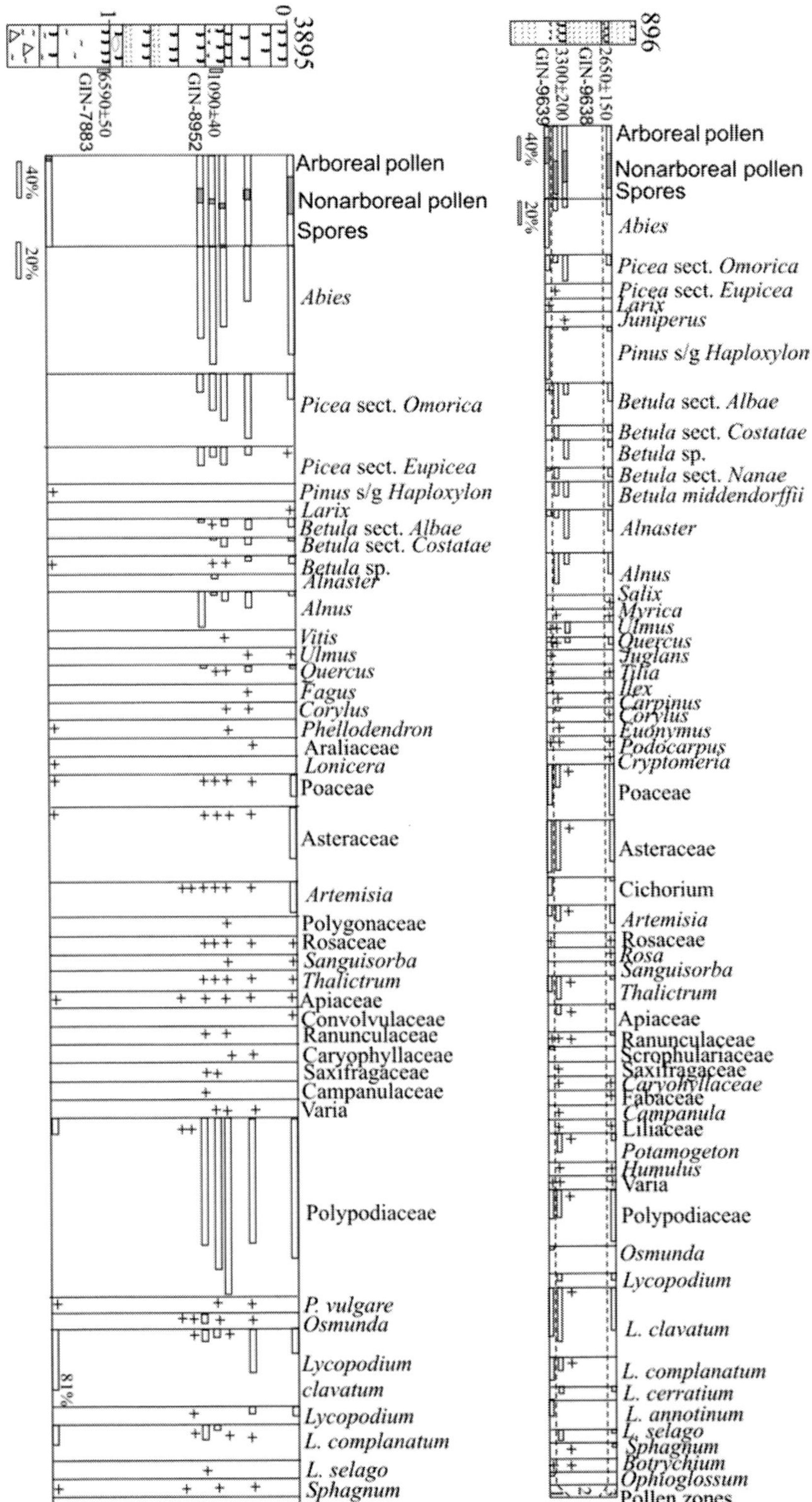

Figure 4. Percentage pollen diagrams for paleosols from dunes near Rogachevka and Saratovka River mouths (section 898), and soil profile near Otradnoe Village (section 3895), Kunashir Island.

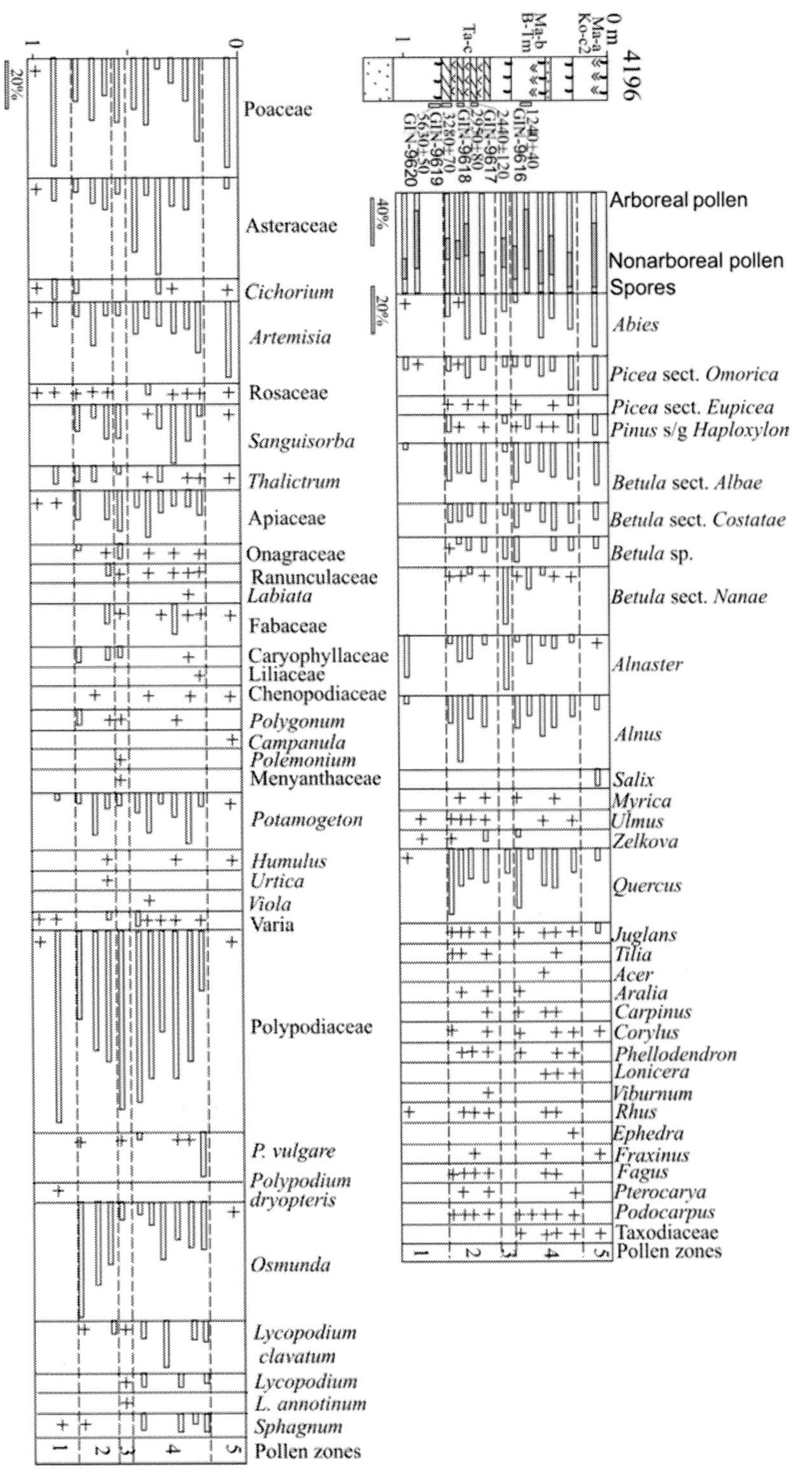

Figure 5. Percentage pollen diagram for soil profile on 2 m terrace, located in 1 km northward Belozerskaya River, Kunashir Island (section 4196).

The structure of the pollen spectra demonstrates that on the coast, South Temperate broad-leaved forests existed at the south end of the island until the beginning of the Late Holocene. Radiocarbon dating obtained for this horizon is for a period from 3280±70 BP, GIN-9619 until 2440±120 BP, GIN-9617 and the sample may be somewhat renewed. The soil includes three volcanic ash seams of rhyolitic and dacitic composition that correspond to volcanic eruptions on Hokkaido Island. Such exotic species as *Podocarpus*, *Taxodiaceae*, which along with *Fagus* and *Pterocarya* pollen may have been brought in from Japanese Islands, were discovered in the Kuril Islands. Redeposition of this pollen is hardly likely since the saturation of the pollen deposition of the Golovnin River glomerate is extremely low [31]. Such an interpretation suggests that predominant southerly winds (Okhotsk-Aleut atmospheric circulation type, according the A. M. Polyakova [30]) were the most typical for the spring and summer seasons in the warm phase of the Holocene. Abundant *Potamogeton* pollen (up to 20%) demonstrates the proximity of nearby lake bodies. Accumulative deposits probably began to form on the southeast portion of Kunashir Island at this time. These were a result of a series of ocean depositions that led to the establishment of marshy areas and numerous small lakes.

The cooling in the second half of the Late Holocene marks a sharp increase in the percent of *Betula* sect. *Nanae* and *Alnaster* pollen. Grassland communities expanded and the percentage of grass pollen in the pollen spectra grew significantly. Ferns, including the genus *Osmunda*, played a large role in grassland associations. The later development of grassland communities on expanding, sandy accumulative deposits is a feature of the lesser Holocene optimum. Aeolian material improved drainage and *Artemisia* (up to 40%) and the forbs (more than 50%) played a large role in grassland communities, despite a general decline in the variety of grassland taxa. It is possible that *Sasa*-mixed herbaceous communities on the slopes of Golovnin Volcano played the determining role in the spectrum represented in the surface soil horizon. The woody vegetation at the south of Kunashir Island is dark coniferous and birch forests with small oak groves.

The age of the lowland Sernovodskii Isthmus is from the Late Holocene. The most ancient portion of the isthmus, formed at the end of the Middle Holocene, is located on the Sea of Okhotsk side of the island and is cut off by dunes formed between the Middle and Late Holocene. Marshy lowlands on the Pacific Ocean side of the isthmus formed in the Late Holocene (^{14}C-data 2220±80 BP, GIN-7895). In the first half of the Late Holocene this portion of the island was covered in mixed coniferous broad-leafed forest [19]. Given supersaturated conditions at the lowland's initial stage of formation, sphagnum moss and alder associations expanded broadly; in the Middle to Late Holocene they were quickly replaced by mixed herbaceous communities that included ferns and forbs. The upper portion of the peatlands were heavy in sand that probably arose from aeolian materials and that show the presence of volcanic ash from Mashu Volcano on Hokkaido Island.

Grasslands on the Vetrovoi Isthmus of Iturup Island belong to relatively young formations. Despite the fact that a strait existed on the isthmus in the inter-glacial period, the surface deposits of the isthmus are of a much younger age. The rise in sea level in the Middle and Late Holocene transgression period [34, 35] did not entirely submerge the isthmus. Dunes, with a series of thin fossil soils on the Sea of Okhotsk side of Vetrovoi Isthmus, contrast to second generation dunes on Kunashir Island [20, 36]. A ridge of dunes along the Pacific Ocean side of the isthmus that arose in the Little Ice Age characteristically lack fossil soils. The surface of the isthmus is universally covered in loom more than 2 meters thick and that is saturated with angular fragments of vulcanite that could possibly have a lahar origin. The age of these deposits is around 1280±110 BP, GIN-10494. Grassland vegetation began to colonize the isthmus during the cooling period that, along with unfavorable strong wind conditions, led to the formation of pseudo-alpine communities. The vegetation on the isthmus has been subjected to serious anthropogenic impact: a landing strip, warehouses and a series of villages were located here, leading to the introduction of weed and exotic species into the association. Grasslands on the coastal lowland near Kurilsk of Iturup Island have natural origin too. These grassland landscape existed last 2000 yr. PB [35].

There are, along with natural climate driven grasslands on the islands, grassland communities that formed as a result of timber harvest that was conducted until the mid 20th century. Intensive harvest took place on the slopes of Mendeleev Volcano [22, 40]. The *Sasa* associations containing thin birch groves on the slopes of Golovnin Volcano are probably also a result of timber harvest. The recent formation of these grassland communities is clearly shown in the pollen spectra for these soil profiles. The grasslands in the area of Otradnoe Peninsula on Kunashir Island may also have their origin in human activity. An abundance of grass pollen (40%) is found only in the surface layer of the soil (Figure 5). During the less Holocene optimum (^{14}C-data 1090±40 BP, GIN-8952) dark coniferous fir-spruce forests with fern understory were spread along this area. Spruce forest also covered dune fields in Golovnin Bay. Spruce roots have been discovered in-situ in the first fossil soil layer.

The grasslands on the shoreline of Kuibyshevsky Bay and the isthmus of the same name that colonized areas where larch and birch-larch forests were harvested on Iturup Island are a result of anthropogenic impact. A high content of grass pollen (37-45%) is noted for this region only in the pollen spectra for fossil soils in dunes located near Maloe Lake. Radiocarbon dating of the age of the upper soil horizon (650±40 BP, GIN-8951) shows that a series of fossil soils formed at the end of the Late Holocene. The composition of the grasses corresponds to the formation of Poaceae-*Artemisia*-mixed herb groups of Poaceae-*Artemisia*-herb assemblages on dune ridges and moist *Carex*-Poaceae associations with *Sphagnum,* Ranunculaceae in marshy areas of the deflation basin. Forest type spectra whose grass pollen content does not exceed 1-3% are characteristic for soils in the central portion of the isthmus. The predominance of wood and shrub pollen species is characteristic of fossil soils in the dune field on the coastline of Kasatka Bay and for the soil profile at 14-16 meters on a terrace like surface in the area of Burevestnik Peninsula. It is interesting to note that the percent of *Pinus* s/g *Haploxylon* pollen increases sharply only in the upper portions of soil profiles. Apparently, as cooling occurred, a series of volcanic ash deposit in the 1600-500 ^{14}C yr BP period led to a much broader distribution of *Pinus pumila* on the Pacific Ocean coastline of the island. The volcanic ash has a primarily basaltic and andesite-basaltic content. Its origin is probably the volcanoes in the Ivan Groznyi group. The regular appearance of *Pinus* s/g *Diploxylon* pollen in the pollen spectra from the soil profile near Burevestnik Peninsula confirms the hypothesis of V. M. Urusov about the existence of a refugium *Pinus parviflora* in the region of Kuibyshevskii isthmus in the Holocene [48].

A polygenic soil profile (Figure 6) has been studied in the western portion of Zeleniy Island on terrace like surface at an elevation of 8-9 meters. The area has mixed grass meadows. The contemporary soil is sod-humic on buried layered ash soil. Two fossil profiles appear at the foundation.

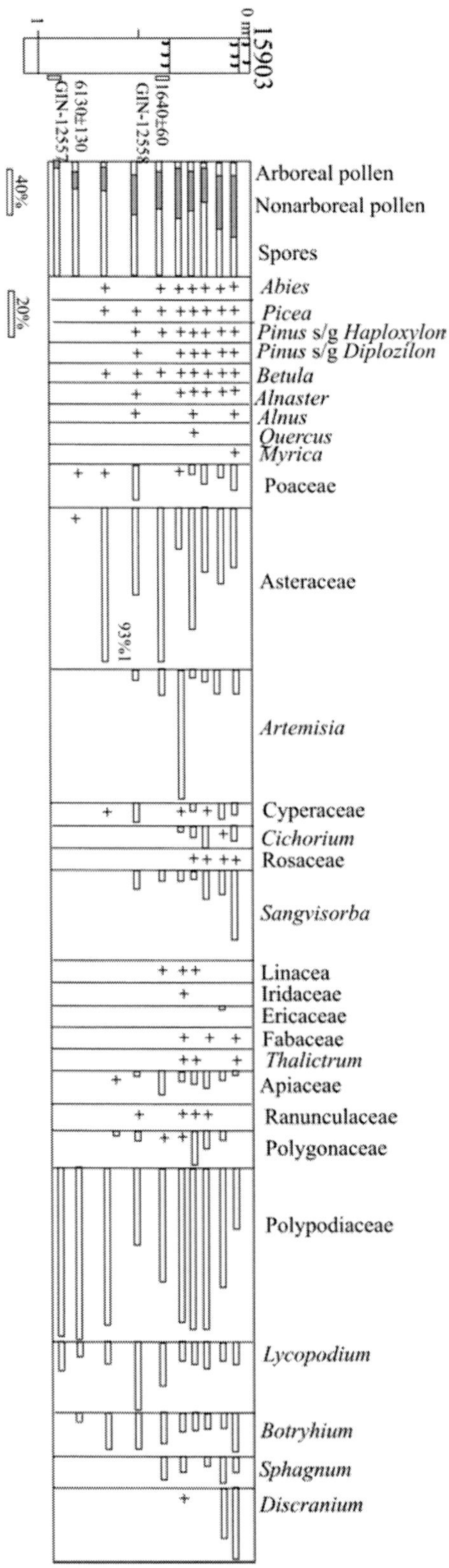

Figure 6. Percentage pollen diagram for soil profiles of Zeleniy Island (section 15903).

From the upper portion of the first fossil soil ^{14}C-data 1640±60 BP, GIN-12557 have been obtained, and from the base of the profile – ^{14}C-data 6130±130 BP, GIN-12558. Grass pollen (up to 56%) and spores (up to 62%) dominate the pollen spectra. The pollen content is diverse and is dominated by Asteraceae (up to 68.6%), *Artemisia* (up to 59%), *Sanguisorba* (up to 33%), Poaceae (up to 12%) that reflects development of mix grass meadows with rich species diversity. Polypodiaceae spores dominate. There is abundant *Lycopodium*. Soil under grassland on Tanfiliev Island has similar profile with Middle Holocene paleosol (^{14}C-data 4240±120 BP, GIN-13461). The pollen spectra show that grassland communities on the island existed at least from the Middle Holocene, and possibly throughout the enter Holocene [33, 37]. Similar data were obtained for Nemuro Peninsula (Hokkaido Island) [15, 24] and Yururi Island near Eastern Hokkaido [25].

Pollen spectra for peatland profiles on Shikotan Island show that forests were well developed in the Middle and Late Holocene [33]. Forests began to decrease at the end of the Late Holocene, possibly in response to ocean impacts in a context of general cooling. Forests were preserved in limited areas, on the western side of the island that were protected by ancient volcanoes from the cooling effects of the Pacific Ocean. Fires and timber harvest may also be reasons for the destruction of small forest areas. The *Sasa* grasslands with dense sod hamper to progress of forest vegetation restoration. Forest vegetation begins to occupy the places with broken soil cover (old roads, fields).

DISCUSSION

Analysis of the paleogeographic material obtained during a study of soil profiles shows that along with anthropogenic promoted grassland communities that have formed following timber harvest, the Southern Kuril Islands are covered by expansive areas of naturally occurring grasslands. The human colonization of the islands began around 3000-4000 ^{14}C yr BP [51]. This colonization, however, does not explain the irreversible impacts on vegetation nor can it serve as the reason for the decline in forest vegetation that over the course of 2000 years has not recovered in some areas.

Development of climate driven grasslands takes a significant amount of time. On windy sections of the islands that are subject to cold, low layers of ocean air, the formation of grassland communities is probably associated with the overall cooling of the climate in the Late Holocene; grasslands, as a landscape element, have existed for around 2000 years. The most extended cooling period in the Southern Kuril Islands took place between 1700-1300 ^{14}C yr BP [19]. The Kofun cooling cycle, which was the most extended in six thousand years, occurred on Japanese islands at that time [41]. The forestless coastal belt on the Southern Kuril Islands is probably associated less with a temperature decrease as with an intensification in the impact of continual winds and fogs that are a product of the cold waters associated with the Oyasio current. The influence of the Oyasio current is especially apparent on flat, terrace like surfaces devoid of mountain barriers that subjected to Pacific Ocean winds. The warm Soya current has a significant warming influence on vegetation on the Sea of Okhotsk coastlines of Kunashir and Iturup Islands [48], declined in the Late Holocene [46].

Marsh-herbaceous groups that are distributed on low marine terraces are various aged and range from the Middle to the Late Holocene, a fact that should be reflected in the floristic composition of communities. These areas may have experienced problems as the climate changed in the Middle and Late Holocene, processes that may have resulted in the formation of new habitat and the introduction of inherited flora. The most diverse herbaceous communities are observed on coastal plains whose formation in the central and southern portions of Kunashir began in the Holocene optimum and where South Temperate forest was developed. In the zone of dark coniferous forests in the north of Kunashir Island the composition is less diverse. On young accumulative forms of the Late Holocene one would expect communities with a poorer floristic composition.

The character of the substratum and the capacity to retain heat are important ecologic factors affecting the vegetation on marine coastlines [43]. The mechanical composition of rocks that determine hydro-physical characteristics and the position of the soil-ground water level are significant features for an assessment of the inertia of geosystems during evolutionary changes that occurred during climate change in the Holocene [11]. In this respect, island territories belong to a group of unique geologic ecosystems where, in contrast to continental regions, the lithogenic component is one of the most dynamic components of the landscape, a component that responds quickly to even the smallest changes in climate as level changes. Based on the example of the development of grassland communities, it has been established that rapid changes in the lithogenic component result in a renewed substratum and, first of all, through aeolian processes that were set in motion during small increases in the regression of the Middle and Late Holocene. Sand buildup changed the hydro-physical characteristics of soils and this significantly influenced the composition of herbaceous communities. With sand buildup, supersaturated sedge and sedge-mixed grass meadows, even in lowland isthmuses, shifted to mix-grass forbic meadows and the role of xerophile species, especially sagebrush, increased.

Volcanic ash accumulations that have varied chemical compositions also have a major impact on the change in the hydro-physical characteristics of the substratum. A loamy composition is characteristic of andesite-basaltic pyroclastic formations; they hinder water exchange. Volcanic ash with rhyolitic and dacitic elements contains, as a rule, aleurite material that is easily saturated. Renewed substratum and a change in the ground water regime can occur in response to slop processes that follow climate rhythms and the dynamics of the lithogenic component on the coastline. They also respond to catastrophic events like tsunamis, mud flows, landslides, and rock fall. Vetrovoi Isthmus, following a mud flow, is an example where vegetation colonized the territory and where the fossil soil layer of loam is at least two meters thick. Unfavorable climatic conditions led to the formation of herbaceous communities with a poor floristic composition.

Ground water levels in coastal lowlands modulate not only in response to the volume of water in various seasons but also in response to erosion dynamics and accumulative processes in stream outlets. Catastrophic events like strong earthquakes, which can increase the amount material available in an outlet, can have an especially strong impact on the formation of outlet spits. Extended spits formed over the course of a year or two following the earthquake in 2004 and this led to the inundation of the outlet portions of streams and to a rise in soil and ground waters in all coastal plains.

If we examine the question of the natural origin of grasslands mainly as a response to a change in climatic conditions in the second half of the Late Holocene, one must note that herbaceous communities over the last 1500-2000 years have been rather stable on the windy island coastlines. The most vulnerable component of island landscapes in modern climatic conditions has been woody vegetation that has poorly recovered on harvest sites, especially on coasts where grassland communities are developing. This trend should be taken into account when exploiting island territories and when planning and executing environmental protection measures.

Conclusions

In addition to the rise of anthropogenic driven grassland communities on windy portions of the coastline of the Southern Kuril Islands, natural grasslands that formed during the cooling period that occurred 1300-1700 years ago are broadly distributed. Grassland community development is probably less associated with drops in temperature than with an intensification of the impact of continual winds and fogs that are, in part, a product of the cold waters associated with the Oyasio current.

Marsh-grassland groups distributed in the lower marine terraces are dated to the Middle to Late Holocene. Phytocenoic complications may have occurred during climatic fluctuations in the Middle and Late Holocene, these associated both with the formation of new habitat during the growth of accumulative forms and as a result of heritage flora.

During small climate shifts in the Middle and Late Holocene that were accompanied by a rise and fall in sea level, the lithogenic component controlling the hydro-physical characteristics of soils that influence the composition of herbaceous communities was one of the most dynamic components of island landscapes. A renewed substratum occurred in response to aeolian processes during small amplitude regressions in the Middle and Late Holocene. Volcanic ash deposits that have varied chemical compositions also played a role.

Grassland communities have over the course of 1500-2000 ^{14}C yr BP been reasonably stable on windy areas of the coastline of islands. The most vulnerable component of island landscapes, given current climatic conditions, is woody vegetation that poorly recovers on harvest sites, especially on the coast where grassland communities develop.

Acknowledgements

We would like to thank N. P. Domra (Institute of Biology and Soil Sciences FEB RAS, Vladivostok) for help in pollen analysis and A. A. Kharlamov (Institute of Oceanology RAS, Moscow) for help with the field work.

The study was supported by RFBR, project 06-05-64033, and field grants of FEB RAS.

References

[1] Aleksandrovskiy, A.L.; Chichagova, O.A. Radiocarbon age of Holocene paleosols in forest-steppe of the East Europe. *Soil science*, 1998, 12. 1414-1422.

[2] Alekseeva, L.M. Flora of Kunashir Island. Vladivostok: Far East Scientific Centre Publication; 1983.

[3] Barkalov, V.Yu. Sketch of vegetation. In: Storozhenko S.Yu, editor. *Vegetation and fauna of Kurils.* Vladivostok: Dalnauka; 2002; 35-66.

[4] Barkalov, V.Yu. Vascular plants of Kuril Reserve (the Sakhalin area). In: Kozhevnikov A.E., editor. *Flora of protected territories of the Russian Far East: Magadanskiy, Bureinskiy and Kurilsky Reserves.* Vladivostok: Dalnauka; 1998; 71-113.

[5] Barkalov, V.Yu.; Eremenko, N.A.. Flora of Natural Reserve "Kurilsky" and Reserve "Minor Kurils" (Sakhalin District). Vladivostok: Dalnauka; 2003.

[6] Belaya, G.A. Structure and functioning of high productive grassy ecosystems. Vladivostok: FEB Academy of Science of the USSR; 1991.

[7] Belaya, G.A.. The structurally-functional organization of natural grassy ecosystems of the Far East. In: Ryanskiy F.N., Yushmanov V.V., editors. *Structural organization and interaction of regulated social-natural systems.* Vladivostok: Dalnauka; 1998; 279-288.

[8] Bronk Ramsey, C. Development of the radiocarbon Program OxCal. *Radiocarbon,* 2001, 43. 2A. 355-363.

[9] Bukreeva, G.F.; Votah, M.R.; Bishaev, A.A. Definition of paleoclimates by pollen data (methods of target iterative classification and regressive analysis). Novosibirsk: Institute of Geology and Geophysics SB Academy of Science of the USSR; 1986.

[10] Chernyaeva, A.M. Flora of Zeleniy Island (Lesser Kuril Arc). *Botanical Journal,* 1977, 62, 1672-1682.

[11] Dyakonov, K.N.; Ivanov, A.N. Stability and inertia of geosystem. *Bulletin of Moscow State University,* 1991, 5. 28-34.

[12] Eremenko, N.A. Season development of plants and plant communities of Southern Kuril Islands. Vladivostok: Biological and Soil Researches Institute FEB RAS Publication; 2004.

[13] Furukawa, R.; Nanayama, F. Holocene Fall Deposits along the Pacific Coastal Region of Eastern Hokkaido. *Bulletin of Volkanological Society of Japan,* 2006, 51. 351-371.

[14] Grabkov, V.K.; Berzan, A.P., Alekseeva, L.M. Distribution of vegetation on a southern part of Kunashir Island. Yuzhno-Sakhalinsk: VINITI Publisher; 1989.

[15] Igarashi, Ya.; Igarashi, T.; Endo, K.; Yamada, O.; Nakagawa, M.; Sumita, M. Vegetation history since the Late Glacial of Habomai Bog and Ochiishi Cape Bog, Nemuro Peninsula, eastern Hokkaido, north Japan. *Japanese Journal Historical Botany,* 2001, 10. 67-79.

[16] Il'iev, A.Ya.; Kaistrenko, V.M.; Gretskaya, E.V.; Tikhonchuk, E.A.; Razjigaeva, N.G.; Grebennikova, T.A.; Ganzey, L.A.; Kharlamov, A.A. Holocene Tsunami Traces on Kunashir Island, Kuril Subduction Zone. In: Satake, K., editor. *Tsunamis: Case Studies and Recent Development.* Netherlands: Springer; 2005; 171-192.

[17] Ivlev, A.M.; Targuljan, V.O.; Kulikov, A.V. Soils of Kunashir Island. In:Kostenkov N.V., Oznobikhin V.I., editors. *Soil cover of the Far East, the problems of its effective using, irrigation, and reservation.* Vladivostok: FEB Academy of Science of the USSR; 1987; 30-49.

[18] Kamenkovich, V.M.; Koshlyakov, M.N.; Monin, A.S. Synoptic Eddies in the Ocean. Leningrad: Gidrometeoizdat; 1987.

[19] Korotky, A.M.; Razjigaeva, N.G.; Grebennikova, T.A.; Ganzey, L.A.; Mokhova, L.M.; Bazarova, V.B.; Sulerzhitsky, L.D.; Lutaenko, K.A. Middle and late-Holocene

environments and vegetation history of Kunashir Island, Kuril Islands, northwestern Pacific. *Holocene*, 2000, 10. 311-331.

[20] Korotky, A.M.; Razjigaeva, N.G.; Mokhova, L.M.; Ganzey, L.A.; Grebennikova, T.A.; Bazarova, V.B. Coastal dunes as indicator of Periods of Global Climatic Deterioration (Kunashiri Island, Kuril). *Geology of Pacific Ocean,* 1996, 13. 73-84.

[21] Liverovsky, Yu. A. Problems of Soil Genesis and Geography. Moscow: Nauka; 1987.

[22] Manko, Yu.I.; Rosenberg V.A. High-altitude zones of vegetation on Mendeleyev volcano. In: Bromley G.F., Butukovsky B.G., Vasiliev N.G., editors. *Biological resources of Sakhalin Island and Kuril Islands*. Vladivostok: FEB Academy of Science of the USSR; 1970; 65-71.

[23] Mokhova, L.M.; Eremenko, N.A. Conformity of structure of pollen rain and subfossil spectra of modern vegetation of Kunashir Island. In: Urusov V.M.; Pshenichnikova, N.F.; Lobanova I.I., editors. *Research and designing of landscapes of the Far East and Siberia.* Vladivostok: Dalnauka; 2001; 41-61.

[24] Morita Y. Vegetation history in Nemuro Peninsula since the Late Pleistocene. *Vegetation Science,* 2001, 19. 39-44.

[25] Morita Y. Vegetation history of Yururi Island in easternmost Hokkaido since the Late Holocene. *Japanese Journal Historical Botany,* 2001, 20. 81-89.

[26] Nakagawa, M.; Ishizuka, Y.; Kudo, T.; Yoshimoto, M.; Hirose, W.;Ishizaki, Y.; Gouchi, N.; Katsui, Y.; Solovyow, A.W.; Steinberg, G.S.; Abdurakhmanov, A.I. Tyatya Volcano, southwestern Kuril arc: Recent eruptive activity inferred from widespread tephra. *The Island Arc,* 2002, 11. 236-254.

[27] Nunn, P.D. Late Quaternary environmental changes on Pacific islands: controversy, certainly and conjecture. *Journal of Quaternary Science*, 1997, 12. 443-450.

[28] Nunn, P.D. Oceanic Islands. Oxford: Blackwell Publication; 1994.

[29] Pokrovskaya, I.M.. Methods of paleopollen studies. In: Pokrovskaya I.M., editor. *Paleopolinology.* Leningrad: Nedra; 1966; 29-60.

[30] Polyakova, A.M. Calendar of types of atmospheric circulation with the account of unstationary of processes above northern part of Pacific ocean and their brief characteristic. Vladivostok: FESU Publishers; 1999.

[31] Pushkar, V.S.; Razzhigaeva, N.G.; Korotky, A.M.; Mokhova, L.M. Sediments and age of late Golovnin transgretion on Kunashir Island (Middle Pleistocene). *Stratigraphy. Geological correlation,* 1998, 6. 50-64.

[32] Razjigaeva, N.G.; Ganzey, L.A. Sedimentary environments on islands in Pleistocene-Holocene. Vladivostok: Dalnauka; 2006.

[33] Razjigaeva, N.G.; Ganzey, L.A.; Belyanina, N.I.; Grebennikova, T.A.; Ganzey, K.S. Paleo- environments and Landscape History of Minor Kuril Islands since Late Glacial. *Quaternary International,* 2008, 179. 83-89.

[34] Razjigaeva, N.G.; Grebennikova, T.A.; Ganzey, L.A.; Mokhova, L.M.; Bazarova, V.B. The role of global and local factors in determining the middle to late Holocene environmental history of the South Kuril and Komandar Islands, northwestern Pacific. *Palaeogeography, Palaeoclimatology, Palaeoecology*, 2004, 209. 313-333.

[35] Razjigaeva, N.G.; Korotky, A.M.; Grebennikova, T.A.; Ganzey, L.A.; Mokhova, L.M.; Bazarova, V.B.; Sulerzhitsky, L.D.; Lutaenko, K. A. Holocene climatic changes and environmental history of Iturup Island, Kuril Islands, northwestern Pacific. *Holocene,* 2002, 12. 469-480.

[36] Razzhigaeva, N.G.; Ganzey, L.A. Development of coastal dunes during sea level changes. *Oceanology*, 2005, 45. 150-160.

[37] Razzhigaeva, N.G.; Ganzey, L.A.; Beljanina, N.I.; Ganzey K.S. Landscapes development on Zeleniy Island (Lesser Kuril Arc) in Late Pleistocene-Holocene. In: Urusov V.M., Lobanova I.I., editors. *Research and designing of landscapes of the Far East and Siberia*. Vladivostok: Dalnauka; 2005; 159-165.

[38] Reference Book of USSR Climate. Leningrad: Gidrometeoizdat; 1968.

[39] Reference Book of USSR Climate. Leningrad: Gidrometeoizdat; 1970.

[40] Rozenberg, V.A.; Manko, J.I.; Vasilyev N.G. Result of studying of arboreal vegetation of Kurils. In: Bromley G.F., Butukovsky B.G., Vasiliev N.G., editors. *Biology Resources of Sakhalin Island and Kurils*. Vladivostok: FEB Academy of Science of the USSR; 1970; 53-58.

[41] Sakaguchi, Y. Warm and cold stages in the past 7600 years in Japan and their global correlation. *Bulletin of the Department of Geography University of Tokyo,* 1983, 15. 1-31.

[42] Seledets, V.P. To botanic-geographical division into districts of Iturup Island (Southern Kurils). In: Vasilieva L.N.; Vorobiev D.P.; Gorovoy P.G.; Probatova, N.S., editors. *Questions of botany on the Far East.* Vladivostok: FEB Academy of Science of the USSR; 1969. P. 181-192.

[43] Seledets, V.P. Method of ecological scales in botanical researches on the Russian Far East. Vladivostok: Dalnauka; 2000.

[44] Sokolov, I. A. *Volcanism and Soil Formation*. Moscow: Nauka; 1973.

[45] Stuiver, M.; Reimer, P.J.; Bard, E. *et al.* INTCAL 98 Radiocarbon Age Calibration, 24000-0 cal BP. *Radiocarbon*, 1998, 40. 1041-1083.

[46] Taira, K.; Lutaenko, K Holocene Palaeoceanographic Changes in the Sea of Japan. *Reports of the Taisetsuzan Institute of Science,* 1993, 28. 65-70.

[47] Tjuremnov, S.N.; Berezina, N.A. About destruction of pollen of tree species in various conditions of a water-mineral regim. *Bulletin of Moscow State University,* 1965, 5. 62-71.

[48] Urusov, V.M.; Chipizubova, M.N. Vegetation of Kurils. Questions of dynamics and origin. Vladivostok: Dalnauka; 2000.

[49] Vascular plants of the Soviet Far East. Editor Kharkevich S.S. Leningrad: Nauka; 1996.

[50] Vorobyov, D.P. Vegetation of Kuril Islands. Moscow: Academy of Science of USSR; 1963.

[51] Zaitseva, G.I.; Popov, S.G.; Krylov, A.P.; Knorozov, Y.V.; Spevakovskiy, A.B. Radiocarbon chronology of archaeological sited of the Kuril Islands. *Radiocarbon,* 1993, 35. 507-510.

In: Grasslands: Ecology, Management and Restoration
Editor: Hans G. Schröder
ISBN 978-1-60692-023-7

Chapter 10

Nitrous Oxide Emissions from Grazed Grasslands: Key Sources and Abatement Strategies (Case Study—Overwintering Areas in the Czech Republic, Central Europe)

Miloslav Šimek

Biology Centre of the Academy of Sciences of the Czech Republic,
Institute of Soil Biology, and University of South Bohemia,
Faculty of Science, Na Sádkách 7,
370 05 České Budějovice, Czech Republic

Abstract

Nitrous oxide (N_2O) emissions from grazed pasture soils represent a significant source of atmospheric N_2O. Nitrous oxide is produced by transformations of nitrogen from cattle excrements, urine and fertilizers entering the soil. Specific conditions are necessary for emissions of N_2O and these emissions are both spatially and temporally highly variable. Spatial, and in some extent temporal variability, is due to the complex effect of nitrogen input, pasture management and environmental conditions (e.g., aeration and water status, pore space and pore size distribution, available carbon and mineral nutrients concentration, pH and other soil chemical properties, as well as microbial communities abundance and diversity) on N_2O emissions and all this makes estimation of emissions very difficult. Several types of environments or events characterized by potentially high N_2O production and emissions (=emission hot spots) can be distinguished in the livestock farming system. These include camping areas, drinking sites, feedlots, shade areas, footpaths, dung and urine patches (where combined effects of nutrients in urine, dung and compaction occur resulting in creating conditions for high nitrification and denitrification rates). Here we propose overwintering areas, that is pasture sites where cattle is located in high stock densities for a relatively long period during winter season (where severe damage of plant cover is common and the effects are typical for the above camping areas other hot spots are even reinforced as another type of hot spots for N_2O emissions. Due to overgrazing effect, damage of vegetation and high stock density, as well as high N-inputs in excrements and lower utilization of deposited N by plants in a cold period, cattle overwintering areas appear to have a large potential for accelerated microbial N transformations and thus gaseous losses. This contribution examines various aspects related to the production of N_2O and its emission from the soil of cattle-grazed pasture sites, aiming to identify its major sources as well as environmental soil conditions favourable for N_2O production (=emission hot spots), as well as to summarize mitigation strategies for N_2O emissions. In particular, it is focused on the cattle overwintering areas, representing specific and often significant emission hot spots. Relevant measures to decrease N_2O fluxes from the grazed grasslands are presented. However, practical options to decrease N_2O emissions seems to be rather limited due to a number of reasons, including a limited knowledge on intrinsic mechanisms regulating N_2O production in the pasture soils.

INTRODUCTION

In highly productive agricultural ecosystems, including arable fields and intensive grasslands, nitrogen inputs often largely exceed nitrogen consumption by crops. In addition to likely direct and indirect effects on edafon, this creates suitable conditions for nitrogen losses. Losses of nitrogen in its various forms from both arable and grassland soils can be considerable and represent the most serious environmental problem of intensive agriculture in many regions. A large proportion of nitrogen losses from soils is in the form of nitrogen gases, namely nitrous oxide (N_2O) and molecular nitrogen (N_2). Over recent decades, nitrogen cycling in terrestrial ecosystems has thus emerged as an important ecological issue resulting in an extensive research focused on many different aspects and levels of the topic. However, the nitrogen cycle is rather complex and there are still many uncertainties concerning, for example, role of individual factors controlling the occurrence and rates of the key nitrogen transformation processes, denitrification and nitrification, both producing nitrous oxide.

N_2O-forming processes and N_2O emissions have been considered in a number of recent reviews (e.g. Bouwman, 1990; Granli and Bockman, 1994; Mosier, 1994; Oenema et al., 1997; Oenema et al., 2005). However, different soil N processes can proceed at the same time, and formation of N_2O in soil tends to have a complicated pattern of response to regulating factors resulting in a well-known extreme spatial heterogeneity and temporal variation of N_2O fluxes (e.g. Parsons et al., 1991; Murray et al., 1995). Particularly N_2O is therefore thought to be often produced in soil microsites ("hot spots"; Parkin, 1987) unevenly distributed in the soil (note: for the purpose of this study, also sites and events characterized by high rates of production and emission of N_2O are called hot spots despite their physical size). Strong temporal pattern of N_2O emissions (e.g. Parsons et al., 1991; Kammann et al., 1998) is related to both changes in environmental conditions (e.g. temperature and rainfall) and management practices (e.g. fertilization, irrigation, ploughing, grazing). Despite the fact that soils are the main sources of N_2O, net N_2O consumption has been found both in natural and agricultural systems too suggesting some soils as a sink for N_2O (Chapuis-Lardy et al., 2007).

Arable (cultivated) soils prevail in the Czech Republic, covering more than 72% of agricultural land. They are common in the whole country; arable farming represents a typical land-use both in lowlands and highlands and occurs also in other marginal parts of the country. However, in less favourite agricultural areas, permanent and temporary grasslands represent a common type of agricultural ecosystems. Permanent grasslands cover about 22% of agricultural land and are used both for hay production (meadows, ca 70% of total area of grasslands) and for grazing (pastures, ca 30%). Although the majority of agricultural production originates in arable agriculture, grasslands make a substantial contribution to the national economy too, through their use in production meat, milk and other animal products. Currently there is a growing tendency to use grasslands for cattle and sheep grazing and the same tendency can be seen in surrounding areas in abroad in Central Europe area.

The purpose of the present paper is to briefly review the soil and environmental conditions supporting N_2O formation, to identify and list the major emission hot spots and to emphasize the importance of spatial and time hot spots for the overall balance of N_2O fluxes. In particular, it is focused on the cattle overwintering areas, representing specific and often significant emission hot spots. In addition to an overview of literature data about the topic, summary of data on N_2O production and emission from the soils of cattle overwintering areas that are located in Central Europe is presented.

CONDITIONS SUPPORTING N_2O FORMATION

When considering N_2O production and its subsequent emission from the soil, it is worth to realize that N_2O is mostly a product of various microbial activities in the soil. Classical idea on N_2O production suggests respiratory denitrification (i.e. conversion of nitrate or nitrite to NO, N_2O and N_2) as the major microbial process responsible for most of N_2O produced in terrestrial ecosystems, although autotrophic nitrification (i.e. conversion of NH_4^+ to NO_3^-) and dissimilatory nitrate reduction to ammonium (i.e. conversion of NO_3^- to NH_4^+) (Fazollari et al., 1990) as a source of N_2O cannot be omitted in various conditions and ecosystems. Moreover, certain amounts of N_2O can be evolved during other reactive N transformations (compare Figure 1), perhaps also due to heterotrophic nitrification carried out by many heterotrophic bacteria and fungi. In this process, reduced nitrogen can be either inorganic (similarly to autotrophic nitrification) and organic. Although there are reports of occurrence and importance of heterotrophic nitrification in various ecosystems including acidic soils, its ecological importance and the proportion of N_2O possibly evolved during the process is largely unknown. Despite many uncertainties, however, it is clear that N_2O production is regulated by a number of factors and conditions controlling either denitrification and nitrification rates and other N-transformations as well.

Atmospheric molecular nitrogen is assimilated by di-nitrogen fixing microorganisms in nitrogen fixation process (f) (or industrially); it has been hypothesized that during nitrogen fixation, certain amounts of N can be evolved as N_2O, although the idea is not widely accepted. Ammonium form of nitrogen NH_4^+ is assimilated (a) into the biomass (R-NH_2), fixed by clay minerals in the soil (j), volatilized as gaseous ammonia (v) and/or nitrified (n) and anaerobically oxidized in anammox reaction to N_2 with NO_2^- as the electron acceptor (am). Nitrification is a source of N_2O and NO as well. Resulting nitrite nitrogen (NO_2^-) can be chemically destroyed to several N gases (at rather low pH values < 4.0). Normally it is further oxidized in the second step of nitrification (n), and nitrate nitrogen (NO_3^-) is formed. Fate of nitrate is various: it can be assimilated (a), leached out of soil (l), or reduced to nitrite (NO_2^-)

due to several different processes and organisms: denitrified in respiratory denitrification by true denitrifiers (d) respired by nitrate respirers (i) or utilized for dissimilative nitrate reduction to ammonium (k).

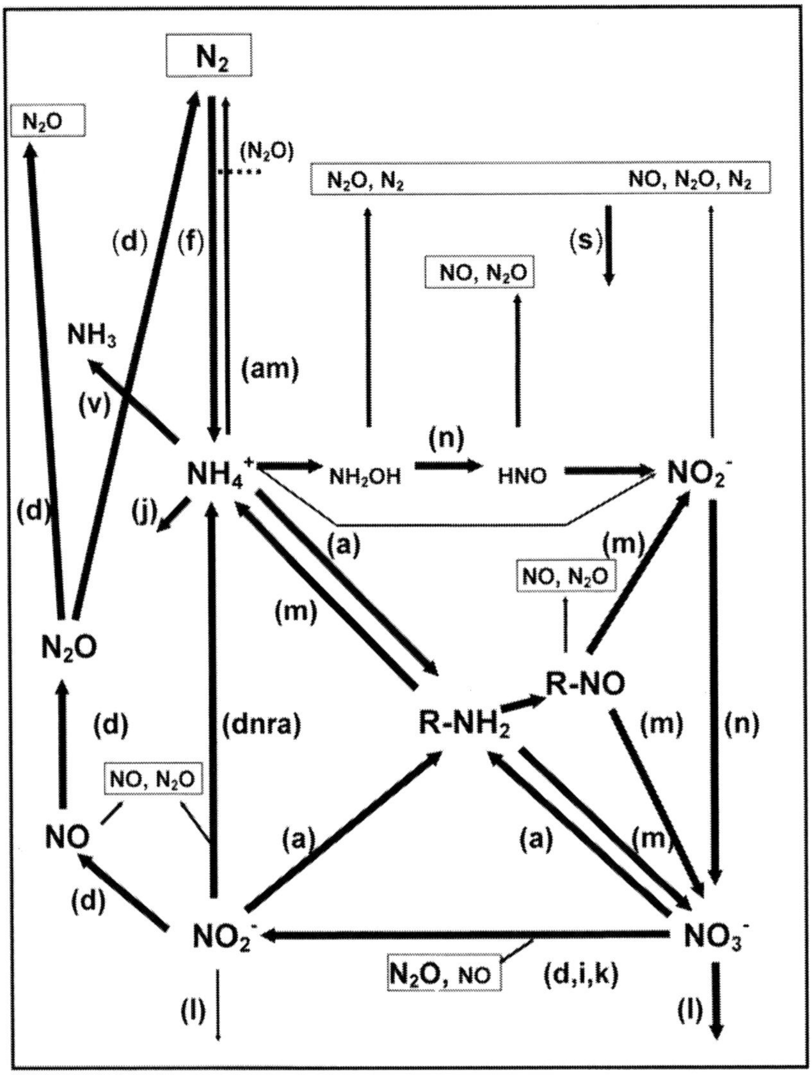

Figure 1. Nitrogen transformations and flows in terrestrial ecosystem; the formation of gases is highlighted in frames.

Both in nitrate and nitrite reductions, NO and N_2O are produced too. In respiratory denitrification first NO, then N_2O and finally N_2 are formed (d), while reduction of NO_2^- can finish in formation of NH_4^+ (dnra). In many ecosystems and environmental conditions, respiratory denitrification is the major source of N_2O (produced via nitrite and nitric oxide reductases), although under specific conditions, other processes can be significant. Nitrogen accumulated in biomass is evolved in mineralization processes (m) producing mostly NH_4^+, but also NO_3^- and NO_2^-, producing small amounts of N gases, too.

Denitrification occurs in general if organic carbon is available and nitrate (or nitrite) is present in a vicinity of denitrifying microorganisms in conditions of lowered partial pressure of molecular oxygen. Therefore increased denitrification can be expected in grazed grasslands receiving relatively unevenly distributed N and C in animal excreta, while soil is impacted by the presence of animals and animal trampling in a way often leading to creation of anoxic microsites in the soil matrix. Although experimental data on the direct effect of soil compaction (caused by the animal treading) on N_2O fluxes are limited, it can be easily hypothesized that such simple relationship exists. For example, Menneer et al. (2005) found that dairy cows treading increased substantially denitrification rate due to reduction of soil aeration through soil physical damage and due to decreased soil N utilization by plants leading to temporary accumulation of mineral nitrogen in the soil. Which is even more important, grazing changes spatial distribution of nitrogen in a grazed ecosystem: nitrogen distributed over the pasture in plant biomass in relatively low concentrations now concentrates in animal excreta. This occurs because grazing ruminants utilize relatively little of the nitrogen consumed, excreting thus most of nitrogen ingested (de Klein and Ledgard, 2005). Nitrogen concentrated in urine and dung patches can not be fully utilized by grass stand and therefore it is available for microbial transformations (which are often characterized by significant "leaks" due to nitrogen gas formation in and subsequent emission from the soil system) as well as it is lost by leaching or via ammonia volatilization. By this mechanism, nitrogen is not only relocated in pasture ecosystem, but its significant proportion is easily lost, which leads to subsequent undesirable impacts of grazing agroecosystems.

BALANCE BETWEEN N_2O PRODUCTION AND CONSUMPTION IN THE SOIL

In addition to their relatively well known role as a source of various atmospheric trace gases, soils can often act as a sink, too. Nitrous oxide consumption in the soil is primarily related to denitrification. It is supposed that most of denitrifiers are equipped with nitrous oxide reductase encoded by *nosZ* gene (Philippot et al., 2007). Thus for example Okereke (1993) found the ability to reduce N_2O to N_2 in 59 out of 71 denitrifying bacteria. In biochemical experiments, N_2O is also substrate of nitrogenase, enzyme responsible for di-nitrogen reduction, which non-specifically reduces molecules with triple bonds. It has not been proved whether this side reaction of nitrogenase plays significant ecological role in any habitat. Other possible N_2O-consuming pathways include nitrifier denitrification and absorption of N_2O in soil liquid phase and its subsequent transportation out of the soil profile. However, these pathways are difficult to quantify and the experimental data are very limited (entrapment of N_2O in the soil and subsoil and associated dissolution of the N_2O into soil water has been extensively reviewed by Clough et al., 2005). Nevertheless reduction of N_2O to N_2 by denitrifying microorganisms represents probably the most important process of N_2O uptake from the soil gaseous phase. Chapuis-Lardy et al. (2007) summarized in their recent excellent review data on negative fluxes of N_2O from various soils. They came to conclusion that soils can often act as a net sink for N_2O. Negative N_2O fluxes were typically connected to low mineral N and low pO_2. Crucial controlling factors for N_2O uptake (either by reduction to N_2 and absorption by water) represented soil properties: in addition to availability of mineral

N and aeration conditions, mainly soil temperature, pH, redox conditions and lability of soil organic C and N may play the role. In general, factors decreasing diffusion of N_2O in soil matrix should increase the probability of N_2O consumption in the soil. The authors also concluded that (logically) in soils with high N_2O production substantial rates of N_2O consumption might take place, but these are masked by larger N_2O production (Chapuis-Lardy et al., 2007). Although it seems probable that in soils exhibiting high denitrification rates and thus producing net positive N_2O fluxes also relatively large N_2O consumption occurs (due to either reduction to N_2 or solubility in soil water), it is not known whether soils with low and/or no production of N_2O could consume significant amounts of atmospheric N_2O. Theoretically any soil containing water represent a net N_2O sink consuming N_2O from the above-ground air due to movement of N_2O molecules first to the soil pores and then to the soil liquid phase. Assuming no reduction, the soil liquid phase becomes a temporary storage body for N_2O (Clough et al., 2005). Moreover the presence of denitrifiers specialized in N_2O reduction to N_2 can not be omitted. Denitrifying microbes can grow with N_2O as sole respiratory electron acceptor when all others such as oxygen, nitrate or nitrite are missing. However, information on these processes possibly occurring in upland soils (analogically to methane oxidation consuming traces of methane from the troposphere) is scarce. As a better understanding of N_2O consumption in the soil is fundamental for better understanding of the whole N_2O budget, more research in this direction is needed.

N_2O Emissions, Hot Spots in Grazed Grasslands and Mitigation Measures

Relatively many papers published in the last decade were dealing with N_2O sources and sites of emissions on grassland livestock systems (see e.g. Oenema et al., 1997; Kammann et al., 1998; Chadwick et al., 1999; Freibauer, 2003; Müller and Sherlock 2004; Saggar et al., 2004; Oenema et al., 2005), thus reflecting the fact that this type of ecosystem is responsible for significant proportion of N_2O emissions worldwide. Recent estimates show that emissions of greenhouse gases from agriculture account for ca 22 % of all global emissions. Livestock production - including land use changes and deforestation, soy-feed production, energy used for grain production, processing and transport, soil organic matter decomposition followed by CO_2 (and CH_4) releases and a number of direct and indirect greenhouse gas sources in animal husbandry - accounts for nearly 80 % of agricultural emissions (McMichael et al., 2007). Emissions from agriculture are dominated by N_2O and CH_4 and agriculture accounts for 52 and 84 % of global anthropogenic emissions of N_2O and CH_4, respectively (Smith et al., 2007). Similar data apply for the Czech Republic; according to national greenhouse gas emission inventory, the major sources of N_2O are agricultural soils and organic fertilizers followed by fossil fuel burning and industry. The estimates for 2005 show that in the country-level sectoral emission budget (expressed as CO_2eq), 58 % emissions in agriculture is N_2O from the soils (both fertilized croplands and grazed pastures), 31 % CH_4 from enteric fermentation, 6 % N_2O and 5 % CH_4 both from organic fertilizers (Anonym, 2007).

The primary source of N_2O emissions from animal production systems is urine and dung, and in smaller extent soils and sometimes also inorganic fertilizers. Therefore improvements in animal waste management are necessary if sustainable reduction of N_2O emissions is required. Several types of environments or events characterized by potentially high N_2O production and thus possible increased emissions (= emission hot spots) can be distinguished in the livestock farming system. These include:

i. hot spots in grazed grasslands (where combine effects of urine, dung, and compaction and other soil structure impacts occur resulting in creating conditions for high nitrification and denitrification rates),
ii. hot spots in manure storage facilities (manure heaps, lagoons and tanks, where either anoxic and oxic conditions occur facilitating microbial transformations in conditions of surplus of nutrients),
iii. hot spots in livestock (ruminants and others, mostly related to methane production by enteric fermentation; however N_2O production can be also expected),
iv. other hot spots (e.g. N_2O emissions from burned dung collected after deposition in paddocks or outside areas)
v. hot spots generated or supported by management practices and policy decisions (decisions influencing occurrence, intensity and timing of grazing, quality of grasslands, numbers of animals, dietary aspects, etc.). For more details see Table 1.

In addition to other known hot spots listed in Table 1 we propose *cattle overwintering areas*, that is pastures where cattle is located in high stock densities for relatively long period during winter season, as potential hot spots of N_2O (and also CH_4) emissions. In these grassland areas severe damage of plant cover is common and the effects typical for the camping and overgrazed areas are even reinforced.

Table 1. N_2O emission hot spots in animal production systems based on grazed grasslands including a newly proposed hot spot - an overwintering area

Area	Hot spots
grazed grasslands	camping areas drinking sites feedlots shade areas footpaths dung (deposited) urine patches **overwintering area**
manure storage, handling and application	manure heaps tanks and lagoons soil after application
livestock	intestine
other direct	dung use as biofuel
indirect: management and policy	grazing intensity grass stand productivity fertilizer management feeding practice dietary additives

Damage of vegetation takes place in the same time of large depositions of nutrients; consequently lower utilization of deposited N by plants in a following period is a prerequisite for higher rates of microbial transformations and subsequent gaseous losses from the system.

There are three mechanisms to lower N_2O emissions from agriculture in general as identified by Smith et al. (2007), which can be adopted to grazed grasslands, too. These include:

i. reduction of existing emissions: Mitigation options include better management of nitrogen flow through the system, increased efficiency of nutrient utilization and thus lowering losses.
ii. enhancement of removals: Nitrous oxide produced in soil, dung, manure and other hot spots can be further reduced to N_2 before escaping the site of origin. Gross production of N_2O can be thus substantially reduced resulting in decrease of net N_2O fluxes.
iii. elimination of N_2O production: Management and policy decisions which can eliminate the possibility for N_2O production in grazed ecosystems.

As pointed out by Smith et al. (2007) many practices were suggested and introduced to mitigate N_2O emissions through above mechanisms, focusing on the hot spots listed in Table 1. A number of publications and extensive reviews on mitigation options for N_2O emissions from agriculture have been published in last years (e.g. Beauchamp, 1997; Oenema et al., 1997; Senevirante, 1999; Senevirante, 2001; Six et al., 2004; de Klein and Ledgard, 2005) including several recent papers (Clough et al., 2007; McMichael et al., 2007; Schils et al., 2007; Smith et al., 2007; Vergé et al., 2007; Noble and Christmas, 2008; van Groenigen et al., 2008). It is therefore not the aim of this brief review to repeat information already widely available. Instead we will focus on the above mentioned newly proposed hot spot, cattle overwintering areas, which represent potentially significant sources of N_2O emissions.

A Case Study of Cattle Overwintering Area and the Significance in N_2O Emissions

Pastures are specific grassland ecosystems used for animal grazing. In the Czech Republic and other Central European regions they are usually located in mountainous and other marginal areas, where environmental and soil conditions are less favourable for field crops, while more precipitation supports growth of grass. Cattle typically graze during the growing season, while, during winter, the animals are located near the animal house on a relatively small plot called an "overwintering area".

In our investigations carried out since 2001, we focused on a cattle overwintering area located in South Bohemia in the Czech Republic (latitude 48°52′ N, longitude 14°13′ E), some 170 km south from Prague (Figure 2). The area was approximately 4 ha, and it had been used for overwintering a cattle herd since the 1994/5 season. The animals were present on the site from the end of October - beginning of November to beginning of May each year. At the beginning, the herd was composed of ca 90 cows and 2 bulls, while new-born suckers, added to the herd in January of each year, increased the total number of animals present on the site.

Figure 2. Cattle overwintering area in Borova farm in the Czech Republic photographed in May 2002.

Always soon after the arrival of animals there was a visible gradient of animal impact from the most impacted areas near the animal house through much less impacted areas in the middle to almost unaffected areas at the opposite side of the overwintering area, where cattle traffic was minimal.

Along this gradient three locations were identified, differing in the presumed rate of animal impact. These included a severely impacted location (S) with totally destroyed plant cover and surface soil, a location with moderate impact (M), where effects of trampling and disturbance of the vegetation were still visible, and a control location (C) with very slight or no impact as judged from soil and vegetation. These differences were most pronounced at the end of the winter season in April, while vegetation regenerated to some extent during summer even at locations S and M. The relative proportion of the size of differently impacted parts of the overwintering area was estimated at 8, 56, and 36 % for locations S, C, and M, respectively. The soil on the site is a sandy loam classified as a Haplic Phaeozem (arenic; WRB system) containing 60-80% sand, 14-32% silt, and 6-14% clay (USDA classification system). Physical and chemical soil properties at the three sampling locations in spring of 2005 are listed in Table 2. The plant cover was a perennial mixture of grasses, clovers and other dicotyledonous plants. The mean annual precipitation in the area is 650 mm and the average annual temperature is 7°C (data from meteorological station located at 7 km distance from the experimental site). More details are available in Šimek et al. (2006a, 2006b) and Hynšt et al. (2007b).

The animal house is located in the left top corner (white roof). The cattle have access either in the animal house and outside on the overwintering pasture. Different rate of impact of animals can be distinguished by different colour. While severely impacted parts are characterized by completely destroyed vegetation cover and have brown colour, much less and none impacted sites are greenish. Footpaths between the animal house and source of water (spring) are visible, too.

Table 2. Selected properties of surface soil (0-20 cm) at three locations of the cattle overwintering area in May 2005. Means of 9 replicates and standard deviations are given. Different letters behind the mean values indicate significant (P<0.05) differences among sites (one way ANOVA, Tukey test)

Location	pH (H_2O)	C_{org}	N_{tot}
S	8.96 (0.16) a	5.8 (0.3) a	1.94 (0.29) a
M	7.25 (0.22) b	4.1 (0.8) b	1.21 (0.45) b
C	5.91 (0.23) c	2.3 (0.3) c	0.65 (0.24) c

Fluxes of N_2O were determined using manual chambers: medium-sized (basal area 0.076 m^2, volume 15 dm^3), non-vented buckets of galvanized steel equipped with a rubber stopper (Figure 3). Gas samples for N_2O analyses (1 ml) were taken through the stopper at time 0 and then 30 and 90 minutes after chamber deployment to 2-4 cm depth. Gas samples were analyzed immediately after transportation to laboratory at the day of sampling. Nitrous oxide was quantified using an HP5890 gas chromatograph equipped with a 3 m Porapak Q column operating at 80°C, and an electron capture detector.

Figure 3. Flux and soil temperature measurements at the cattle overwintering area in 2004. The figure shows the chamber deployed to the severely impacted part of the area. The chamber covers ca 0.076 m^2 and its volume is ca 15 dm^3. Notice the soil surface damaged by the animals.

We found that soil from the cattle overwintering area had a high potential for production of N_2O as a consequence of strong animal impact and accumulation of nutrients from their excrements (Šimek et al., 2006a; Hynšt et al., 2007b). The outburst of N_2O occurred at the end of winter and in spring, often during a few short periods with favourable conditions following rainfall when there was enough nitrogen available from the animal excrements and biological activities including intensive nitrogen transformations (mineralization and nitrification) were reinforced. The emissions of N_2O (and probably N_2) increased with the degree of animal impact and were positively correlated with soil nitrate, although on some occasions N_2O production from less impacted parts of the overwintering area exceeded that from the most impacted locations. Maximum N_2O fluxes of up to 8 mg N_2O-N m^{-2} h^{-1} were recorded at the most impacted location near the animal house, where the highest concentrations of soil mineral nitrogen (up to 450 mg . kg^{-1} nitrate + ammonium N) also occurred. The peaks of N_2O emissions were thus relatively large in comparison with other studies of grazed pasture soils, where usually fluxes did not exceed 0.5 mg N_2O-N m^{-2} h^{-1} (e.g. Velthof and Oenema, 1995; Kammann et al., 1998; Rudaz et al., 1999; Williams et al., 1999; van Groenigen et al., 2005). Total N_2O-N emission during winter season was estimated to be 10.5, 6.6, and 1.5 kg N ha^{-1} from locations S, M and C, respectively (means from two seasons, see Figure 4). As the flux measurement frequency was 1-2 weeks only, it is however likely that these cumulative flux data are significantly underestimated. Despite this, the results show that cattle overwintering areas are important sources of greenhouse gases, including N_2O (Šimek et al., 2006a; Hynšt et al., 2007b).

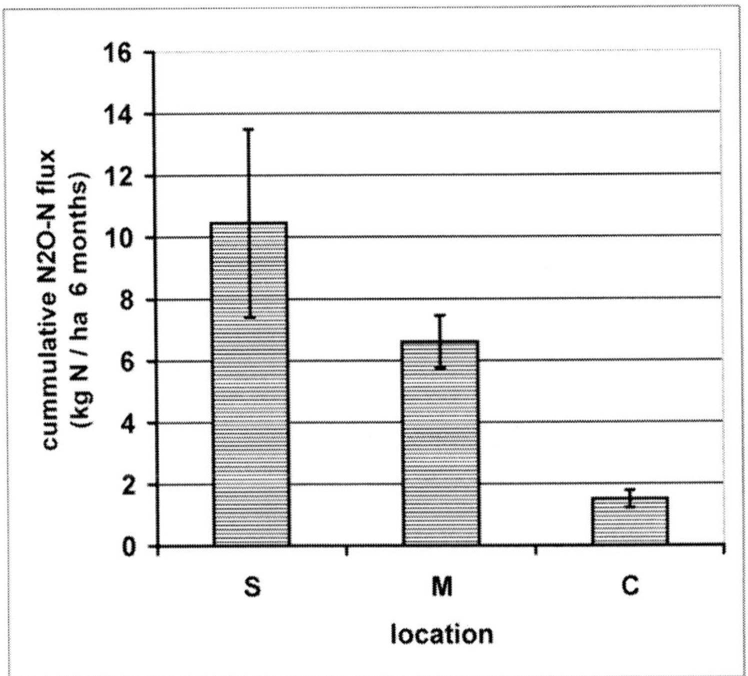

Figure 4. Total N_2O-N emission during winter season accounting for 6 months from locations S, M and C, respectively. Means for two consequtive winter seasons 2002/3 and 2003/4 and standard deviations are shown.

In another study we investigated the response of soils under study to amendments with nitrate nitrogen in doses corresponding to 10 - 1 500 kg N and ha^{-1}, covering the range of nutrient inputs occurring in real field conditions (Hynšt et al., 2007a). Peak N_2O emissions were up to 94 mg N_2O-N m^{-2} h^{-1} 5-8 hours after amendment. No upper limit of N_2O emissions was detected as the emissions were directly related to the dose of nutrients in the whole range of amendments used, but the fluxes reflected the soil and environmental conditions, too.

In three independent experiments, total cumulative losses of N_2O-N ranged from 0.2 to 5.6% of the applied 500 kg NO_3^--N ha^{-1}. Splitting of high nutrient doses lowered the rate of N_2O fluxes following the first amendment, but the effect of splitting on the total amount of N_2O-N released from the soil was insignificant, as the initial lower values of emissions in the split variants were compensated for by a longer duration of gas fluxes. The results suggest that the cattle impacted soil has the potential to metabolize large inputs of mineral nitrogen over short periods (~days). The emission factors for NO_3^--N did not exceed values reported in literature.

In a more detailed investigation carried out in May 2005, N_2O fluxes in the field were compared with potential denitrifying enzyme activity in soil sampled at the same time and the rates of gas production were related to bacterial *nirK*, *nirS* and *nosZ* genes used as functional markers of the denitrifying communities (Chronakova et al., unpublished results). Field N_2O emissions were the highest in location M, while total potential denitrification rates were the highest in soil from the location S (Figure 5a,b), where the highest abundances of all genes of interest were found, too. However, significant proportion of N_2 was produced from the soil M during laboratory incubations and N_2 was major denitrification product in the soil S (Figure 5b).

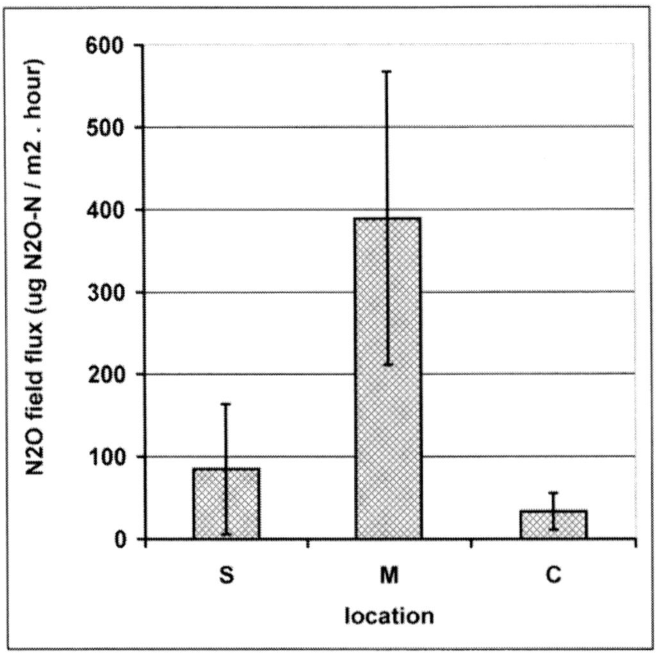

Figure 5A. N_2O emission from locations S, M and C in May 2005. Means of 9 replicates and standard deviations are shown.

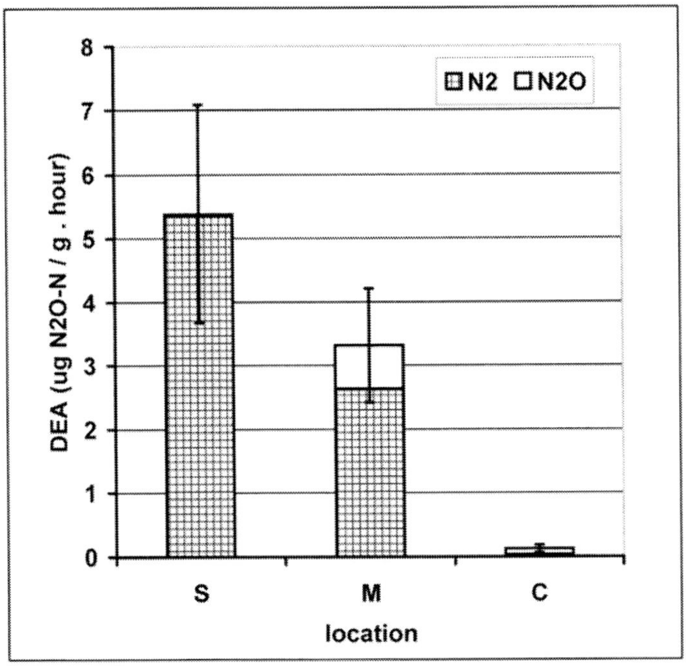

Figure 5. B. Potential denitrifying enzyme activity (DEA) in soils from locations S, M, and C in May 2005. Both N_2O and N_2 productions are shown (means of 9 replicates and standard deviations).

We therefore suggest that the major denitrification product in soil S is N_2, presumably due to significant shifts in soil pH towards alkaline regions and increase in available carbon (compare Table 2). This assumption is supported by the results received in 2004 by Šimek et al. (2006b). They found that after amendments of soils with ^{15}N-labeled NO_3^-, most of field gaseous loss was N_2 from both S and M soils, while N_2O was the main gas from C treatment. The mole fraction of evolved N_2O was 0.04, 0.15 and 0.75 for the S, M, and C treatments, respectively (Šimek et al., 2006b).

In a summary, cattle overwintering area investigated extensively by Šimek and co-workers since 2001 represents a significant "point" source of N_2O emissions (= emission hot spot). The list of obvious hot spots for N_2O emissions from animal production systems has to be therefore extended and the animal overwintering areas included. It is likely that to mitigate N_2O fluxes from the animal overwintering areas, the same mitigation option as in the case of other hot spots related to grazed grasslands, could be widely adopted. However, more research is necessary to solve this problem.

Conclusion

Grazed grasslands as a part of livestock production systems produce large quantities of greenhouse gases including nitrous oxide. In addition to other significant sources (e.g. manure heaps and liquid manure tanks), several emission hot spots can be distinguished in the pastures. These include camping areas, drinking sites, feedlots, shade areas, footpaths and dung and urine patches (where combine effects of nutrients in urine, dung and compaction occur resulting in creating conditions for high nitrification and denitrification rates and, consequently, high N_2O fluxes). In this chapter we proposed overwintering areas, that is pasture sites where cattle (and/or other livestock) is located in high stock densities for relatively long periods during winter season as another type of hot spots for N_2O emissions. Due to overgrazing effect, damage of vegetation and high stock density accompanied with high N-inputs in excrements and lower utilization of deposited N by plants in a cold period, cattle overwintering areas appear to have a large potential for accelerated microbial N transformations and thus gaseous losses from the system, including nitrogen gas fluxes.

Acknowledgements

This study was supported by the project of Ministry of Education of the Czech Republic (LC06066), the Research Plans (AV 0Z60660521 and MSM 6007665801), and the grant of the Grant Agency of the Academy of Sciences of Czech Republic (IAA600660605). J. Hynšt, J. Čuhel, A. Chroňáková, P. Brůček, D. Elhottová, A. Němcová, V. Krištůfek, V. Šlajchrtová, E. Zadáková and L. Jíšová are greatly acknowledged for collaboration in research at the cattle overwintering area, field works and laboratory analyses. Mr and Mrs Kamír are thanked for allowing access to the experimental plot and for their significant support of field work.

References

Anonym (2007). National greenhouse gas inventory report of the Czech Republic, NIR. Reported inventory 2005. Prague: Czech Hydrometeorological Institute.

Beauchamp, E.G. (1997). Nitrous oxide emission from agricultural soils. *Canadian Journal of Soil Science*, 77, 113-123.

Bouwman, A. F., 1990. Exchange of greenhouse gases between terrestrial ecosystems and the atmosphere. In A.F. Bouwman (Ed.), *Soils and the Greenhouse Effect* (pp. 61-127). Chichester: John Wiley and Sons.

Chadwick, D.R., Sneath, R.W., Phillips, V.R., Pain, B.F. (1999). A UK inventory of nitrous oxide emissions from farmed livestock. *Atmospheric Environment*, 33, 3345-3354.

Chapuis – Lardy, L., Wrage, N., Metay, A., Chotte, J.-L., Bernoux, M. (2007). Soils, a sink for N_2O? A review. *Global Change Biology*, 13, 1-17.

Clough., T.J., Sherlock, R.R., Rolston, D.E. (2005). A review of the movement and fate of N_2O in the subsoil. *Nutrient Cycling in Agroecosystems*, 72, 3-11.

Clough., T.J., Di, H.J., Cameron, K.C., Sherlock, R.R., Metherell, A.K., Clark, H., Rys, G. (2007). Accounting for the utilization of a N_2O mitigation tool in the IPCC inventory methodology for agricultural soils. *Nutrient Cycling in Agroecosystems*, 78, 1-14.

de Klein, C.A.M., Ledgard, S.F. (2005). Nitrous oxide emissions from New Zealand agriculture – key sources and mitigation strategies. *Nutrient Cycling in Agroecosystems*, 72, 77-85.

Fazzolari, E., Mariotti, A., Germon, J.C. (1990). Nitrate reduction to ammonia: a dissimilatory process in Enterobacter amnigenus. *Canadian Journal of Microbiology*, 36, 779-785.

Freibauer, A. (2003). Regionalised inventory of biogenic greenhouse gas emissions from European agriculture. *European Journal of Agronomy*, 19, 135-160.

Granli, T., Bockman, O.C. (1994). Nitrous oxide from agriculture. *Norvegian Journal of agricultural Sciences*, 12, 7-129.

Hynšt, J., Brůček, P., Šimek, M. (2007a). Nitrous oxide emissions from cattle-impacted pasture soil amended with nitrate and glucose. *Biology and Fertility of Soils*, 43, 853-859.

Hynšt, J., Šimek, M., Brůček, P., Petersen, S.O. (2007b). High fluxes but different patterns of nitrous oxide and carbon dioxide emissions from soil in a cattle overwintering area. *Agriculture, Ecosystems and Environment*, 120, 269-279.

Kammann, C.,Grünhage, L., Müller, Ch., Jacobi, S., Jäger, H.-J. (1998). Seasonal variability and mitigation options for N_2O emissions from differently managed grasslands. *Environmental Pollution*, 102, 179-186.

McMichael, A., Powles, J.W., Butler, C.D., Uauy, R. (2007). Food, livestock production, energy, climate change, and health. *Lancet,* 370, 1253-1263.

Menneer, J.C., Ledgard, S., McLay, Ch., Silvester, W. (2005). Animal treading stimulates denitrification in soil under pasture. *Soil Biology and Biochemistry*, 37, 1625-1629.

Mosier, A.R. (1994). Nitrous oxide emissions from agricultural soils. *Fertilizer Research* , 37, 191-200.

Müller, Ch., Sherlock, R.R. (2004). Nitrous oxide emissions from temperate grassland ecosystems in the Northern and Southern Hemispheres. *Global Biogeochemical Cycles*, 18, 1-11.

Murray, R.E., Feig, Y.S., Tiedje, J M. (1995). Spatial heterogeneity in the distribution of denitrifying bacteria associated with denitrification activity zones. *Applied and Environmental Microbiology,* 61, 2791-2793.

Noble, B.F., Christmas, L.M. (2008). Strategic environmental assessment of greenhouse gas mitigation options in the Canadian agricultural sector. *Environmental Management*, 41, 64-78.

Oenema, O., Velthof, G.L., Yamulki, S., Jarvis S.C. (1997). Nitrous oxide emissions from grazed grassland. *Soil Use and Management*, 13, 288-295.

Oenema, O., Wrage, N., Velthof, G.L., van Groenigen, J.W., Dolfing, J., Kuikman, P.J. (2005). Trends in global nitrous oxide emissions from animal production systems. *Nutrient Cycling in Agroecosystems*, 72, 51-65.

Okereke, G.U. (1993). Growth yield of denitrifiers using nitrous oxide as a terminal electron acceptor. *World Journal of Microbiology and Biotechnology*, 9, 59-62.

Parkin, T.B. (1987). Soil microsites as a source of denitrification variability. *Soil Science Society of America Journal*, 51, 1492-1501.

Parsons, L.L., Murray, R.E., Smith, M.S. (1991). Soil denitrification dynamics: spatial and temporal variations of enzyme activity, populations, and nitrogen gas loss. *Soil Science Society of America Journal,* 55, 90-95.

Philippot, L., Hallin, S., Schloter, M. (2007). Ecology of denitrifying prokaryotes in agricultural soil. *Advances in Agronomy*, 96, 249-305.

Rudaz, A.O., Wälti, G., Lehmann, P., Fuhrer, J. (1999). Temporal variation in N_2O and N_2 fluxes from a permanent pasture in Switzerland in relation to management, soil water content and soil temperature. *Agriculture, Ecosystems and Environment*, 73, 83-91.

Saggar, S., Andrew, R.M., Tate, K.R., Hedley, C.B., Rodda, N.J., Townsend, J.A. (2004). Modelling nitrous oxide emissions from dairy – grazed pastures. *Nutrient Cycling in Agroecosystems,* 68, 243-255.

Schils, R.L.M., Olesen, J.E., del Prado, A., Soussana, J.F. (2007). A reviw of farm level modelling approaches for mitigating greenhouse gas emissions from ruminant livestock systems. *Livestock Science*, 112, 240-251.

Seneviratne, G. (1999). A cheap way to control N_2O in agriculture. *Ambio*, 28, 378.

Seneviratne, G. (2001). Mitigating nitrous oxide emission in tropical agriculture: Myths and realities. *Current Science*, 80, 117-118.

Šimek, M., Brůček, P., Hynšt, J., Uhlířová, E., Petersen, S.O. (2006a). Effects of excretal returns and soil compaction on nitrous oxide emissions from a cattle overwintering area. *Agriculture, Ecosystems and Environment* 112, 186-191.

Šimek, M., Stevens, R.J., Laughlin, R.J., Hynšt, J., Brůček, P., Čuhel, J., Pietola, L. (2006b). Gaseous nitrogen losses from a grassland area used for overwintering cattle. In: C.R. Soliva, J. Takahashi, M. Kreuzer, M. (Eds.), *Greenhouse gases and animal agriculture: an update*. (International Congress Series, 1293, pp. 343-346). Amsterdam: Elsevier.

Six, J., Ogle, S.M., Breidt, F.J., Conant, R.T., Mosier, A.R., Paustian, K. (2004). The potential to mitigate global warming with no-tillage management is only realized when practised in the long term. *Global Change Biology*, 10, 155-160.

Smith, P., Martino, D., Cai, Z., Gwary, D., Janzen, H., Kumar, P., McCarl, B., Ogle, S., O´Mara, F., Rice, Ch., Scholes, B., Sirotenko, O., Howden, M., McAllister, T., Pan, G., Romanenkov, V., Schneider, U., Towprayoon, S., Wattenbach, M., Smith, J. (2008). Greenhouse gas mitigation in agriculture. *Philosophical Transactions of the Royal Society B*, 363, 789-813.

Van Groenigen, J.W., Schils, R.L.M., Velthof, G.L., Kuikman, P.J., Oudendag, D.A., Oenema, O. (2008). Mitigation strategies for greenhouse gas emissions from animal production systems: synergy between measuring and modelling at different scales. *Australian Journal of Experimental Agriculture*, 48, 46-53.

Van Groenigen, J.W., Velthof, G.L., van der Bolt, F.J.E., Vos, A., Kuikman, P.J. (2005). Seasonal variation in N_2O emissions from urine patches: Effects of urine concentration, soil compaction and dung. *Plant and Soil*, 273, 15-27.

Velthof, G.L., Oenema, O. (1995). Nitrous oxide fluxes from grassland in the Netherlands: II. Effects of soil type, nitrogen fertilizer application and grazing. *European Journal of Soil Science*, 46, 541-549.

Vergé, X.P.C., de Kimpe, C., Desjardins, R.L. (2007). Agricultural production, greenhouse gas emissions and mitigation potential. *Agricultural and Forest Meteorology*, 142, 255-269.

Williams, D.L., Ineson. P., Coward, P.A. (1999). Temporal variations in nitrous oxide fluxes from urine – affected grassland. *Soil Biology and Biochemistry*, 31, 779-788.

Reviewed by

Prof. Hermann Bothe
University of Cologne
Gyrhofstrasse 15
D-50923 Cologne
GERMANY

INDEX

A

B

C

D

E

F

G

H

I

J

K

L

M

N

O

P

Q

R

S

T

U

Z